ASPIRE
SUCCEED
PROGRESS

Cambridge IGCSE® & O Level

Essential
Biology

Third Edition

Gareth Williams

Richard Fosbery

Editor: Lawrie Ryan

Choong Yao-Jin

Alice Robinson

OXFORD
UNIVERSITY PRESS

Great Clarendon Street, Oxford, OX2 6DP, United Kingdom

Oxford University Press is a department of the University of Oxford. It furthers the University's objective of excellence in research, scholarship, and education by publishing worldwide. Oxford is a registered trade mark of Oxford University Press in the UK and in certain other countries

First published in 2021

British Library Cataloguing in Publication Data
Data available

978-1-38-200603-3 (standard)
10 9

978-1-38-200602-6 (enhanced)
10 9 8 7 6 5 4

The manufacturing process conforms to the environmental regulations of the country of origin.

Printed and bound in Great Britain by Bell & Bain Ltd, Glasgow

Acknowledgements

The publisher and authors would like to thank the following for permission to use photographs and other copyright material:

Cover: Dirk Ercken/Shutterstock. Photos: Footer: Shanvood/Shutterstock; **p2:** megasquib/iStock/Getty Images; **p3:** Markanja/iStock/Getty Images; **p4**(t): Rinek/iStock/Getty Images; **p4**(c): Shadow_Hawk/iStock/Getty Images; **p4**(b): konmesa/iStock/Gett Images; **p5:** Javier Larrea/age fotostock/Alamy Stock Photo; **p6**(c): David Scharf/Science Photo Library; **p6**(b): Gregory Dimijian/Science Photo Library; **p7:** M.I. Walker/Science Photo Library; **p8**(t): Natural Visions/Alamy Stock Photo; **p8**(c): Mark Kostich/iStock/Getty Images; **p8**(b): M.H. Sharp/Science Photo Library; **p9**(t): Juanvi Carrasco/Alamy Stock Photo; **p9**(c): Abeselom Zerit/Shutterstock; **p10**(t): Ted Kinsman/Science Photo Library; **p10**(c): Lupos/iStock/Getty Images; **p10**(bl): Stuart Wilson/Science Photo Library; **p10**(br): David Aubrey/Science Photo Library; **p11**(t): M.H. Sharp/Science Photo Library; **p11**(c): johnaudrey/iStock/Getty Images; **p12**(t): Adrian Bicker/Science Photo Library; **p12**(bl): Sinclair Stammers/Science Photo Library; **p12**(br): Bjorn Svensson/Science Photo Library; **p18**: Kateryna Kon/123RF; **p19**: Power And Syred/Science Photo Library; **p21**(t): Dr Gopal Murti/Science Photo Library; **p21**(c): David Aubrey/Science Photo Library; **p21**(b): CNRI/Science Photo Library; **p33**(ct): Mike Samworth; **p33**(cb): Mike Samworth; **p43**(t): GoGo Images Corporation/Alamy Stock Photo; **p43**(c): A. Barrington Brown, © Gonville & Caius College/Science Photo Library; **p46**: J.C. Revy, ISM/Science Photo Library; **p51**: Laguna Design/Science Photo Library; **p54**: Biophoto Associates/Science Photo Library; **p56**(cl): Ecoimages/Dreamstime; **p56**(br): Cordelia Molloy/Science Photo Library; **p58**: Ecoimages/Adobe Stock; **p59**: IslandLeigh/iStock/Getty Images; **p63**(t): Rosenfeld Images LTD/Science Photo Library; **p63**(c): Nigel Cattlin/Alamy Stock Photo; **p66**: Geoff Kidd/Science Photo Library; **p70**: Shutterstock; **p72**: geogphotos/Alamy Stock Photo; **p73**(t): filo/iStock/Getty Images; **p73**(c): Marilyn Barbone/Alamy Stock Photo; **p74**(c): Biophoto Associates/Science Photo Library; **p74**(b): Biophoto Associates/Science Photo Library; **p75**: Biophoto Associates/Science Photo Library; **p78**: Ian Boddy/Science Photo Library; **p79**(t): Science Photo Library; **p79**(c): Alex Bartel/Science Photo Library; **p83**: Eye Of Science/Science Photo Library; **p86**(c): National Geographic Image Collection/Alamy Stock Photo; **p86**(b): Louisa Preston/Science Photo Library; **p87**(t): Chuck Brown/Science Photo Library; **p87**(c): claudiodivizia/123RF; **p88**(br): Rattiya Thongdumhyu/Shutterstock; **p88**(bl): Power And Syred/Science Photo Library; **p89**(t): Martyn Chillmaid; **p89**(c): Nigel Cattlin/Alamy Stock Photo; **p90**(c): Dr Jeremy Burgess/Science Photo Library; **p90**(bc): Biophoto Associates/Science Photo Library; **p90**(br): Biophoto Associates/Science Photo Library; **p94**: Biophoto Associates/Science Photo Library; **p95**: Clouds Hill Imaging Ltd/Science Photo Library; **p97**: Dr David Furness, Keele University/Science Photo Library; **p102**: Sputnik/Science Photo Library; **p103**: Photo Researchers, Inc./Science Photo Library; **p106**(c): CNRI/Science Photo Library; **p106**(b): Biophoto Associates/Science Photo Library; **p107**: PDPics from Pixabay; **p108**: Kateryna Kon/123RF; **p109**: royaltystockphoto.com/Shutterstock; **p110**(c): Steve Gschmeissner/Science Photo Library; **p110**(b): Biology Media/Science Photo Library; **p111**: Biophoto Associates/Science Photo Library; **p114**(c): Adam Hart-Davis/Science Photo Library; **p114**(b): satit_srihin/iStock/Getty Images; **p115**: gregepperson/iStock/Getty Images; **p118**: JPC-PROD/Shutterstock; **p119**: Irfan/Xinhua/Alamy Stock Photo; **p120**: lostinbids/iStock/Getty Images; **p121**(t): Ton Bangkeaw/Shutterstock; **p121**(c): Phanie/RGB Ventures/SuperStock/Alamy Stock Photo; **p122**(c): Ariel Skelley/DigitalVision/Getty Images; **p122**(b): turk_stock_photographer/iStock/Getty Images; **p123**(t): Michael Dwyer/Alamy Stock Photo; **p123**(c): Universal Images Group/Getty Images; **p124**: Steve Turner/Alamy Stock Photo; **p125**(t): Eric Audras/ONOKY/Getty Images; **p125**(c): John Durham/Science Photo Library; **p128**(lt): CNRI/Science Photo Library; **p134**: indykb/iStock/Getty Images; **p137**: Ed Reschke/Stone/Getty Images; **p138**: Pseudolongino/Dreamstime; **p140**(c): Power And Syred/Science Photo Library; **p140**(b): moodboard/Alamy Stock Photo; **p141**: Cultura Creative/Alamy Stock Photo; **p147**(t): magicmine/Alamy Stock Photo; **p147**(c): Jose Luis Calvo/Shutterstock; **p148**: Jubal Harshaw/Shutterstock; **p154**(c): Biophoto Associates/Science Photo Library; **p154**(b): Biophoto Associates/Science Photo Library; **p156**: Prof S. Cinti/Science Photo Library; **p158**: Michael Freeman/Alamy Stock Photo; **p159**(tc): Szymon Mucha/Alamy Stock Photo; **p159**(bc): WhyNotChannel/Shutterstock; **p162**: Johnny Habell/Shutterstock; **p165**: Astrid & Hanns-Frieder Michler/Science Photo Library; **p166**: JohnPitcher/iStock/Getty Images; **p169**(t): Katrina Leigh/Shutterstock; **p169**(cr): JEROME WEXLER/Science photo Library; **p170**: GH Photos/Alamy Stock Photo; **p174**: Svetlana Kuznetsova/123RF; **p175**: Dr Jeremy Burgess/Science Photo Library; **p176**(c): AMI Images/Science Photo Library; **p176**(b): Frank Walker/Alamy Stock Photo; **p177**(cl): blickwinkel/Jagel/Alamy Stock Photo; **p177**(cr): Eye Of Science/Science Photo Library; **p178**: Darwin Dale/Science Photo Library; **p179**: Dr Jeremy Burgess/Science Photo Library; **p180**(c): Ed Reschke/Getty Images; **p180**(b): Susumu Nishinaga/Science Photo Library; **p183**(tc): Steve Gschmeissner/Science Photo Library; **p183**(bc): D. Phillips/Science Photo Library; **p185**: Prof. P. Motta/Dept. of Anatomy/University/Science Photo Library; **p189**(tr): Anatomical Travelogue/Science Photo Library; **p189**(utc): Anatomical Travelogue/Science Photo Library; **p189**(ltc): James Stevenson/Science Photo Library; **p190**: Monkey Business Images/Shutterstock; **p192**: DR J. P. Abeille/Science Photo Library; **p194**(c): NIAID/National Institutes Of Health/Science Photo Library; **p194**(b): Sean Sprague/Alamy Stock Photo; **p197**: Frank Fox/Science Photo Library; **p198**(t): video1/iStock/Getty Images; **p198**(c): CNRI/Science Photo Library; **p204**: Michel Delarue, ISM/Science Photo Library; **p205**(bl): M.I. Walker/Science Photo Library; **p205**(br): Dimarion/Shutterstock; **p206**: D. Phillips/Science Photo Library; **p207**: Pascal Goetgheluck/Science Photo Library; **p209**: Martin Shields/Alamy Stock Photo; **p210**: PeterAustin/iStock/Getty Images; **p212**: Revy, ISM/Science Photo Library; **p213**: Alfio Scisetti/Alamy Stock Photo; **p215**(t): Biophoto Associates/Science Photo Library; **p215**(c): David Nicholls/Science Photo Library; **p218**: Ian Shaw/Alamy Stock Photo; **p219**: sirisris/Shutterstock; **p220**: Chris Howes/Wild Places Photography/Alamy Stock Photo; **p221**: iperl/123RF; **p224**(cl): maxaltamor/123RF; **p224**(c): WaterFrame/Alamy Stock Photo; **p224**(cr): karlumbriaco/Getty Images; **p224**(b): Shutterstock; **p225**(t): Mauro Fermariello/Science Photo Library; **p225**(c): Steve Allen/Science Photo Library; **p226**(c): W K Fletcher/Science Photo Library; **p226**(b): Tony Camacho/Science Photo Library; **p228**(t): Mark Conlin/Alamy Stock Photo; **p228**(b): Bruce Anderson; **p229**(t): Esa Hiltula/Alamy Stock Photo; **p229**(c): Andrew Gardner/Alamy Stock Photo; **p230**(tc): John Moss/Science Photo Library; **p230**(bc): John Giustina/The Image Bank/Getty Images; **p230**(b): Michele Westmorland/Corbis Documentary/Getty Images; **p231**: driftlessstudio/iStock/Getty Images; **p234**: Elliot Nichol/Alamy Stock Photo; **p243**(t): Hugh Spencer/Science Photo Library; **p243**(tc): Power And Syred/Science Photo Library; **p244**: amrishwad/iStock/Getty Images; **p248**(c): frei-impuls/iStock/Getty Images; **p248**(b): Adek Berry/AFP/Getty Images; **p249**(t): Kurupin_Kp2/Shutterstock; **p249**(tc): Dennis Cox/Alamy Stock Photo; **p249**(c): Sukpaiboonwat/Shutterstock; **p250**(c): osiris59/iStock/Getty Images; **p250**(b): Weldon Schloneger/123RF; **p251**: jeanro/iStock/Getty Images; **p252**(t): John Potter/Alamy Stock Photo; **p252**(bl): Pexels/Pixabay; **p252**(br): Sasin Tipchai/Pixabay; **p253**: chiarito/iStock/Getty Images; **p254**(c): David Nunuk/Science Photo Library; **p254**(b): blickwinkel/Schuetz/Alamy Stock Photo; **p255**: Dr Jeremy Burgess/Science Photo Library; **p256**: Joe Sohm/Visions Of America/Science Photo Library; **p258**(c): brianafrica/Alamy Stock Photo; **p258**(b): georgeclerk/iStock/Getty Images; **p259**(t): Design Pics Inc/Alamy Stock Photo; **p259**(c): Peter Scoones/Science Photo Library; **p260**(t): John Carnemolla/Corbis/Getty Images; **p260**(c): Michael & Patricia Fogden/Nature Picture Libraray; **p260**(b): Ievgenii Fesenko/123RF; **p261**(t): Enjoylife2/iStock/Getty Images; **p261**(c): GoodOlga/iStock/Getty Images; **p262**(c): Jeffrey Wiles/123RF; **p262**(b): KeithSzafranski/E+/Getty Images; **p263**(t): Joe McDonald/Shutterstock; **p263**(tc): Tony Karumba/ARF/Getty Images; **p263**(c): Dante Fenolio/Science Photo Library; **p264**(t): Erick Margarita Images/Shutterstock; **p264**(c): mkos83/iStock/Getty Images; **p264**(b): Juniors Bildarchiv GmbH/F279/Alamy Stock Photo; **p265**(t): Scubazoo/Science Photo Library; **p265**(c): Jim Richardson/National Geographic Image Collection/Alamy Stock Photo; **p268**(c): margouillat/123RF; **p268**(b): SM Rafiq Photography/Moment Open/Getty Images; **p269**(t): Adrian Muttitt/Alamy Stock Photo; **p269**(c): Rosenfeld Images Ltd/Science Photo Library; **p270**: Power And Syred/Science Photo Library; **p272**: Science Photo Library/Alamy Stock Photo; **p273**: Maximilian Stock Ltd/Science Photo Library; **p274**: chas53/iStock/Getty Images; **p281**(t): Georgette Douwma/Science Photo Library; **p281**(c): AndreyTTL/iStock/Getty Images; **p283**: Luis César Tejo/Shutterstock; **p300**: Dennis Kunkel Microscopy/Science Photo Library; **p303**: Biophoto Associates/Science Photo Library; **p306**(bl): Bruce Anderson; **p306**(bc): Cameron Ewart Smith; **p306**(br): Cameron Ewart Smith.

Artwork by Q2A Media Services Pvt. Ltd, Aptara, Six Red Marbles, Greengate Publishing Services, Gareth Williams, Thomson Digital, OUP, Barking Dog, Mike Gordon, Susan Harrison, Jane Cope, IFA Design Ltd, Jordan Publishing Design, Oxford Designers & Illustrators, Harry Venning, Peters & Zabransky Ltd and Tony Wilkins.

Every effort has been made to contact copyright holders of material reproduced in this book. Any omissions will be rectified in subsequent printings if notice is given to the publisher.

This Student Book refers to the Cambridge IGCSE Biology (0610) Syllabus published by Cambridge Assessment International Education.

This work has been developed independently from and is not endorsed by or otherwise connected with Cambridge Assessment International Education.

Contents

Contents

Introduction

This book is designed specifically for Cambridge IGCSE® Biology 0610. Experienced teachers have been involved in all aspects of the book, including detailed planning to ensure that the content gives the best match possible to the syllabus.

Using this book will ensure that you are well prepared for studies beyond the IGCSE level in pure sciences, in applied sciences or in science-dependent vocational courses. The features of the book outlined below are designed to make learning as interesting and effective as possible:

LEARNING OUTCOMES

- These are at the start of each spread and will tell you what you should be able to do at the end of the spread.

S
- Some outcomes will be needed only if you are taking a supplement paper and these are clearly labelled, as is any content in the spread that goes beyond the syllabus.

EXAM TIP

Experienced teachers give you suggestions on how to avoid common errors or give useful advice on how to tackle questions.

KEY POINTS

These summarise the most important things to learn from the spread.

PRACTICAL

These show the opportunities for practical work. The results are included to help you if you do not actually tackle the experiment or are studying at home.

SUMMARY QUESTIONS

These questions are at the end of each spread and allow you to test your understanding of the work covered in the spread.

At the end of each unit there is a double page of examination-style questions written by the author.

At the end of the book you will also find:

'Alternative to practical' section – this provides guidance if you are doing this examination paper instead of coursework or the practical examination.

Assessment structure

Paper 1: Multiple Choice (Core)

Paper 2: Multiple Choice (Supplement)

Paper 3: Theory (Core)

Paper 4: Theory (Supplement)

Paper 5: Practical Test

Paper 6: Alternative to Practical

Extra resources, including **answers**, and a Revision Checklist:

www.oxfordsecondary.com/essential-igcse-science

Syllabus matching grid

Topic number in this book	Topic title	IGCSE Syllabus section
1.1	Characteristics	1.1
1.2	Classification	1.2
1.3	Features of organisms	1.3
1.4	Vertebrates	1.3
1.5	Invertebrates	1.3
1.6	Ferns and flowering plants	1.3
1.7	Dichotomous keys	1.3
2.1	Structure of cells	2.1
2.2	Cell organelles	2.1/2.2
2.3	Different types of cell	2.1/2.2
2.4	Levels of organisation	2.1
3.1	Diffusion	3.1
3.2	Osmosis	3.2
3.3	Osmosis in plant cells	3.2
3.4	Active transport	3.3
4.1	Biological molecules	4.1
4.2	Chemical tests for biological molecules	4.1
4.3	DNA	4.1
5.1	Structure and action of enzymes	5.1
5.2	Factors affecting enzyme action: temperature	5.1
5.3	Enzymes and pH	5.1
6.1	Photosynthesis	6.1
6.2	What is needed for photosynthesis?	6.1
6.3	Products of photosynthesis	6.1
6.4	Rate of photosynthesis	6.1
6.5	Limiting factors	6.1
6.6	Leaves	6.2
6.7	Mineral requirements	6.1
7.1	A balanced diet	7.1

Topic number in this book	Topic title	IGCSE Syllabus section
7.2	Sources of nutrients	7.1
7.3	Nutrient deficiency	7.1
7.4	Digestion	7.2/7.3/7.4
7.5	Teeth	7.3
7.6	Mouth, oesophagus and stomach	7.2
7.7	Small intestine and absorption	7.4/7.5
8.1	Transport systems	8.1
8.2	Water uptake	8.2
8.3	Transpiration	8.3
8.4	Measuring transpiration	8.3
8.5	Translocation	8.4
9.1	Circulation	9.1
9.2	The heart	9.2
9.3	Heart and exercise	9.2
9.4	Blood vessels	9.3
9.5	Coronary heart disease (CHD)	9.2
9.6	Blood	9.4
9.7	Blood in defence	9.4
10.1	Disease	10.1
10.2	Defence against disease	10.1
10.3	Vaccination	10.1
10.4	Aspects of immunity	10.1
10.5	Controlling the spread of disease	10.1
10.6	Drugs	15.1
11.1	The gas exchange system	11.1
11.2	Gas exchange	11.1
11.3	Breathing	11.1
11.4	Rate and depth of breathing	11.1
12.1	Aerobic respiration	12.1/12.2
12.2	Anaerobic respiration	12.1/12.3
13.1	Excretion	13.1

Topic number in this book	Topic title	IGCSE Syllabus section
13.2	Kidney structure	13.1
13.3	Kidney function	13.1
14.1	Nervous control in humans	14.1
14.2	Neurones and reflex arcs	14.1
14.3	Synapses	14.1
14.4	Sense organs	14.2
14.5	More about the eye	14.2
14.6	Hormones	14.3/14.4
14.7	Controlling conditions	14.4
14.8	Controlling body temperature	14.4
14.9	Tropic responses	14.5
14.10	Investigating tropic responses	14.5
15.1	Asexual and sexual reproduction	16.1/16.2
15.2	Flower structure	16.3
15.3	Pollination	16.3
15.4	Fertilisation and seed formation	16.3
15.5	The male reproductive system	16.4
15.6	The female reproductive system	16.4
15.7	Fertilisation and implantation	16.4
15.8	Pregnancy	16.4
15.9	Sex hormones	16.5
15.10	The menstrual cycle	16.5
15.11	Sexually transmitted infections (STIs)	16.6
16.1	Chromosomes, genes and DNA	17.1
16.2	Protein synthesis 1	17.1
16.3	Protein synthesis 2	17.1
16.4	Mitosis	17.2
16.5	Meiosis	17.3
16.6	Inheritance and genes	17.4
16.7	Monohybrid inheritance	17.4

Topic number in this book	Topic title	IGCSE Syllabus section
16.8	Codominance	17.4
16.9	Sex linkage	17.4
17.1	Variation	18.1
17.2	Investigations into variation	18.1
17.3	Mutations	18.1
17.4	Adaptive features	18.2
17.5	Natural selection	18.3
17.6	Natural selection in action	18.3
17.7	Selective breeding	18.3
18.1	Energy flow	19.1
18.2	Pyramids of numbers and biomass	19.2
18.3	Shortening the food chain	19.2
18.4	Nutrient cycles	19.3
18.5	The nitrogen cycle	19.3
18.6	Populations, communities and ecosystems	19.4
19.1	Food supply	20.1
19.2	Habitat destruction	20.2
19.3	Pollution	20.3
19.4	Water pollution	20.3
19.5	The greenhouse effect	20.3
19.6	Sustainable resources	20.4
19.7	Endangered species	20.4
19.8	Captive breeding	20.4
19.9	Conservation	20.4
20.1	Microorganisms and biotechnology	21.1/21.2
20.2	Enzymes and biotechnology	21.2
20.3	Fermenters	21.2
20.4	Genetic modification	21.1/21.3

1.1 Characteristics

Figure 1.1.1 Salmon provide nutrition for bears.

Figure 1.1.2 This plant tendril moves slowly as it grows round a support.

Biology is the study of living things or living organisms, which are classified into five major groups called **kingdoms**:

animals, plants, fungi, protoctists and prokaryotes (bacteria)

There are seven characteristics of living organisms. These characteristics are often described as life processes:

nutrition, respiration, movement, growth, excretion, sensitivity and reproduction.

Nutrition

Nutrition is the obtaining of food to provide energy and substances needed for growth, development and repair. Nutrients are compounds that may be large and complex (like carbohydrates, proteins and vitamins) or simple (like mineral ions).

Nutrition in green plants involves **photosynthesis**, in which the energy from sunlight is absorbed and used to turn carbon dioxide and water into simple sugars. Plants then convert these simple sugars into complex compounds, such as cellulose and proteins. They need mineral ions from the soil to make proteins.

Animals cannot make their own food like plants, so they have to eat plants or other animals to gain energy and nutrients. The process of taking in food is called **ingestion**. The food is digested, absorbed into the blood and then assimilated by cells for growth and repair. Food which is not digested and absorbed is egested in faeces.

Respiration

All living organisms respire because they all need energy. Respiration involves chemical reactions that occur in cells to break down nutrients, such as glucose, to release energy. Oxygen is usually needed for respiration to happen. This is the word equation for respiration involving oxygen:

glucose + oxygen $\longrightarrow$ carbon dioxide + water (+ energy released)

Living things use this energy for movement, growth, repair and reproduction.

Movement

Organisms move themselves or move parts of themselves into new areas or to change position.

Plants move slowly when they grow. Their roots move down into the soil and their leaves and stems move up towards the light. Leaves can move to face the Sun so they can absorb as much light as possible.

Most animals are able to move their whole bodies. They move to obtain their food or to avoid being caught by predators. Some animals remain fixed to one place throughout their lives, but they are able to move parts, such as the tentacles on a sea anemone.

Growth

Growth is a permanent increase in size of an organism. This involves an increase in cell number, cell size or both. It always involves making more complex chemicals, such as proteins, which is why the dry mass increases. Plants carry on growing throughout their lives. Animals stop growing when they reach a certain size.

Excretion

All living organisms produce toxic (poisonous) waste substances as a result of metabolism. **Metabolism** is all the chemical reactions that occur in an organism. Respiration is a major part of metabolism.

Excretion is the removal of these waste materials and substances in excess of requirements from the body. Plants store waste substances in their leaves, so the waste chemicals are removed when the leaves fall off. Animals breathe out carbon dioxide; other waste substances leave the body in the urine.

Sensitivity

Living organisms are able to detect or sense changes in their internal and external environments. A change like this is a **stimulus** (plural: stimuli). Sensitivity is the ability to detect these stimuli and **respond** to them.

Plants respond to movement of the Sun by moving their leaves to face the light. The flowers of some plants open in the morning and close at night. Animals have sensory cells and sense organs for detecting light, sound, touch, pressure and chemicals in the air and in food.

Reproduction

Organisms reproduce to make new individuals. **Asexual reproduction** involves one parent producing offspring that are identical to each other and to the parent. **Sexual reproduction** involves two parent organisms producing **gametes** (sex cells), which fuse to produce to the next generation. The offspring show **variation**. They are not identical to each other or to their parents.

Figure 1.1.3 Sea anemone.

KEY POINTS

The mnemonic 'Mrs Gren' can help you remember the seven characteristics of living organisms:

Movement causes an organism (or part of an organism) to change its position or place.

Respiration involves chemical reactions that release energy in cells.

Sensitivity is being able to detect and respond to internal and external stimuli.

Growth is a permanent increase in size and dry mass.

Reproduction results in the formation of new individuals.

Excretion is the removal of waste chemicals made in the cells during metabolism.

Nutrition involves the use of food for energy and growth.

SUMMARY QUESTIONS

1 A visitor from outer space lands on Earth. The first thing that it sees is a motor car passing by.
   a Give two reasons why the visitor thinks that the car is alive.
   b Give two reasons why you think that the visitor is wrong.

2 Plants and animals are two groups of living organisms. Find out and then describe how plants differ from animals in the ways in which they:
   a feed   b move   c grow   d use their senses

1.2 Classification

LEARNING OUTCOMES

- Describe the binomial system of naming organisms
- Describe the importance of classification to the study of evolution
- **S** Describe the use of biochemistry in classification.

We classify organisms into groups. The largest grouping is the **kingdom** (Topic 1.1) and the smallest is the **species**. The organisms in a kingdom share some similar features. For example, within the plant kingdom, all plants are green and carry out photosynthesis. Each kingdom is subdivided into groups known as **phyla** (singular: **phylum**). But first we must explain how organisms are named.

The binomial system

Meerkat, suricate and Sun angel are all names for a type of mongoose that lives in the Kalahari Desert in southern Africa. To avoid confusion, organisms are given scientific names using the **binomial system** (binomial means 'two names').

A **species** is a group of individuals that look alike. They live in the same habitat and breed together to give offspring that are fertile and can also breed together.

Each species is given two names: the first name is for the **genus** and the trivial name applies to one species within the genus. The trivial name should never be used on its own.

A genus is a group of species that are closely related, but do not interbreed with each other. Some genera (plural of genus) consist of only one species, as is the case with meerkats. This may be because other species in that genus are extinct.

The binomial system is used by biologists all over the world as it is an international language for naming organisms. The table has some examples.

Figure 1.2.1 A meerkat, *Suricata suricatta*.

Table 1.2.1 Examples of the binomial system for naming living organisms.

common name	scientific name
meerkat	*Suricata suricatta*
human	*Homo sapiens*
baobab tree	*Adansonia digitata*
cholera bacterium	*Vibrio cholerae*
malarial parasite	*Plasmodium falciparum*
oyster mushroom	*Pleurotus ostreatus*

Figure 1.2.2 The great white pelican, *Pelecanus onocrotalus*.

Figure 1.2.3 The spot-billed (grey) pelican, *Pelecanus philippensis*.

Supplement

Classification using biochemistry

New techniques in biochemistry have helped determine how closely one species is related to another. Large biological molecules such as nucleic acids and proteins are found in all living cells, but they are not identical in all organisms. These differences can be used to reflect their evolutionary relationships.

Classification using DNA and amino acids

DNA is a nucleic acid found in all living organisms. New techniques enable molecular scientists to extract DNA from cells. They can then use sequencing machines to find out the sequence of bases in the extracted DNA.

This makes it possible to compare the base sequences of DNA from different organisms to see how similar or different they are. The more similar the base sequence, the more closely related the species.

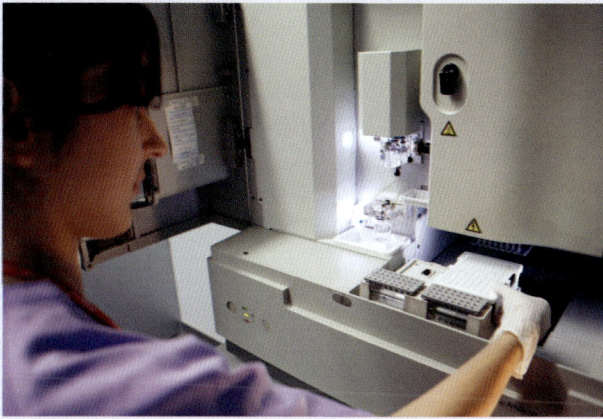

Figure 1.2.4 Loading samples into a DNA sequencer.

DNA base sequences code for the amino acid sequences of proteins. A similar technique is used to work out the amino acid sequences of proteins.

Haemoglobin is a protein that transports oxygen in animals. Human haemoglobin has 574 amino acids. Each amino acid is coded by three bases, so there are at least 1722 bases in the DNA that codes for haemoglobin.

The DNA sequences are compared with sequences for the same protein in different animals. The more similarities there are between amino acid sequences for any two animals, the more closely related they are likely to be. For example, haemoglobin in a gorilla has 572 amino acids in common with the 574 amino acids found in human haemoglobin. A horse, on the other hand, has only 557 of its amino acids in haemoglobin in common with humans. So, the evolutionary relationship between a gorilla and a human is closer than that between a human and a horse.

SUMMARY QUESTIONS

1 Copy and complete the sentences using these words:

fertile genus shared trivial classification binomial breed

The system of putting organisms into groups is called _____ . The _____ system is used for naming organisms. Each organism has two names. The first is called the _____ name and the second is the _____ name. A species is a very similar group of individuals that can _____ together and produce _____ offspring. We classify living things on the basis of their _____ features.

2 a Explain the term *binomial system*.

   b Find the scientific names of two plant species and two animal species not mentioned in this topic and write them alongside their common names.

3 In the 18th century, Carl Linnaeus worked out the binomial system of naming living things. Find out about his life and work and then write a brief biography in your own words.

1.3 Features of organisms

LEARNING OUTCOMES

- State the features of cells in living organisms
- List the features used to place organisms into one of the five kingdoms.

The cells of all living organisms are made up of a **cell membrane**, **cytoplasm** and **DNA**. They also have small structures called **ribosomes** present in the cytoplasm, which carry out protein synthesis, along with enzymes that are involved in processes such as respiration.

There are two main types of cell. Bacteria have a simple cell structure that you can see in Figure 1.3.1. Cells of bacteria do not have a nucleus and are known as **prokaryotic** cells. The cells of all other organisms have nuclei and these cells are known as **eukaryotic**.

Biologists use the features that organisms share to classify them into five large groups known as **kingdoms**. For example, all bacteria are classified together into the same kingdom because they have the same cell structure. There are five kingdoms, including the plant kingdom and the animal kingdom.

Plant kingdom Plants are multicellular organisms. The bodies of plants are not compact like those of animals. Their roots spread out through the soil to gain water and ions and their stems grow many leaves to absorb light. Growth occurs in special places such as the tips of roots and stems. This kingdom includes mosses and ferns, which do not reproduce using seeds, and the seed-bearing plants, such as conifers and the flowering plants. All plants have chlorophyll and carry out photosynthesis, using light to make their own food. Plant cells have chloroplasts full of chlorophyll, cellulose cell walls and a large sap-filled vacuole.

Animal kingdom Animals are also multicellular organisms. Most animals have more compact bodies than plants. Growth occurs throughout the body. Animals do not photosynthesise and obtain their food by eating plants and/or other animals. Animal cells do not have cell walls, chloroplasts or large vacuoles. All animals have nervous systems that coordinate their responses to stimuli and their movement.

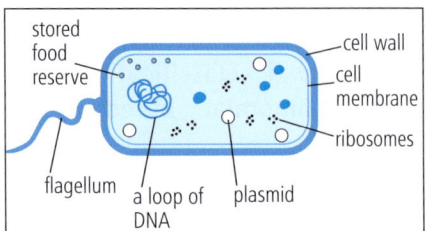

Figure 1.3.1 A typical bacterium.

stored food reserve · cell wall · cell membrane · ribosomes · flagellum · a loop of DNA · plasmid

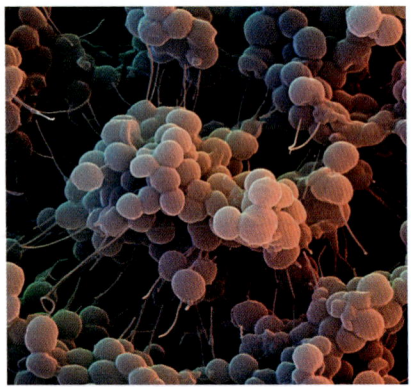

Figure 1.3.2 *Staphylococcus*, a type of bacterium found on the human skin (×4000).

Figure 1.3.3 Bread mould, *Rhizopus nigricans*, is a fungus. The black structures contain spores.

Supplement

Prokaryote kingdom Bacteria have a simple cell structure that you can see in Figure 1.3.1. Some bacteria are spherical and some are rod-shaped. Many exist in short chains of cells. Most bacteria are a few micrometres in length and can only be seen with light and electron microscopes. (One micrometre is one-thousandth of a millimetre.)

Bacterial cells are surrounded by cell walls. The cells of some bacterial species are also surrounded by slime capsules.

There is no nucleus, just a loop of DNA within the cytoplasm. Bacteria often have additional loops of DNA inside their cytoplasm called **plasmids** (see page 268). The structure of bacteria differs from animal and plant cells as they do not have chloroplasts or mitochondria. Some bacteria have extensions called flagella (singular: flagellum) for moving through water or other fluids.

Protoctists A diverse group of organisms that are classified in this kingdom as they do not belong in any of the other four kingdoms. All protoctists have cells with nuclei; many are unicellular and some are multicellular. Unicellular protoctists include organisms commonly known as protozoans, including *Plasmodium*, which is the parasitic organism that causes malaria. Algae are classified as protoctists and they vary in size from tiny unicellular organisms to the kelps, which are giant seaweeds that form underwater 'forests' in parts of the sea where there is plenty of light so they can photosynthesise.

Fungi kingdom Fungi are visible with a light microscope. Many are also visible to the naked eye. Most fungi are multicellular although yeasts are single-celled. Each cell has a nucleus and a cell wall that is made out of **chitin**, not cellulose as in plants. Fungi do not have chlorophyll and cannot carry out photosynthesis.

The main fungus body is called the **mycelium**. It consists of a branching network of threads or **hyphae** that grow over the surface of its food source, releasing **enzymes** that digest the food outside the fungus. The digested food is then absorbed by the hyphae.

Fungi reproduce by making spores that can be carried by the wind. Most fungi are **saprotrophs**, which means they feed on dead or decaying matter, but some are **parasites**.

Viruses

Viruses are not cells. They are particles made up of genetic material (DNA or RNA) surrounded by a protein coat. Viruses are **parasites** that enter the cells of another organism (the **host**) in order to multiply. Viruses take over the host cell and direct it to make new viruses.

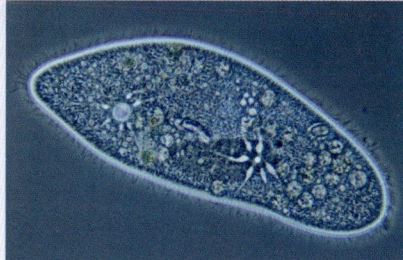

Figure 1.3.4 Paramecium, a single-celled protoctist (×206).

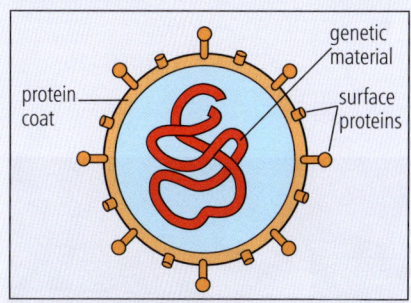

Figure 1.3.5 Structure of a human virus. Viruses are extremely small. This virus that causes influenza is about 120 nanometres in size. (One nanometre is one-thousandth of a micrometre.) Viruses are visible only using an electron microscope.

KEY POINTS

1 The cells of all living organisms have cytoplasm, a cell membrane and DNA.

S 2 There are five kingdoms: Plant, Animal, Prokaryote, Protoctist and Fungi.

3 Prokaryotes have cells without nuclei. They can be seen only using a light microscope.

4 Protoctists are a group of organisms with many features of eukaryotic cells.

5 Fungi are made up of threads called hyphae. These grow over the food supply, digest it externally and then absorb it.

6 Viruses can be seen only with an electron microscope. They are not cells but a few genes inside a protein coat.

SUMMARY QUESTIONS

1 State the features present in all cells.

2 The plant kingdom and the animal kingdom contain many species that are familiar to us. Make a table to compare plants with animals. You should include similarities as well as differences.

S 3 Compare the structure of bacterial cells and plant cells.

4 Make a table to compare the features of the five kingdoms.

5 Why are viruses not classified in one of the five kingdoms?

6 How many nanometres are there in a metre?

1.4 Vertebrates

- Define the term *vertebrate*
- Describe the five main groups of vertebrates using visible, external characteristic features
- Describe how these vertebrates are adapted to their environment

Figure 1.4.1 This dace has fins, a tail and a streamlined shape – adaptations for swimming.

Figure 1.4.2 The bright red eyes of this tree frog are thought to startle predators, giving time for the frog to escape.

Figure 1.4.3 American crocodile, *Crocodylus acutus*.

Animals that have a **vertebral column** or backbone are called **vertebrates**. All vertebrates have an internal skeleton made of either bone or cartilage. They all belong to the phylum *chordata,* which includes some invertebrates that share common features with the vertebrates.

There are five main groups of vertebrates:

fish, amphibians, reptiles, birds and mammals.

Fish

Most fish live in water permanently, but there are some species, such as the mud skipper, that can survive out of water for varying lengths of time.

Typically, fish are streamlined and have fins for swimming and for balance. They have eyes and a lateral line for detecting pressure changes in water. They breathe dissolved oxygen from the water using their gills. Their skin is covered with scales. Examples include tuna, herring, shark, catfish and cod.

Unlike fish, the other classes of vertebrates have ears for detecting sound and four limbs, although some have evolved into legless forms like snakes and some lizards.

Amphibians

These vertebrates have smooth, moist skin. Although most amphibians live on land, they return to water to breed. Fertilisation is external because sperm and eggs are released into the water (not inside a female body). Development is external: the fertilised eggs hatch into swimming tadpoles, which have gills for breathing.

On land, the adult amphibians breathe using lungs. However, when they are in water, they can breathe through their skin. Examples of amphibians include frogs, toads and salamanders.

Reptiles

Reptiles have dry, scaly skin to cut down water loss. They can live in dry regions as they do not have to return to water to breed.

Fertilisation takes place inside the female's body. However, development is external as they lay eggs with leathery, waterproof shells that stop them from drying out. Reptiles have lungs to breathe air. Crocodiles, lizards, snakes, turtles and tortoises are all reptiles.

Birds

Birds have feathers and their front limbs are modified as wings. Most of them are able to fly but some, like penguins and ostriches, cannot. Birds have no teeth but different species have beaks adapted to deal with different types of food.

Fertilisation is internal and development is external as the females lay eggs that are protected by hard shells.

Birds are **homeothermic** (warm-blooded). This means that they are able to regulate their body temperature. They can keep it constant even though the outside temperature changes. Examples of birds include hawks, eagles, sparrows, parrots and starlings.

Mammals

Mammals are the vertebrates that have hair or fur. Fertilisation is internal and so is development. The young develop in a womb and they are born already well developed. Female mammals suckle their young on milk from mammary glands. All mammals, even aquatic ones like whales and dolphins, use lungs for breathing.

Like birds, mammals are also homeothermic, maintaining a constant internal temperature. Leopards, bats, dolphins, bears, lemurs and wolves are all mammals.

Figure 1.4.4 Grey heron, *Ardea cinerea* – a bird with a long beak for catching fish.

EXAM TIP

Make a table to help you learn the features of these five groups. Making the table will help you learn this topic.

Figure 1.4.5 A chimpanzee has hands and feet adapted to living in trees.

SUMMARY QUESTIONS

1 Copy and complete the sentences using these words:

development   five   wings   scaly   moist   land   water   classes   breed   fly   fish   fur   feathers

There are _____ different groups of vertebrate. Sharks belong to the group called _____ and spend all their time in _____. Amphibians have _____ skin and return to water to _____. Reptiles have _____ skin and lay their eggs on _____. Birds have front limbs that are modified to form _____. They also have _____ and most are able to _____. Mammals have _____ or hair and give birth to their young in an advanced state of _____.

2 a State the features that are exhibited by all vertebrates.

  b Name the five groups of vertebrates.

  c Give an example of each group.

3 To which group of vertebrates does each of the following belong?

  a turtle   b bat   c whale   d salamander

  e snake   f hawk   g shark

KEY POINTS

1 Vertebrates are animals that have a vertebral column or backbone.

2 The vertebrates are divided into five groups: fish, amphibians, reptiles, birds and mammals.

1.5 Invertebrates

LEARNING OUTCOMES

- Describe the external features of arthropods
- Describe the features of insects, crustaceans, arachnids and myriapods that are used to classify them into groups
- Describe how the animals in these groups are adapted to their environment

Invertebrates are animals that do not have a vertebral column or backbone.

Arthropods form the largest of the groups in the animal kingdom containing the largest number of species. Each arthropod species has a segmented body, an external skeleton (**exoskeleton**) and jointed legs.

The hard exoskeleton allows arthropods to live on dry land. When they grow too big for their exoskeleton, they moult and grow a new one. Some moult all through their lives, others moult only during the early stages of their life.

Four of the different types of arthropod are:

crustaceans, myriapods, insects and arachnids

Crustaceans

Crustaceans have a body divided into a **cephalothorax** (head-thorax) and abdomen. Many have a chalky exoskeleton that provides a very hard and effective protection against predators. Crustaceans have two pairs of **antennae** and **compound eyes** that are made of many tiny individual components. They have between five and twenty pairs of legs. They breathe using **gills**.

Nearly all crustaceans live in water. Some crabs live on land but return to water to breed. Woodlice (also known as slaters) and some land crabs are exceptions as they do not use water for breeding. Examples of crustaceans include crabs, shrimps, crayfish and lobsters.

Figure 1.5.1 Woodlouse or slater. Notice its antennae for detecting stimuli.

Myriapods

These are the **centipedes** and the **millipedes**. They have long bodies made up of many segments.

Their bodies are not divided into separate regions such as the thorax and abdomen. Centipedes have one pair of legs on each segment so that the total number of legs depends upon how many segments there are. Centipedes are fast-moving carnivores. They have powerful jaws and can paralyse their prey.

Millipedes have two pairs of legs on each body segment. They are slow-moving herbivores. You can often find them feeding in leaf litter.

Figure 1.5.2 Sally Lightfoot crab, *Grapsus grapsus.* Notice its large claws for feeding and defence.

Figure 1.5.3 Millipede – with its many legs for moving efficiently.

Figure 1.5.4 Centipede – a fast-moving carnivore.

Insects

Insects have bodies that are divided into three parts: head, thorax and abdomen.

On the thorax there are three pairs of legs and many species have two pairs of wings. They have one pair of antennae on the head and compound eyes. They breathe through holes in the sides of the thorax and abdomen called **spiracles**.

Insects have colonised most habitats in the world, although there are very few species that live in the sea. Two reasons why they are so successful on land is that they are covered by a waterproof **cuticle** that stops them losing too much water, and they can fly!

This is the largest group within the arthropods. Examples include beetles, flies, locusts, cockroaches, dragonflies, butterflies, moths, bees and wasps.

Figure 1.5.5 Monarch butterfly, *Danaus plexippus*. Notice its wings for its migration over long distances.

Arachnids

These arthropods have bodies divided into two parts, the cephalothorax and abdomen.

Arachnids have four pairs of legs and no wings. They have no antennae but do have several pairs of simple, not compound, eyes. They paralyse their prey with poison fangs. Some spiders are able to weave silken webs with their spinnerets. Scorpions, ticks and mites are also classified as arachnids.

Figure 1.5.6 Tarantula – a large spider with venom to paralyse its prey.

SUMMARY QUESTIONS

1 a Name four groups of arthropod.
  b List the key features of insects.
  c State three ways in which arachnids look different from insects.

2 Distinguish between the following on the basis of visible features:
  a centipedes and millipedes
  b crustaceans and arachnids

3 Describe the features of insects that adapt them for living successfully on land.

4 Copy and complete the table:

feature	myriapods	crustaceans	insects	arachnids
number of pairs of legs				
body regions				
number of pairs of antennae				
type of eyes				
wings				

KEY POINTS

1 Arthropods are segmented animals with jointed legs and an exoskeleton.

2 Arthropods are classified into different groups including:

crustaceans, myriapods, insects and arachnids

1.6 Ferns and flowering plants

Supplement

Like all plants, ferns and flowering plants are multicellular. They are green in colour because many of their cells contain **chloroplasts**. These chloroplasts contain the green pigment **chlorophyll**, which absorbs light for photosynthesis. Each cell is surrounded by a cell wall made of cellulose.

Both ferns and flowering plants have transport systems consisting of tiny tubes. These are called **xylem vessels**, which carry water and mineral ions, and **phloem tubes**, which transport dissolved substances such as sugars. Both ferns and flowering plants are well adapted to live on land.

Ferns

Ferns are a group of plants that have become well adapted for life on land. Ferns have strong stems, roots and leaves. Their leaves have a waxy layer (the cuticle) that helps to reduce water loss. Most of them have leaves resembling those of the species in Figure 1.6.1. Ferns live in many different habitats including some that are quite dry. The largest are the tree ferns that have a thick stem that supports a crown of leaves at heights of several metres. Many ferns grow from a thick underground stem called a rhizome.

Unlike conifers and flowering plants, ferns do not produce seeds. If you look under the leaves of ferns at certain times of the year you will see structures that make and release microscopic spores. These spores are carried by the wind to form new plants.

Figure 1.6.1 The fern *Dryopteris filix-mas*.

Features of flowering plants

Flowering plants have true stems, roots and leaves. They reproduce by means of flowers, which make seeds. The seeds are produced inside the ovary within the flower.

The **shoot** is the part of the plant above ground. The shoot is made up of a **stem** bearing leaves, buds and flowers. The **apical bud** is the part where the stem grows new leaves. The stem supports the

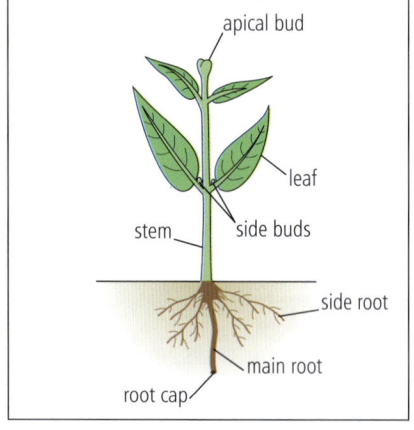

Figure 1.6.2 Structure of a typical dicotyledonous plant.

Figure 1.6.3 *Magnolia* is a dicotyledon.

Figure 1.6.4 Maize, *Zea mays*, is a monocotyledon.

structures of the shoot, and spaces out the leaves so that they can receive adequate light and air (Topic 6.6) It holds the flowers in a position that enables **pollination** to take place.

The stem allows transport of water from the soil up to the leaves and food from the leaves to other parts of the plant, such as the roots.

The **roots** are the parts of flowering plants found below ground. Roots are usually white since they do not contain chlorophyll. Roots anchor the plant firmly in the ground and prevent it from being blown over by the wind. Roots also absorb water and mineral ions from the soil (Topic 8.2).

Dicotyledons and monocotyledons

Flowering plants can be divided into two main groups:

dicotyledons and **monocotyledons**

Dicotyledons look like the plant in Figure 1.6.2.

Their leaves are often broad with a network of branching veins. The parts of the flower, for example the male parts known as stamens, are in multiples of four or five in each flower. Dicotyledons have *two* **cotyledons** (seed leaves) in a seed.

Grasses and cereals, like the maize plant at the bottom of page 12 are monocotyledons.

The leaves of most monocotyledons have parallel veins. Grasses and cereals have long, narrow leaves. Other monocotyledons have leaves with a variety of shapes, such as those of palm trees. The parts of the flower are in multiples of three. Monocotyledons have *one* cotyledon inside each seed.

clusters of flowers

leaf

Figure 1.6.5 Structure of a typical monocotyledon plant.

EXAM TIP

Be aware of the differences between flowering plants and animals. Note and learn the differences in cell structure (Topic 2.1) and reproduction (from Unit 16).

KEY POINTS

1 Flowering plants are multicellular – each cell is surrounded by a cellulose cell wall; those in leaves and some stems contain chloroplasts.

2 Monocotyledons have one cotyledon in their seeds and leaves with parallel veins.

3 Dicotyledons have two cotyledons and broad leaves with a network of branching veins.

4 Ferns are plants well adapted to life on land. They have xylem and reproduce by means of spores.

SUMMARY QUESTIONS

1 Copy and complete the sentences using these words:

> parallel   two   narrow   flowering
> broad   network   one

Monocotyledons are _____ plants that often have _____ leaves with _____ veins. They have _____ cotyledon inside the seed. Dicotyledons often have _____ leaves with a _____ of veins. They have _____ cotyledons inside each seed.

2 Find the names of five dicotyledonous plants and five monocotyledonous plants that grow where you live. Construct a dichotomous key (Topic 1.7) for the ten plants you have identified using their visible features.

3 Make a table of differences between ferns and flowering plants.

1.7 Dichotomous keys

Sorting things out

To identify the name of a plant or animal you could look through the pictures in a book until you found the right one. However, that would take a lot of time and effort.

Scientists use **dichotomous keys** to identify living things. Dichotomous means dividing into two.

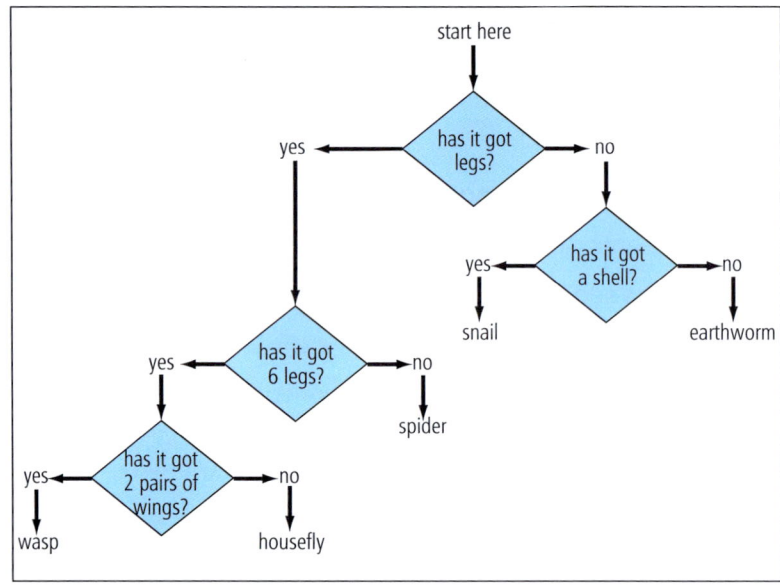

Figure 1.7.1 A branching dichotomous key.

A key has a number of steps – you can see them in the branching key and the numbered key on this page. At each step in the branching key you find a question or statement. Start at the beginning and answer 'yes' or 'no' to the first question or statement. This takes you to another question or to an identification. Use the branching key shown in Figure 1.7.1 to identify the animals in Figure 1.7.2

Use the numbered key below to identify the same animals. It is set out differently from the first key, but it works in the same way. Start at the beginning and answer the questions at each stage.

1	Has legs	Go to 2
	Has no legs	Go to 4
2	Has 6 legs	Go to 3
	Has 8 legs	Spider
3	Has 1 pair of wings	Housefly
	Has 2 pairs of wings	Wasp
4	Has a shell	Snail
	Has no shell	Earthworm

Figure 1.7.2 Five invertebrates.

Making a leaf key

Now try making a key of your own.

1 Take six different leaves and label them A to F.

2 Put the six leaves out in front of you.

3 Think of a question that will divide them into two groups.

4 Write down the question.

5 Now think up questions to divide each group into two.

6 Write these down.

7 Carry on until you come to the last pair of leaves.

8 Write out your key as a branching key or as a numbered key.

Pond animals key

Now try making a key of these pond animals below.

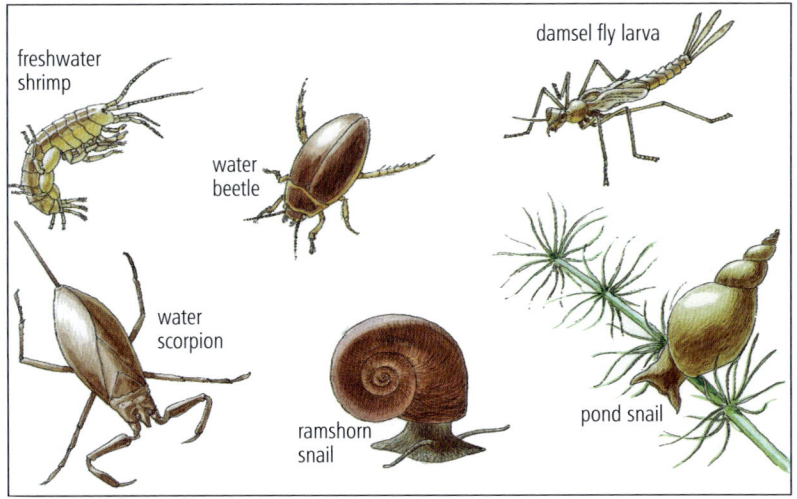

Figure 1.7.3 Pond animals.

1 When a scientist visited an island she discovered some insects.

She made some drawings and brought them back to the laboratory. Her drawings are shown on the right.

  a Give each insect a suitable common name and a suitable scientific name.

  b Make a dichotomous key to identify them. Present your key either as a branching key or as a numbered key.

2 Write down the features that would enable you to classify a fish, an amphibian, a reptile, a bird and a mammal. (You can use the photos on pages 8 and 9 to help you.) Now make a key that you could use to identify each of these vertebrate groups.

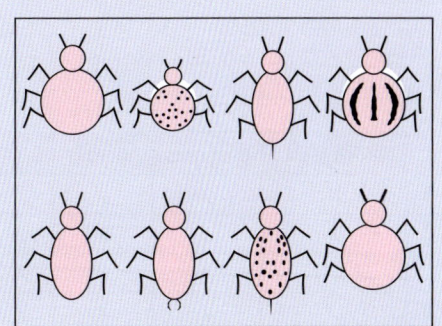

Practice questions

1 Which process is carried out by all the organisms that live in a forest?

  A moulting

  B photosynthesis

  C pollination

  D respiration

  (Paper 1)   [1]

2 What are characteristics of all organisms?

  A excretion and respiration

  B ingestion and growth

  C photosynthesis and egestion

  D respiration and photosynthesis

  (Paper 1)   [1]

3 Which group of animals includes those with a segmented body, an exoskeleton and jointed limbs?

  A amphibians

  B arthropods

  C reptiles

  D vertebrates

  (Paper 1)   [1]

4 Which group of vertebrates includes those with dry scaly skin and four legs?

  A amphibians

  B fish

  C mammals

  D reptiles

  (Paper 1)   [1]

5 Each cell is surrounded by a cell membrane and contains cytoplasm.

  The table shows five other features of cells. Which row shows the features shown by cells from organisms in all five kingdoms?

	chloroplasts	enzymes	nucleus	DNA	ribosomes
A	✗	✓	✓	✓	✗
B	✗	✓	✗	✓	✓
C	✓	✗	✓	✗	✓
D	✓	✓	✗	✓	✗

  (Paper 2)   [1]

6 Which is not associated with excretion from a mammal?

  A exhaling carbon dioxide from the lungs

  B release of heat from the skin

  C removal of toxic waste products of metabolism

  D removal of water that the body does not need

  (Paper 2)   [1]

7 The evolutionary relationships between five species are shown in the diagram.

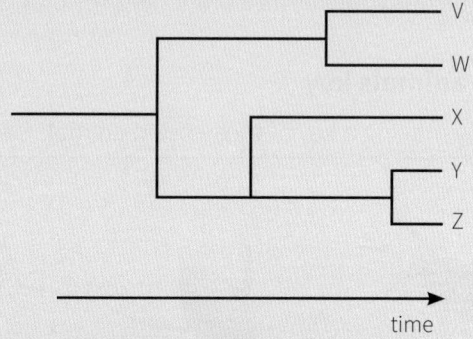

time

  An analysis of base sequences from the DNA of these five species was carried out. Which two species will have the most similar sequences?

  A V and W

  B W and X

  C X and Z

  D Y and Z

  (Paper 2)   [1]

8 Which is an example of a feature used in classifying vertebrates?

  A egg laying

  B type of body covering

  C type of diet

  D pattern of bones in the front limb

  (Paper 2)   [1]

9 Which are features of viruses?

  1 protein coat

  2 cell wall

  3 genetic material   [1]

A 1, 2 and 3

B 1 and 2 only

C 1 and 3 only

D 2 and 3 only

(Paper 2)

10 The drawings show five arthropods (not drawn to the same scale).

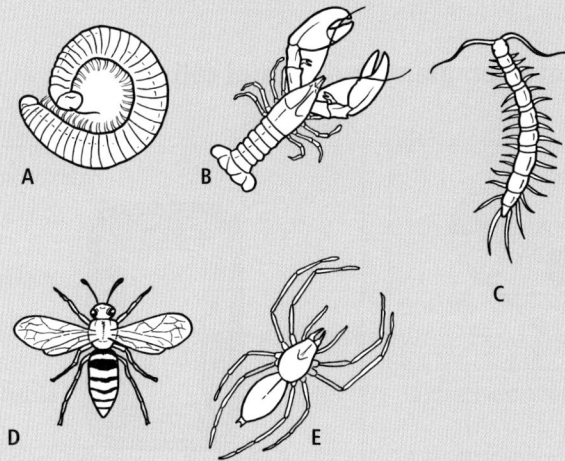

A   B   C

D   E

(a) Make a table to compare the features of the five arthropods that you see in the drawings. [6]

(b) Use the features that you have used in your table to make a key to the five arthropods. (The key may be either a branching key or a numbered key.) [6]

(Paper 3)

11 Copy and complete the table using ticks and crosses to indicate whether the four groups show the features or not.

feature	fish	amphibian	bird	mammal
backbone				
fins				
wings				
fur				
homeothermic				

(Paper 3) [5]

12 **(a)** Five of the features of living things and their definitions are listed.

Match each of the features with its definition by pairing a letter and a number.

Features

A movement   **B** sensitivity

C respiration   **D** nutrition   **E** growth

Definitions

1 A permanent increase in size

2 The chemical reactions that occur in cells to break down nutrient molecules with the release of energy

3 An action by an organism to cause a change of position or place

4 The ability to detect and respond to changes in the environment

5 The taking in of materials from the environment for energy, growth and development [3]

(b) Explain what is meant by the term species. [2]

(Paper 3)

13 The drawings show leaves from three species of flowering plant. They are not drawn to scale.

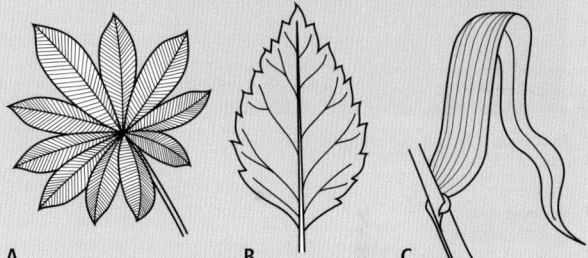

A   B   C

(a) (i) Identify the group of flowering plants to which each species belongs. [2]

(ii) State the features that you used to make your identifications. [5]

(b) List five features shown by the leaves that would be useful in devising a dichotomous key to identify species of flowering plants. [5]

(c) (i) State three structural features that ferns have in common with flowering plants. [3]

(ii) Explain why fungi, which have cell walls, are not classified in the same kingdom with plants. [3]

(Paper 4)

2.1 Structure of cells

LEARNING OUTCOMES

- Describe the structures of plant and animal cells as seen using a light microscope
- Describe the differences in structure between plant and animal cells
- State the functions of the structures seen using the light microscope in the plant cell and in the animal cell

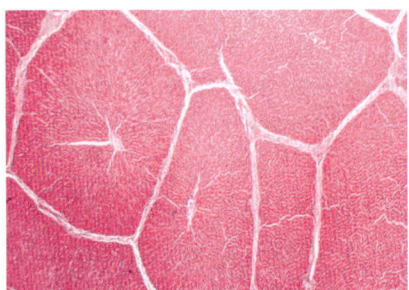

Figure 2.1.3 Liver cells magnified by ×40.

EXAM TIP

When you have to state differences between plant and animal cells, make sure you write something about both types of cell.

Cells

Cells are the small building blocks that make up all living organisms. Very small living things such as bacteria are made of only one cell.

An insect such as a fly may contain millions of cells. No one knows for certain how many cells there are in a human being – estimates vary between 10×10^{12} and 50×10^{12}.

Differences between plant and animal cells

Animal cells

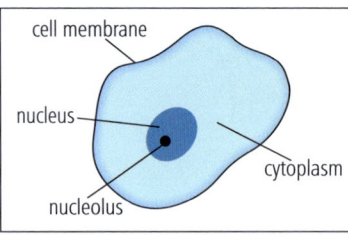

Figure 2.1.1 A cheek cell (stained with a blue dye).

Plant cells

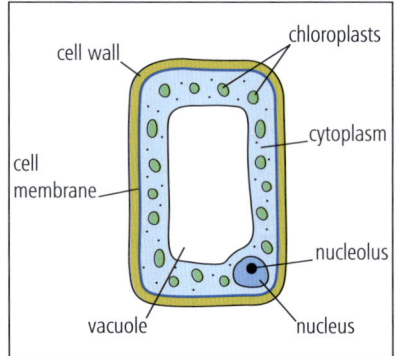

Figure 2.1.2 A palisade mesophyll cell from a leaf.

The differences between plant and animal cells are summarised in this table.

Table 2.1.1 A summary of differences between plant cells and animal cells. All the structures in Table 2.1.1 except mitochondria and ribosomes are visible using a light microscope. Mitochondria and ribosomes are visible using an electron microscope (see page 20).

feature	plant cell	animal cell
cellulose cell wall	present	absent
cell membrane	present; surrounded by cell wall	present
shape	permanent shape determined by the cell wall; shapes can be nearly spherical, box-like or cylindrical	shapes vary as there is no cell wall
chloroplasts	present in some cells	absent
vacuole	large permanent vacuole in cytoplasm containing cell sap	small vacuoles in cytoplasm; do not contain cell sap
nucleus	present (often at the side of the cell close to the cell wall)	present (found anywhere within the cell)
cytoplasm	present	present
mitochondria	present	present
ribosomes	present	present

Functions of cell structures

Table 2.1.2 The functions of cell structures.

cell structure	functions
cell membrane	• forms a barrier between the cell and its surroundings • keeps contents of cell inside • allows simple substances to enter and leave the cell, e.g., oxygen, carbon dioxide and water • controls movement of other substances into and out of the cell, e.g., glucose • often described as partially permeable (see page 28)
nucleus	• controls all activities in the cell • controls how cells develop
cytoplasm	• place where many chemical reactions take place, e.g., respiration and making proteins for the cell
chloroplast (plant cells only)	• photosynthesis • stores starch
cell wall (plant cells only)	• stops cells from bursting when they fill with water • gives shape to cells • allows water and dissolved substances to pass through freely (often described as freely or fully permeable)
sap vacuole (plant cells only)	• full of water to maintain shape and 'firmness' of cell • stores ions and sugars
mitochondria	• release energy during aerobic respiration
ribosomes	• make proteins for the cell

You can see the structure of a bacterial cell on page 6.

You can see the structure of a bacterial cell on page 6.

Figure 2.1.4 Cells of onion epidermis magnified by ×250.

SUMMARY QUESTIONS

1 Copy and complete this table by using ticks and crosses to indicate if the structures are present or not.

cell structures	cheek cell (animal)	bacterial cell	leaf cell (plant)
nucleus			
cell wall			
chloroplasts			
large vacuole			
cytoplasm			

2 State the functions of the following parts of a cell:
   a chloroplast **b** cell membrane **c** cell wall **d** nucleus
   e cytoplasm **f** vacuoles

KEY POINTS

1 A cell membrane, cytoplasm and a nucleus are found in both plant and animal cells.

2 All plant cells have a cellulose cell wall; some have chloroplasts and a vacuole containing cell sap. Animal cells do not have these parts.

3 Refer to Table 2.1.2 for the functions of the cell structures.

2.2 Cell organelles

LEARNING OUTCOMES

- State that the cytoplasm of all cells contains vesicles and ribosomes

- Identify mitochondria in diagrams and photographs of cells

- State that aerobic respiration occurs in mitochondria

- **S** Calculate the magnification and actual size of biological specimens using micrometres as units

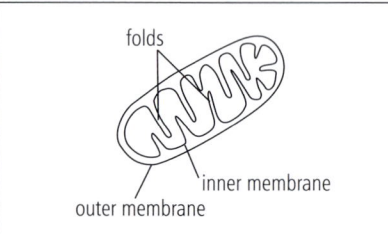

Figure 2.2.1 Structure of a mitochondrion.

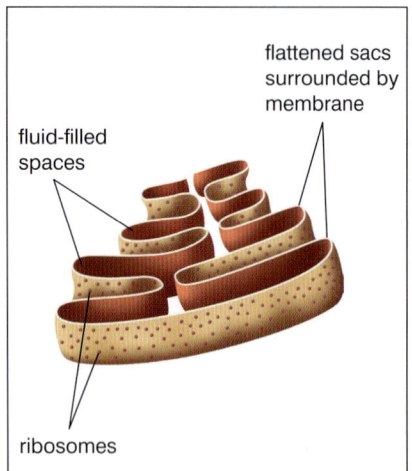

Figure 2.2.2 Most of the ribosomes in eukaryotic cells are found on the surface of rough endoplasmic reticulum.

As you know, plant and animal cells contain a number of small structures present in the cytoplasm of the cell. These are called **cell organelles** and include **mitochondria**, **ribosomes** and **vesicles**.

Mitochondria

Mitochondria are relatively large organelles found in all eukaryotic cells. They are often visible using the electron microscope as structures about 1 μm wide and 5 μm long.

Each mitochondrion has a double membrane, the outer one of which controls the entry and exit of materials. The inner membrane forms many folds on which some of the chemical reactions of aerobic respiration take place.

Mitochondria provide most of the energy that cells require. This is because aerobic respiration occurs in mitochondria.

Cells with high rates of respiration have many mitochondria to provide sufficient energy. For example, human liver cells and insect flight muscle cells contain vast numbers of mitochondria.

Ribosomes

Ribosomes are small organelles often found in large numbers in all cells. They are about 20 nanometres (nm) in diameter in eukaryotic cells but smaller in prokaryotic cells.

The function of ribosomes is to synthesise proteins, such as the enzymes involved in respiration.

The **endoplasmic reticulum (ER)** is a complex system of flattened sacs surrounded by membranes and filled with fluid. Most of the ribosomes in plant and animal cells, as well as the cells of protoctists and fungi, are found on the surface of ER.

Where ribosomes are present on their outer surface, the membranes are called **rough endoplasmic reticulum**. The main function of rough ER is to package and transport proteins made by the ribosomes.

Cells that produce a lot of protein, for instance those making digestive enzymes in the alimentary canal, have huge numbers of ribosomes attached to a very extensive system of rough ER (RER). RER forms a network in the cytoplasm allowing molecules to be transported within the cell.

Vesicles

Small pieces of RER may be pinched off at the ends to form small vesicles, which can travel through the cytoplasm, attach to cell membrane and release their contents outside the cell. Vesicles have a similar function at synapses at the ends of neurones (see page 156).

Size of cells and specimens

Look at the photograph of some human cheek cells. The cells are 1000 times larger than in real life. This means that they have been *magnified* 1000 times.

To work out the *actual size* of a cell, measure the length of one magnified cell in millimetres. Let's say that it measures 13 mm. Now we can use this formula to work out actual size:

$$\text{actual size} = \frac{\text{image size}}{\text{magnification}}$$

So the actual size of our cell $= \frac{13\,\text{mm}}{1000} = 0.013\,\text{mm}$

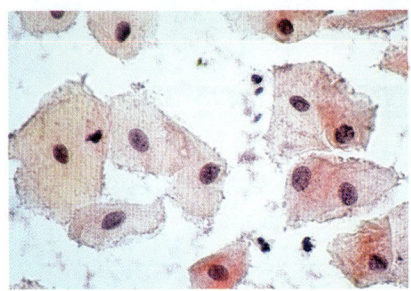

Figure 2.2.3 Human cheek cells (×1000).

Supplement

There are 1000 micrometres (μm) in a millimetre. We can convert the answer to micrometres by multiplying by 1000 to give 13 μm.

We can use a similar technique when the image size has been reduced, as in this photograph of a goldfish. This time there has been a *reduction* of actual size in the photograph. If you calculate the fish's actual size using the same formula you will find it is 90 mm long.

Figure 2.2.4 The image of this goldfish is one third (×0.33) actual size.

To calculate the magnification of an image we reorganise the formula:

$$\text{magnification} = \frac{\text{image size}}{\text{actual size}}$$

SUMMARY QUESTIONS

1 Which organelles carry out the following functions?
   a manufacture proteins
   b convert light energy into chemical energy
   c control the substances that pass into and out of a cell
   d carry out reactions in aerobic respiration

2 Copy and complete the table by using ticks and crosses.

feature	bacterial cell	liver cell	palisade cell
cell wall			
chloroplast			
mitochondrion			
vesicle			

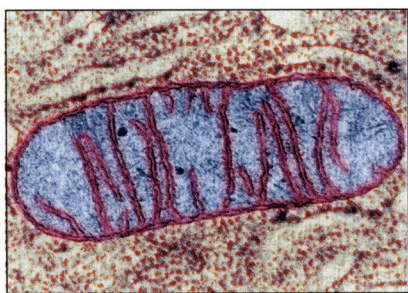

Figure 2.2.5 An electron micrograph of a mitochondrion (×80 000).

S 3 a Measure the length of the mitochondrion in the electron micrograph. It has been magnified 80 000 times (×80 000).

Calculate the actual length of the mitochondrion in micrometres (μm).

b Calculate the actual size of the liver cells and the onion cells on page 19.

KEY POINTS

1 Ribosomes in the cytoplasm make proteins.

2 Aerobic respiration occurs in mitochondria.

3 Cells with high rates of respiration have many mitochondria.

4 Multiply measurements in millimetres by 1000 to convert to micrometres.

2.3 Different types of cell

LEARNING OUTCOMES

- Identify different types of cell from diagrams and photographs
- Relate the features of these cells to their functions
- Calculate magnification and actual size of biological specimens using millimetres as units

To function efficiently, many-celled organisms have cells that are specialised to carry out certain functions. This means that the functions of the body are divided between different groups of cells. As an organism develops from a fertilised egg, new cells are produced. These new cells are formed by the division of existing cells. These cells grow and change to become specialised for certain functions. In this topic we look at some examples of specialised cells.

Specialised cells

Ciliated cells are found in the air passages in the lungs (trachea and bronchi) and in the oviducts in the female reproductive system. These cells have **cilia** on their surfaces. Cilia beat back and forth to create a current in the fluid next to the cell surfaces.

In the airways, cilia move the mucus that traps dust and pathogens up to the nose and throat. In the oviducts, cilia move the egg from the ovary to the uterus.

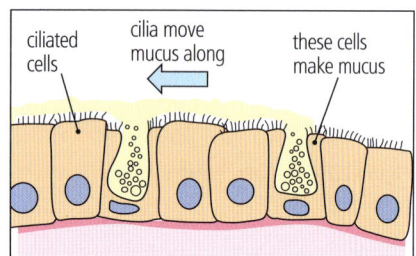

Figure 2.3.1 Ciliated cells line the airways

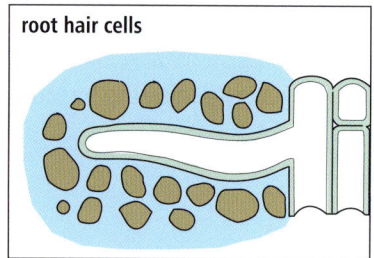

Figure 2.3.2 Root hair cells are long and thin to absorb water from the soil.

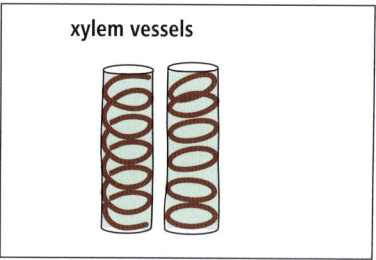

Figure 2.3.3 Xylem vessels are small tubes that carry water up the stem.

Root hair cells in a plant have long extensions that give them a large surface area to absorb water and ions from the soil.

Xylem vessels are cylindrical and hollow. They are arranged into columns like pipes. The cell walls are thickened with bands or spirals of cellulose and a waterproof material called **lignin**. These cells allow water and ions to move from the roots to the rest of the plant. They also help to support the stem and leaves.

Neurones (nerve cells) are highly specialised cells. They have thin extensions of the cytoplasm like wires. They are able to transmit information in the form of electrical impulses around the body.

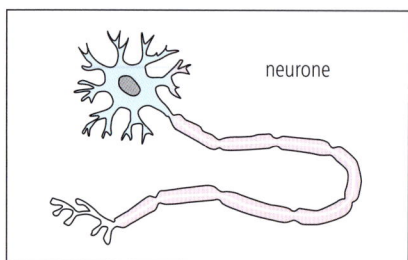

Figure 2.3.4 Neurones carry impulses around the body.

Red blood cells contain the protein **haemoglobin** that carries oxygen. They are shaped like flattened discs. This shape provides a large surface area compared with their volume, which makes for efficient absorption of oxygen.

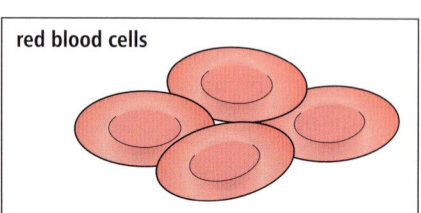

Figure 2.3.5 Red blood cells have a substance called haemoglobin that carries oxygen.

Palisade mesophyll cells are plant cells with numerous **chloroplasts** in the cytoplasm. The chloroplasts trap light energy for **photosynthesis**. Palisade cells have a cell wall made up of tough **cellulose**, which strengthens the cell. In the cytoplasm there is a large **vacuole**, which is filled with cell sap. Starch grains are found in the cytoplasm. These are formed by photosynthesis and are a temporary store of energy.

Sperm cells have a tail and are adapted for swimming. The head of the sperm carries genetic information from the male parent to the female parent. Genes from the father are present in the sperm nucleus.

Egg cells are much bigger than sperm cells. They contain yolk as a store of energy. Genes of the mother are found inside the nucleus of the egg.

Look carefully at Figure 2.3.6. The cells are 500 times larger than in real life. This means that the cells have been magnified 500 times. To work out the actual size of these cells, measure the length of one of the cells in millimetres. Let's say it measures 20 mm. Now we can use a formula to work out the actual size of the cell:

$$\text{actual size} = \frac{\text{image size}}{\text{magnification}}$$

So the actual size of our cell is 20 mm divided by 500 = 0.04 mm

To calculate the magnification of an object we reorganise the formula:

$$\text{magnification} = \frac{\text{image size}}{\text{actual size}}$$

The actual diameter of an egg cell is 0.1 mm. To calculate its magnification in Figure 2.3.8 we divided the diameter of the drawing by 0.1 mm to give ×180.

SUMMARY QUESTIONS

1 Draw a table matching the type of cell in the first column with its correct shape and function:

type of cell	shape	function
red blood cells	hollow tube	keep the air passages free from dust
ciliated cells	like wires	transport water and ions
root hair cells	flat discs	transmit electrical impulses
nerve cells	long and thin	transport oxygen
xylem vessels	have cilia that move	absorb water from the soil

2 Here are some other specialised cells. Use this book to find out the functions of these cells and how they are adapted to carry out their functions.
   a   muscle cell   b   goblet cell   c   white blood cell
   d   palisade mesophyll cell

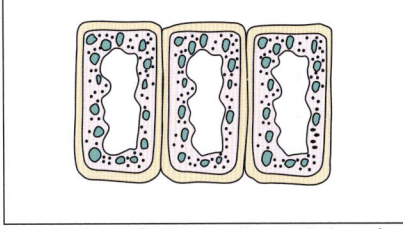

Figure 2.3.6 Leaf palisade cells contain lots of chloroplasts for photosynthesis (×500).

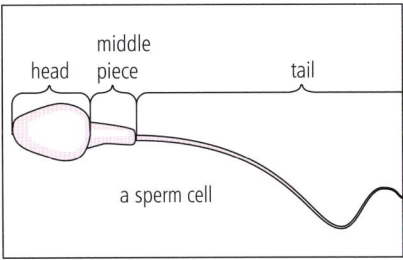

Figure 2.3.7 Sperm cells have a tail for swimming (×2000).

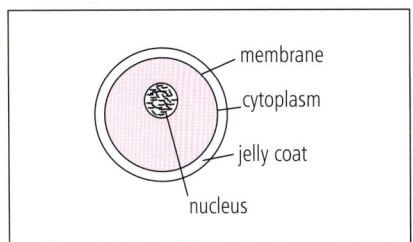

Figure 2.3.8 Egg cells contain an energy store (×180).

2.4 Levels of organisation

LEARNING OUTCOMES

- Define the terms *tissue,*
  organ and *organ system*

- Describe examples of the
  above that occur in plants
  and animals

Tissues and organs

A group of similar cells is called a **tissue**. All the cells in a tissue look the same and they work together to carry out a shared function. Muscle tissue is made up of identical **muscle** cells. These cells work together and so the muscle tissue contracts.

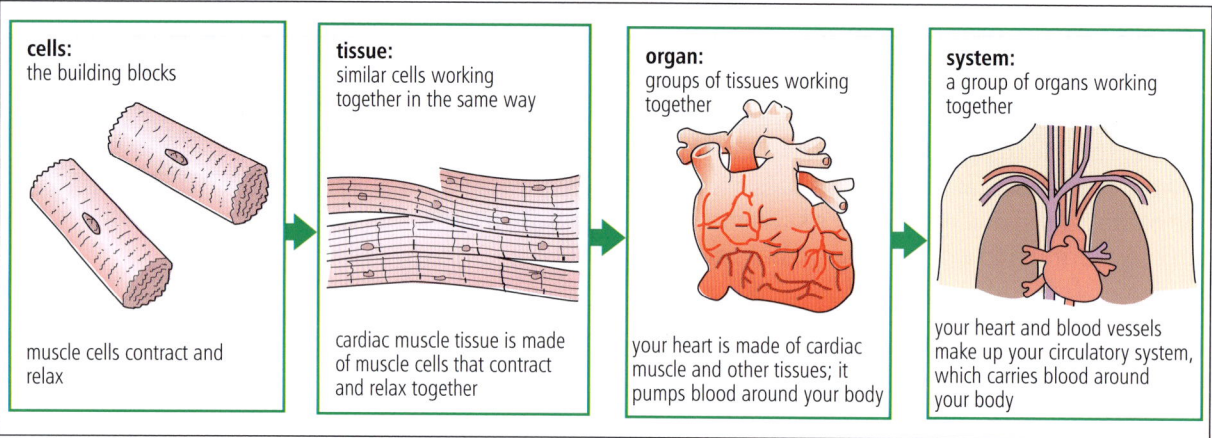

Figure 2.4.1 Levels of organisation in the human circulatory system.

An **organ** is made up of a group of different tissues that work together to perform specific functions.

The heart is an example of an organ. It is made up of different tissues such as cardiac muscle, nervous tissue, fibrous tissue and blood that work together to pump blood around the body.

The stomach, lungs, brain and kidneys are all organs.

Different organs work together as part of an **organ system**. Organ systems consist of a group of organs with related functions, working together to perform body functions. For example, the heart and blood vessels work together as part of the circulatory system. Here are some other organ systems:

- The digestive system is made up of the gullet (oesophagus), stomach, pancreas, liver and intestines.
- The excretory system is made up of the kidneys, ureters and bladder.
- The nervous system is made up of the brain, spinal cord and nerves.
- The reproductive system in females is the ovaries, oviducts, uterus and vagina; in males it is the testes, sperm ducts, prostate gland and penis.

All the different organ systems make up a living **organism**.

Plant tissues and organs

The diagrams show the tissues in a leaf. The tissue that carries out photosynthesis in leaves is called **mesophyll**. The cells making up the upper layer of the mesophyll are called palisade cells. These cells are closely packed and full of chloroplasts so that they are well adapted to absorb lots of light. The palisade cells make up the **palisade mesophyll tissue**. All the cells making up this tissue look alike and do the same function – they absorb light for photosynthesis.

A leaf is an organ. Other plant organs are roots and stems. Other structures, such as flowers and fruits, are modified leaves.

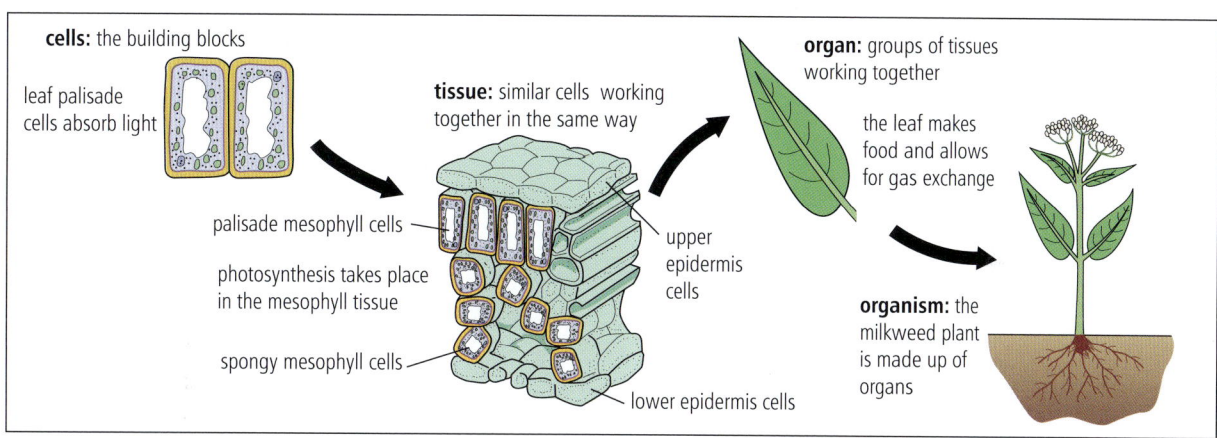

cells: the building blocks

leaf palisade cells absorb light

tissue: similar cells working together in the same way

palisade mesophyll cells

photosynthesis takes place in the mesophyll tissue

spongy mesophyll cells

upper epidermis cells

lower epidermis cells

organ: groups of tissues working together

the leaf makes food and allows for gas exchange

organism: the milkweed plant is made up of organs

Figure 2.4.2 Levels of organisation in a flowering plant.

SUMMARY QUESTIONS

1 Copy and complete the sentences using these words:

   organ system   **cells**   **tissues**   **function**

   A tissue is made up of _____ that carry out the same _____.
   An organ is formed from a group of _____ working together.
   An _____ is a group of organs working together to perform several body functions.

2 Copy out the organs listed on the left. Match each with the correct system from those listed on the right.

organs	system
lungs and trachea	digestive
heart and blood vessels	nervous
brain and spinal cord	gas exchange
ovaries, oviducts and uterus	excretory
gullet, stomach and intestines	reproductive
kidneys and bladder	circulatory

3 Arrange the following words into the correct sequence, starting with the smallest and ending with the largest:

   organ   **organ system**   **tissue**   **organism**   **cell**

EXAM TIP

Make sure you can explain how a palisade mesophyll cell is adapted to carry out photosynthesis by listing the features and explaining how each one helps.

KEY POINTS

1 Cells that have the same function are grouped together to form tissues.

2 Different tissues make up organs which work together to do a particular function.

3 Different organs work together as organ systems.

Practice questions

1 Which structure is found only in plant cells?

  A cell membrane

  B chloroplast

  C cytoplasm

  D nucleus

  (Paper 1)   [1]

2 Which structure is not found in animal cells?

  A cell membrane

  B cell wall

  C cytoplasm

  D nucleus

  (Paper 1)   [1]

3 Which is the correct sequence, starting with the smallest and ending with the largest?

  A nucleus, chloroplast, mitochondrion, ribosome

  B chloroplast, ribosome, nucleus, mitochondrion

  C mitochondrion, nucleus, ribosome, chloroplast

  D ribosome, mitochondrion, chloroplast, nucleus

  (Paper 1)   [1]

4 A student makes a drawing of a biological specimen. The length of the specimen in the drawing is 140 mm. The magnification is ×40. What is the actual size of the specimen?

  A 5600 mm   **B** 56 mm

  C 35 mm   **D** 3.5 mm

  (Paper 1)   [1]

5 Which is not found in an animal cell?

  A nucleus

  B permanent vacuole

  C ribosomes

  D mitochondria

  (Paper 2)   [1]

6 What is the function of root hair cells?

  A absorption of water

  B growth of roots

  C protection

  D transport of sugars

  (Paper 1)

7 The diagram shows a mitochondrion from a liver cell.

1.0 µm

The magnification of the diagram is:

  A ×40   **B** ×400

  C ×4000   **D** ×40 000

  (Paper 2)   [1]

8 The substances that move in and out of mitochondria are:

  A CO_2 in, O_2 out, nutrients in

  B CO_2 out, O_2 in, nutrients in

  C CO_2 in, O_2 in, nutrients out

  D CO_2 out, O_2 out, nutrients out

  (Paper 2)   [1]

9 Cells of organisms classified in the prokaryote kingdom contain which of the following?

  A cell membrane, enzymes, nucleus, vesicles

  B cell wall, cell membrane, cytoplasm, ribosomes

  C cytoplasm, DNA, mitochondria, ribosomes

  D cytoplasm, DNA, enzymes, mitochondria

  (Paper 2)   [1]

10 The palisade mesophyll cell is a type of plant cell. A liver cell is a type of animal cell. Copy and complete the table to compare these two cells. Put a tick (✔) if you think the structure is present and a cross (✗) if you think it is absent.

cell structure	palisade mesophyll cell	liver cell
cell wall		
cell membrane		

cytoplasm		
nucleus		
chloroplast		
large vacuole		

(Paper 3) [6]

11 The diagram shows three animal cells.

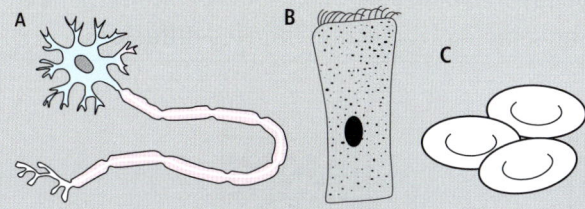

(a) Identify the cells **A**, **B** and **C**. [3]

(b) State the functions of the three cells. [3]

(c) State where in the body these cells are found. [3]

(d) The magnification of cell **B** is ×1000. Calculate its actual size in millimetres. [3]

(Paper 3)

12 The diagram shows three plant cells.

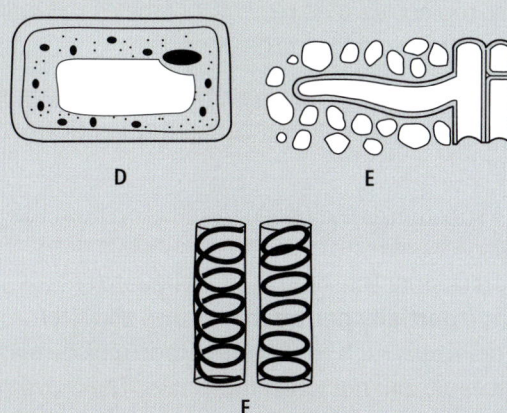

(a) Identify the cells **D**, **E** and **F**. [3]

(b) State the functions of the three cells. [3]

(c) For each cell, **D**, **E** and **F**, state a plant organ where it is found. In each case give a different organ. [3]

(d) The actual length of cell **D** is 0.05 mm. Calculate the magnification of the drawing. [2]

(Paper 3)

13 (a) Name:

   (i) an animal cell that does not have a nucleus; [1]

   (ii) a plant cell that does not have a nucleus; [1]

(b) State the differences between the following pairs of terms.

   (i) Cytoplasm and nucleus [2]

   (ii) Cell membrane and cell wall [2]

(Paper 3)

14 (a) Six cell structures and their functions are listed below.

Cell structures

A   nucleus

B   cell membrane

C   cell wall

D   large vacuole

E   ribosome

F   mitochondrion

Functions

1   controls the movement of substances into and out of the cell

2   stores water and ions

3   makes proteins

4   withstands pressure of water inside the cell

5   carries out aerobic respiration

6   stores DNA and controls the activities of the cell

Match each of the cell structures with its function by writing a letter and a number. [5]

(b) A student looked at a cell using a light microscope. She made a drawing of the cell and showed the diameter of the nucleus in her drawing as 70 mm. She calculated the magnification of her drawing as ×10 000. What is the actual size of the nucleus in micrometres? Show your working. [2]

(Paper 4)

3.1 Diffusion

Molecules in gases move about in a random way. They have kinetic energy and bump into one another and spread out to fill up all the space available. Molecules and ions in a liquid do this as well, although it takes longer for them to fill the space. Movement in a gas is faster as the molecules are more spread out. The difference in speed between movement in gases and liquids is important for organisms. This movement of molecules is called **diffusion**.

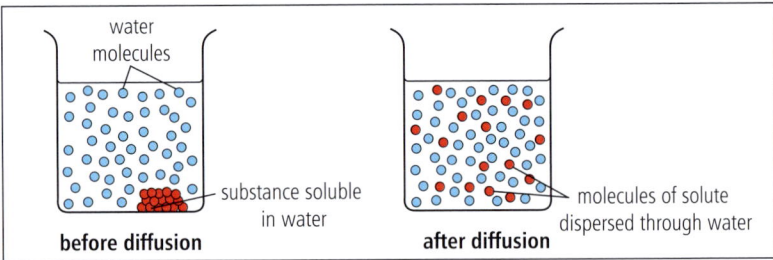

Figure 3.1.2 Diffusion in a liquid.

When molecules or ions diffuse they spread out from where there are lots of them in a given volume (a high concentration) to where there are not as many of them (a low concentration). The difference between the concentration of molecules in two places is a **concentration gradient**. Molecules carry on diffusing until they are spread out evenly. When this happens the molecules keep moving, but there is no longer a difference in concentrations so diffusion has stopped.

Diffusion is the net movement of molecules or ions from a region of high concentration to a region of lower concentration down a concentration gradient.

Cells gain some of the substances they need by diffusion from their surroundings. They also lose some of their waste substances to their surroundings by diffusion. These substances have to cross cell membranes that are **partially permeable** as they allow the movement of small molecules such as oxygen, carbon dioxide and water to pass through easily, but not larger molecules. The movement of molecules by diffusion across cell membranes is **passive movement** as cells do not need to use energy to move the molecules.

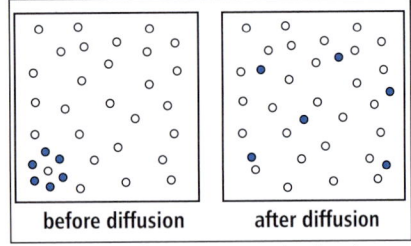

Figure 3.1.1 Diffusion in a gas. Molecules of a coloured gas spread out through the container.

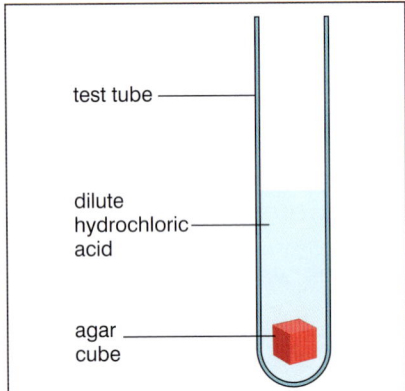

Figure 3.1.3 The test tube contains a cube of agar containing the indicator cresol red. This indicator is red in alkali but changes to yellow in acidic conditions. The test tube is half filled with a dilute solution of hydrochloric acid.

PRACTICAL

What factors affect diffusion?

- **Surface area:** The effect of surface area can be investigated by cutting three cubes of agar containing the pH indicator cresol red. The agar is cut to make one cube 10 × 10 × 10 mm and two cubes that are both 5 × 5 × 5 mm. The larger cube is placed in a test tube containing dilute hydrochloric acid. The two smaller cubes are placed in another test tube containing the same volume of dilute hydrochloric acid.

- The smaller cubes take less than half the time for the acid to diffuse into the agar and change the colour of the indicator from red to yellow. The smaller cubes have a larger surface area for the acid to diffuse through.

- **Temperature:** The effect of temperature on diffusion can be investigated by having two test tubes of dilute hydrochloric acid with the same size cubes of agar (e.g., 10 × 10 × 10 mm). One test tube is put in a water bath at 20 °C and the other in a water bath at 30 °C. The cube of agar at 30 °C changes colour from red to yellow more quickly than the cube at 20 °C.

- **Concentration:** The effect of concentration can be investigated by using two different concentrations of hydrochloric acid. The cube of agar in the higher concentration of acid changes colour from red to yellow more quickly than the cube in the lower concentration as there is a steeper concentration gradient.

Gas versus liquid

Animals and plants exchange the gases oxygen and carbon dioxide with their surroundings at gas exchange surfaces. In mammals, the gas exchange surface is formed of the alveoli in the lungs (see page 131). Blood transports these two gases between the lungs and all the cells in the body. In the alveoli, oxygen diffuses across a very thin layer of cells into the blood. Carbon dioxide diffuses in the opposite direction. Breathing constantly refreshes the air in the alveoli and blood constantly removes oxygen and brings carbon dioxide, so the concentration gradients are always steep. There are many alveoli to give a very large surface area for gas exchange.

In plants, gas exchange occurs inside the leaves. The spongy mesophyll cells provide a large surface area for the exchange of gases. There are air spaces between the cells in a plant and each cell exchanges gases with this air (see page 64). This is efficient because diffusion through the air is 300 000 times faster than through water.

(see page 131)
(see page 64)

EXAM TIP

Remember as you read about diffusion that the energy for diffusion comes from the kinetic energy of the random movement of molecules and ions.

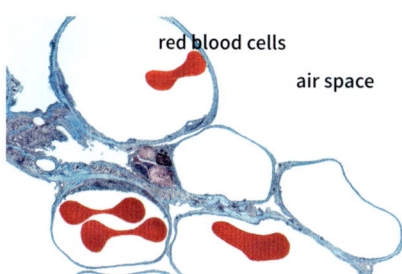

red blood cells

air space

Figure 3.1.4 There is a short distance between the red blood cells and the air in the alveoli. Magnification × 1500.

SUMMARY QUESTIONS

1. Copy and complete the sentences using these words:

   **liquid   random   diffusion
   low   gas   high**

   _____ is the net movement of molecules of a _____ or a _____ from an area of _____ to an area of _____ concentration as a result of the _____ movement of molecules.

2. State how the following factors affect diffusion into cells:
   a distance
   b size of molecule
   c surface area
   d concentration gradient
   e temperature

3. What do the terms *solution* and *concentration gradient* mean?

KEY POINTS

1. Diffusion is the net movement of molecules or ions from a region of high concentration to a region of low concentration down a concentration gradient.

2. Factors that affect diffusion are: size of molecule, distance, surface area, temperature and the steepness of the concentration gradient.

3.2 Osmosis

LEARNING OUTCOMES

- Define the term *osmosis*
- Describe the effect of osmosis on plant and animal tissues
- Investigate osmosis using dialysis tubing
- **S** Explain the movement of water into and out of cells using the term *water potential*

Water as a solvent

A **solution** is made up of two parts, the **solute** and the **solvent**. The solute dissolves in the solvent. If you dissolve sugar in water you make a sugar solution. The sugar is the solute and the water is the solvent. The solute is not always a solid like sugar. Liquids and gases can be solutes as they can dissolve in solvents too. Something that dissolves in a solvent is described as being **soluble**.

Water is sometimes called the universal solvent. About 75% of cytoplasm is water and it is the main component of transport fluids like blood, and xylem sap and phloem sap in plants. Everything transported in plants and animals has to dissolve in water and most of the chemical reactions that occur in cells happen in water. Also, water is needed for digestion and excretion to take place.

Each cell is surrounded by a cell membrane. It separates the contents of the cell from the outside. The cell membrane has tiny holes in it that allow small molecules to pass through but not large ones. The cell membrane is described as being **partially permeable**.

Osmosis is a special kind of diffusion involving water molecules. It occurs when two solutions are separated by a **partially permeable membrane**.

Osmosis is the diffusion of water from a dilute solution into a more concentrated solution through a partially permeable membrane.

The tiny holes in the membrane allow small water molecules to pass through, but the large solute molecules are too big to pass through the partially permeable membrane. Water is diffusing from a place where there is a dilute solution with a **high concentration of water** to a place where there is a concentrated solution with a **lower concentration of water**.

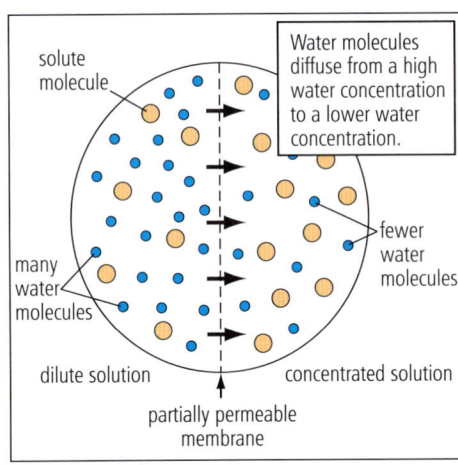

Figure 3.2.1 Osmosis.

Supplement

Water potential is a way of thinking about the ability of water to move by osmosis. This is influenced by how much water is available, but also by other factors such as the pressure exerted on water in plant cells by the cell wall. It is more accurate to say that a dilute solution (containing a lot of water molecules) has a **high water potential**. A concentrated solution (containing fewer water molecules) has a **low water potential**.

In Figure 3.2.1, there is a **water potential gradient** between the two sides of the membrane. The water molecules diffuse *down* this water potential gradient, from a region of high water potential to a region of lower water potential through a partially permeable membrane.

Water molecules are free to move through the membrane in both directions by kinetic energy. However, since there are many more water molecules present on the left hand side, there will be a **net** movement of water molecules through the membrane from left to right, down the water potential gradient.

PRACTICAL

1 Cut two pieces of dialysis tubing, each 12 cm long. Tie one end of each with cotton.

2 Fill one model cell with a dilute sugar solution (cell A). Fill the other model cell with water (cell B).

3 Tie the other end of both model cells and weigh them on a balance. Put cell A into a beaker of water and put cell B into a beaker containing a concentrated solution of sugar. After 30 minutes take out the model cells and weigh them again.

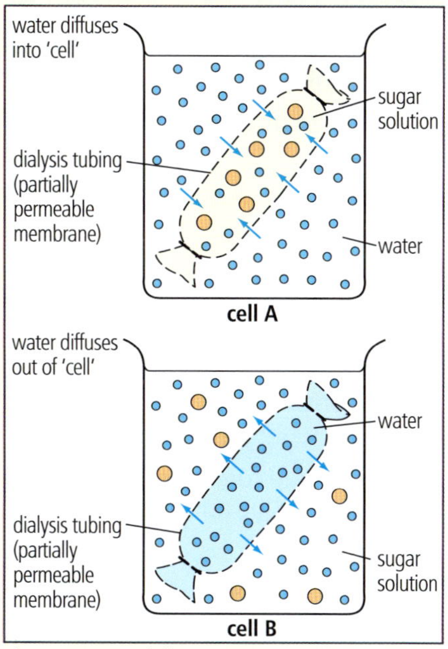

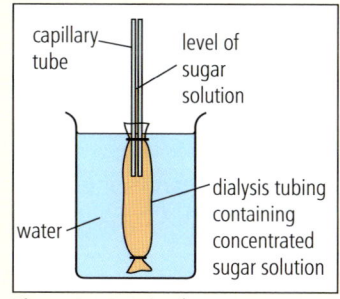

Figure 3.2.2 Modelling osmosis in cells.

Model cell A increases in mass because water has diffused *into* the 'cell' by osmosis. Model cell B decreases in mass because water has diffused *out of* the 'cell' by osmosis.

An osmometer You can see the effects of osmosis if you set up the apparatus in Figure 3.2.3.

Figure 3.2.3 A simple osmometer.

Fill the partially permeable membrane with a very concentrated solution of sugar.

Tie it to a capillary tube and stand it in water.

Very quickly you will see the liquid moving up the tube. You can measure how fast it is moving using a ruler and a stopwatch. Use your ideas about osmosis to explain why the liquid rises in the tube.

SUMMARY QUESTIONS

1 Define the term *diffusion*.

2 Describe how you can find out how fast water diffuses by osmosis into a sugar solution. Remember to include all practical details.

3 Explain why water is an important solvent for animals and plants.

S 4 Define the terms *partially permeable membrane* and *osmosis*.

5 Explain in terms of water potential how water passes into plant cells placed in distilled water.

A model cell

Dialysis tubing (Visking tubing) is partially permeable. We can use dialysis tubing to represent the cell membrane and the sugar solution to represent the cytoplasm.

EXAM TIP

You may be given the results of a practical demonstration of osmosis and be expected to explain the results. Make sure you remember the definition of osmosis and apply it to the results you can see. Always explain that the results are due to water molecules moving by osmosis.

KEY POINTS

1 Osmosis is the diffusion of water molecules from a region of their higher concentration (dilute solution) to a region of their lower concentration (concentrated solution) through a partially permeable membrane.

2 A partially permeable membrane allows small molecules such as water to pass through but not large solute molecules.

3 Water is important because all transport and most chemical reactions in cells take place in water.

4 Water molecules diffuse down a water potential gradient from a region of higher water potential to a region of lower water potential.

3.3 Osmosis in plant cells

LEARNING OUTCOMES

- Investigate the effect of immersing plant tissues in different solutions
- Describe how water can enter and leave plant cells by osmosis
- **S** Explain the effects of osmosis on plant cells

EXAM TIP

When describing the cell membrane, always say that it is <u>partially</u> permeable, not semi-permeable.

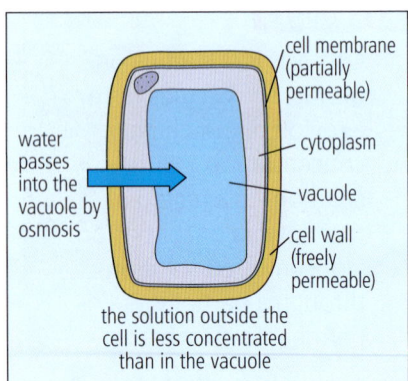

Figure 3.3.1 Osmosis in a plant cell.

EXAM TIP

S In Paper 4 you may be expected to use the term 'water potential' in writing about osmosis in plant or animal cells and tissues. You can also refer to 'water potential gradients' to explain water movement into and out of cells and tissues.

PRACTICAL

Osmosis in potato cells

1. Cut nine cores from a potato so that they are exactly the same length. Record this length.

2. Feel the cores to see how firm and 'bendy' they are.

3. Set up the following test tubes:

   A – distilled water

   B – dilute sugar solution

   C – concentrated sugar solution.

4. Place three cores of potato into each test tube and leave them for 60 minutes.

5. Remove the cores and measure their lengths. Calculate the average length of the cores in each test tube.

   Compared to the start, cores will be:

   A – longer and firmer

   B – about the same length and firmness

   C – shorter and softer and 'bendier'.

   These results are explained in the text.

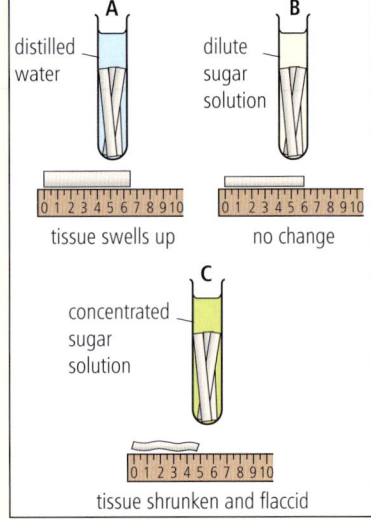

Figure 3.3.2 Osmosis with potato cores.

Cells that contain the maximum volume of water provide support for the stems and leaves of plants. The pressure of water within each cell presses against the cell wall, keeping the cell firm. The cells keep the stems of many plants upright and leaves at right angles to the stem. However, when these cells lose water, they are no longer firm and plants wilt (see page 90).

Supplement

Turgidity

The cell membrane of the plant cell is partially permeable and the cell sap inside the vacuole is a solution of ions and sugars. When plant cells are placed in water, some water enters the cells. This is because there is a **water potential gradient** so that water molecules diffuse into the cells by osmosis.

As water enters it makes the cell swell up. The water pushes against the cell wall developing a **turgor pressure**. Eventually the cell contains as much water as it can hold. It's like a blown-up balloon. The strong cell wall stops the cell bursting and the cell is **turgid**. This is what has happened to the cells in the potato cores in tube A in Figure 3.3.2. The cells have absorbed some water by osmosis, become turgid and caused the core to get slightly longer.

Supplement

Plasmolysis

When plant cells are placed into a concentrated sugar or salt solution water passes *out* of the cells by osmosis. As water passes out, the sap vacuole starts to shrink. These cells are no longer firm, and become limp. We say that they are **flaccid**. As more water leaves the cells the cytoplasm starts to move away from the cell wall. These cells are now **plasmolysed**. This is what has happened to the cells in the potato cores in test tube **C**. The cells have decreased in volume so the whole core is shorter than at the start.

The cores in test tube **B** did not change in length very much because the water potential of the sugar solution was about the same as the water potential of the cell sap in the potato cells. There has been no overall diffusion of water into or out of the cells so they have stayed about the same length.

Importance of water potential and osmosis

Water potential gradients are important to living organisms. When water passes into a root hair it moves from an area of high water potential (in the soil water), into an area of lower water potential (in the cell sap of the root hair).

We have seen the importance of osmosis in supporting herbaceous stems by maintaining turgor in their cells.

Amoeba is a single-celled organism that lives in fresh water. Water is continually entering the cell from an area of high water potential into an area of lower water potential. *Amoeba* has an organelle called a **contractile vacuole**. The star-shaped structure you can see in the photograph of *Paramecium* on page 7 (Figure 1.3.4) is a contractile vacuole. Water gradually accumulates inside the vacuole and when it is full, the water is released to the outside.

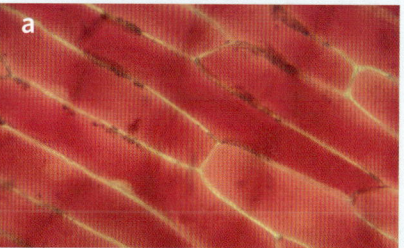

the solution outside the cell is more concentrated than in the vacuole

water passes out of the cell by osmosis

as cytoplasm is pulled away from the cell wall, the cell becomes plasmolysed

the vacuole shrinks

Figure 3.3.3 Water passes out of a plant cell by osmosis.

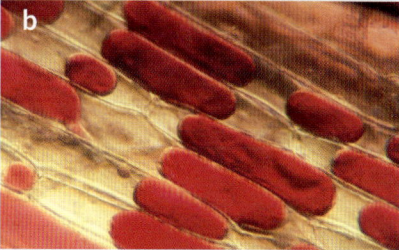

Figure 3.3.4 These epidermal cells from a red onion are (a) turgid and (b) plasmolysed.

SUMMARY QUESTIONS

1 Explain what is meant by each of these terms:

turgor pressure turgid flaccid plasmolysis

2 Some potato cores were weighed and then placed into a dilute sugar solution. After 2 hours, they were taken out of the solution, dried on a paper towel and weighed again.

The mass of the potato cores remained unchanged.

  a What does this tell you about the concentration of the sugar solution?

  b Explain your answer to part **a** in terms of osmosis.

s 3 Explain, using the term *water potential*, what happens to cores of potato when they are placed into distilled water for 60 minutes.

KEY POINTS

1 Water passes into plant cells by osmosis. A plant cell that is full of water is turgid. Turgid cells provide support for leaves and young stems.

2 If plant cells are placed into a concentrated sugar solution, water passes out by osmosis. These cells are no longer firm, they are flaccid. As the vacuole shrinks, the cell membrane moves away from the cell wall – the cell is now plasmolysed.

3.4 Active transport

LEARNING OUTCOMES

- Describe movement of molecules and ions across cell membranes by active transport
- Explain the importance of active transport as a process that requires energy
- Describe the active transport of ions in plant roots
- Describe the active uptake of glucose by epithelial cells in villi and in kidney tubes

Active transport

Cells take up molecules and ions and keep them in high concentrations.

Look at the concentration of magnesium ions in the root hair cell and the concentration of magnesium ions in the soil solution.

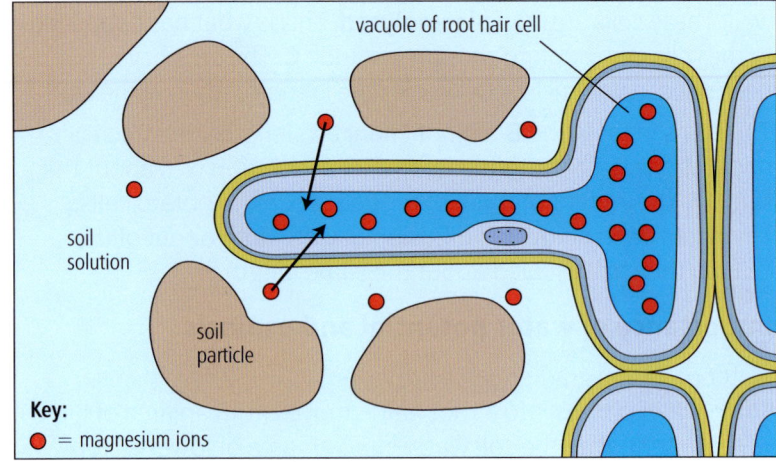

Figure 3.4.1 Root hair cells absorb ions by active transport.

The concentration of magnesium ions is far greater inside the vacuole of the root hair cell than it is in the water in the soil. We might expect magnesium ions to diffuse out of the root hair cell into the soil water down a diffusion gradient. The magnesium ions are maintained at a high concentration inside the root hair cell by a process called **active transport**.

Active transport is the movement of ions or molecules in or out of a cell through the cell membrane against a concentration gradient, using energy released during respiration.

EXAM TIP

Notice the use of the term 'concentration gradient' in the definition of active transport. 'Against a concentration gradient' means from a low concentration to a high concentration.

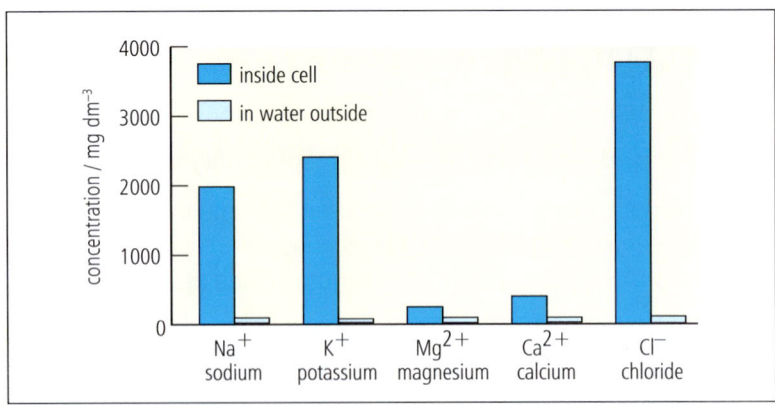

Figure 3.4.2 Ion concentration due to active transport.

The bar chart shows the concentrations of some ions inside the cells of a freshwater plant and in the water in which it lives. These ions cannot have been taken into the plant by diffusion. They are taken in against a concentration gradient by active transport.

Supplement

Active transport needs energy

The cell membrane contains **carrier proteins**. These carrier proteins span the cell membrane and provide means by which ions and molecules can enter or leave a cell by active transport. First the molecule or ion combines with a carrier protein. Energy from respiration enables the carrier protein to change its shape to carry the ion or molecule through the membrane. The molecule or ion is released to the other side of the membrane and the carrier protein returns to its original shape.

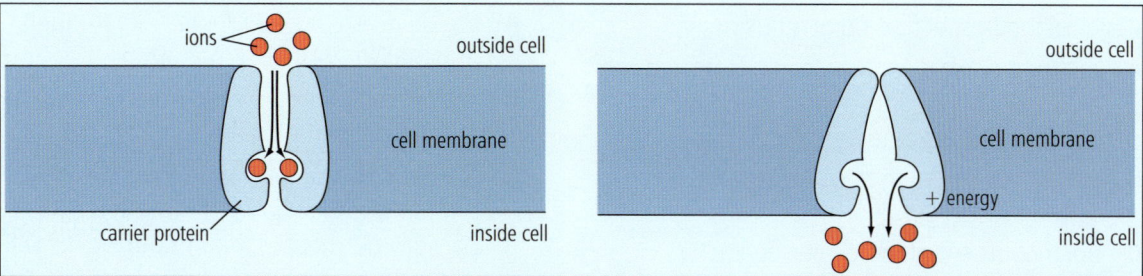

Figure 3.4.3 Carrier proteins in membranes carry out active transport.

Epithelial cells lining the **villi** in the small intestine and the kidney tubules absorb glucose by active transport (Topic 7.7). These cells have high rates of respiration to provide energy for this active transport.

Active transport relies upon respiration to take up ions or molecules against a concentration gradient. Any factor that affects the rate of respiration will also affect the rate of active transport. So a lack of oxygen would reduce respiration rate and active transport.

An increase in temperature would increase the rate of respiration, up to a point, so would also have the same effect on active transport. The presence of poisons such as cyanide can stop respiration, so active transport would stop altogether.

EXAM TIP

Root hair cells and epithelial cells of villi are adapted for active transport by having many carrier proteins in their cell membranes and a high rate of respiration to provide energy.

SUMMARY QUESTIONS

1 Explain what is meant by active transport.

S 2 a Describe the role of carrier proteins and respiration in active transport.

   b Give two examples of active transport taking place.

3 Make a table to compare diffusion with active transport.

   Make sure that you have three columns headed 'features', 'diffusion' and 'active transport'. For features you can include 'needs energy from the cell', 'concentration gradient'. You may be able to think of some other features to use in your table.

4 State what effect each of the following would have on active transport, in each case give reasons for your answer:

   a a lack of oxygen

   b an increase in temperature

   c the presence of a substance, such as cyanide, that inhibits respiration.

KEY POINTS

1 Active transport is the movement of ions or molecules across the cell membrane, against a concentration gradient, using energy from respiration.

2 Active transport enables root hair cells to take up ions, and epithelial cells of the villi to take up glucose. **S**

Practice questions

1 Which is *not* partially permeable?

  A cell membrane of palisade mesophyll cell

  B cell membrane of red blood cell

  C cell wall

  D dialysis tubing

(Paper 1) [1]

2 Which is the *best* definition of diffusion?

  A movement of solvent molecules through a partially permeable membrane down a concentration gradient

  B net movement of molecules down a concentration gradient

  C net movement of molecules against a concentration gradient

  D random movement of molecules in a gas or a liquid

(Paper 1) [1]

3 Root hair cells are surrounded by soil water, which is a very dilute solution of mineral ions, such as nitrate ions and magnesium ions. Plant cells have higher concentrations of these ions. How do root hair cells absorb ions from soil water?

  A active transport

  B diffusion

  C osmosis

  D random movement

(Paper 1) [1]

4 A student cut up a potato into pieces that looked like chips. The student described these pieces as 'fairly firm'. The pieces were put into three different liquids. Which row shows the results that the student obtained?

	very concentrated salt solution	very dilute salt solution	water
A	fairly firm	swollen and very firm	soft
B	soft	fairly firm	swollen and very firm
C	soft	soft	fairly firm
D	swollen and very firm	soft	fairly firm

(Paper 1) [1]

5 Water is an important molecule in organisms because it is a:

  A solute   **B** solution

  C solvent   **D** suspension

(Paper 1) [1]

6 Which is the *best* definition of osmosis?

  A the diffusion of water molecules through a partially permeable membrane

  B the net diffusion of water molecules down a water potential gradient

  C the net diffusion of water molecules down a water potential gradient through a partially permeable membrane

  D the net diffusion of water molecules through a partially permeable membrane from a solution with a low water potential to a solution with a high water potential

(Paper 2) [1]

7 Some fresh plant tissue was put into a concentrated salt solution for 60 minutes. Which *best* explains why the tissue became softer?

  A water diffused down a water potential gradient from the cells to their surroundings

  B water diffused into the cells so that they became turgid

  C water diffused out of the cells so that they became plasmolysed

  D the cells lost turgor pressure and became flaccid

(Paper 2) [1]

8 (a) Explain the importance to humans of the following:

    (i) diffusion of oxygen in the alveoli [2]

    (ii) diffusion of carbon dioxide in the alveoli [2]

    (iii) absorption of glucose by diffusion in the small intestine. [2]

(b) Explain the importance to plants of the following:

    (i) diffusion of carbon dioxide into leaves [2]

(ii) absorption of magnesium ions by root hair cells. [2]

(c) Osmosis is a type of diffusion. Explain how osmosis differs from diffusion. [3]

(Paper 3)

9 (a) State two ways in which active transport differs from diffusion. [2]

(b) Plants are supported by the pressure of water inside their cells.

(i) State the cell structure that stores water inside plant cells. [1]

(ii) State the cell structure that withstands the pressure inside plant cells, preventing them from bursting. [1]

(iii) Explain how water is absorbed by plant cells. [2]

(Paper 3)

10 A group of students investigated osmosis. They began by peeling 50 small onions. They divided the peeled onions into five batches of 10 onions and weighed them. Each batch was placed into a solution of different concentrations of salt (sodium chloride). After immersion for two hours each batch was surface dried and reweighed. The students calculated the percentage change in mass. The table shows their results.

conc. of salt / g dm^{-3}	mean mass of onions / g		percentage change in mass
	before immersion	after 2 hours immersion	
0	147	173	+ 18.0
25	153	165	+ 8.0
50	176	172	−2.0
100	154	149	
150	149	142	−4.5
200	183	175	−4.5

Information and data used to compile the table from Practical osmosis in vegetable pickling, Ray W James. Journal of Biological Education (1993) 27 (2), pages 90–91

(a) Calculate the percentage change in mass for the onions kept in the 100 g dm^{-3} salt solution. Show your working. [2]

(b) State why the students calculated the percentage change in mass. [1]

(c) Plot a graph of the results. [5]

(d) Use your graph to find the salt solution in which there is no change in mass. [1]

(Paper 6)

11 An experiment was set up to investigate the factors influencing the uptake of ions by plant roots. Some roots were cut from a plant, washed and placed in three solutions, **A**, **B** and **C**, containing potassium ions. A mixture of gases was bubbled through each of the solutions. A gas mixture rich in oxygen was bubbled through solution **A**; solution **B** received a gas mixture with a very low concentration of oxygen; solution **C** received no oxygen in the gas mixture. The roots were left for 24 hours and then the rate of uptake of potassium ions was determined.

The rate of uptake of potassium ions was highest in solution **A** and lowest in solution **C**.

(a) Explain how the supply of oxygen to roots influences their uptake of potassium ions. [3]

The experiment was repeated with solutions containing roots kept at different temperatures. They were all provided with the gas mixture that had been given to solution **A**. The results are shown in the table.

temperature / °C	rate of uptake of potassium ions / arbitrary units
5	3
10	5
20	10
30	20
40	15

(b) Draw a graph of the results. [6]

(c) Describe the results shown in your graph. [4]

(d) Explain the effect of temperature on the uptake of potassium ions by the roots. [3]

(Paper 4)

4.1 Biological molecules

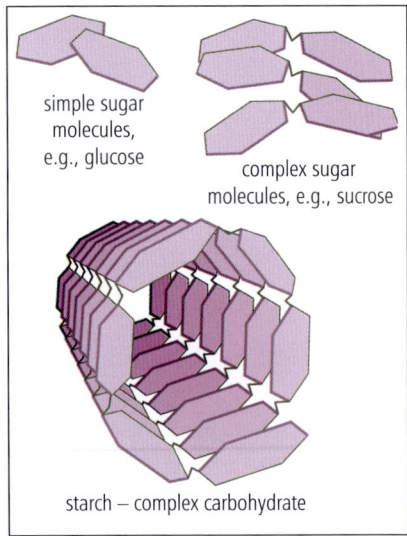

simple sugar molecules, e.g., glucose

complex sugar molecules, e.g., sucrose

starch – complex carbohydrate

Figure 4.1.1 Different types of carbohydrate.

Biological molecules are complex chemicals like carbohydrates, proteins and fats. They are useful chemicals that are needed by living organisms for **metabolism**. By metabolism we mean all the chemical reactions taking place in the cells of the body. These reactions include the release of energy in respiration, protein synthesis, and the growth and repair of cells.

Green plants make the complex chemical compounds that they need from simple raw materials. Carbon dioxide and water are the raw materials for **photosynthesis**. The simple sugars produced in photosynthesis are used to make a wide range of other compounds. Plants need minerals to make some of these complex compounds.

Animals eat plants and/or other animals that feed on plants. Biological molecules that animals require are present in their diet. The different biological molecules needed for a balanced diet in humans are carbohydrates, proteins, fats, vitamins, minerals, fibre and water.

Carbohydrates

These contain the elements carbon (C), hydrogen (H) and oxygen (O). Carbohydrates include sugars and starches.

Glucose is a simple sugar that is made in photosynthesis, used in respiration and transported in the blood. It consists of six carbon atoms arranged into a ring.

Sucrose is a double-sugar molecule made up of two molecules of simple sugars joined by chemical bonds. Complex carbohydrates are made by joining many simple sugar molecules together by chemical bonds. Plants store **starch** as an energy store. They have enzymes to catalyse the reactions that join glucose together into long chains. **Glycogen** is another complex carbohydrate made from glucose by animals as a store of energy. It is stored in the liver and muscles.

Starch and glycogen, unlike sugars, are insoluble and do not taste sweet. **Cellulose** is a complex, structural carbohydrate made up of thousands of glucose units. The glucose units are held together by bonds forming long and unbranched chains. These cellulose molecules are linked together to form fibres, which give plant cell walls their strength and rigidity.

EXAM TIP

Biological molecules are made by organisms. They only become food molecules when eaten by an animal. The term nutrient (or nutrient molecule) is often used to refer to the biological molecules described in Topics 4.1 and 4.2. Beware – it is also used to refer to ions, such as nitrate ions, sodium ions and magnesium ions, which organisms need from their environment.

Proteins

Proteins are complex molecules made up of carbon, hydrogen and oxygen, but they also contain nitrogen (N) and many have sulfur (S). Proteins are long-chain molecules made up of smaller molecules called **amino acids**. After formation they are either folded into different shapes (Topic 5.1) or become arranged into long fibres.

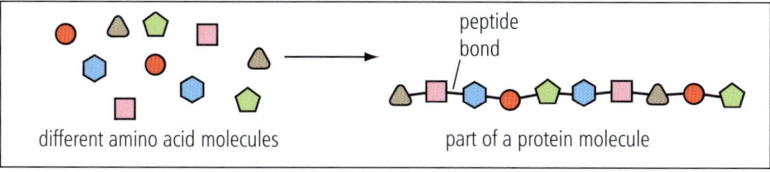

Figure 4.1.2 How a protein is made from amino acids.

There are about 20 different types of amino acid. Molecules of amino acids are made into chains as you can see in the diagram. It is the sequence of the different amino acids in the chain that determines the type of protein that is formed. Each individual amino acid joins the chain by means of a chemical bond called a peptide bond.

The different sequences of amino acids give different shapes to protein molecules. These different shapes of protein molecules can be related to their function. Enzymes are proteins that provide a surface for reactions to take place called the active site (see page 47). Antibodies are proteins with a structure that has binding sites on its surface. This enables them to bind to the surface of pathogens (see page 108).

Fats

Fats and oils are made up of the elements carbon, hydrogen and oxygen. Each fat molecule is made up of one molecule of **glycerol** attached to three **fatty acids**. There are different types of fatty acid and these can form different fats with different properties. Fats are used for energy storage and thermal insulation in the body.

EXAM TIP

You may be asked to compare these three groups of biological molecules. Make a table to compare them using headings such as 'Name of biological molecules group' and 'Elements'. You will be able to add to this table in Topic 7.2. When naming the elements always use their full names, not their symbols.

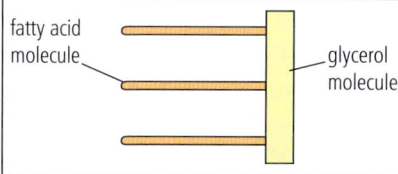

Figure 4.1.3 A molecule of fat.

KEY POINTS

1 Carbohydrates and fats are made up of the elements carbon, hydrogen and oxygen. Proteins are made up of the same elements plus nitrogen and sometimes sulfur.

2 Carbohydrates are simple sugars (e.g., glucose), and complex carbohydrates such as starch, glycogen and cellulose.

3 Small molecules are joined together to make long chain molecules. Sugars are made into starch, glycogen and cellulose. Amino acids are made into proteins.

4 A molecule of a fat is made by combining three fatty acid molecules with a molecule of glycerol.

SUMMARY QUESTIONS

1 a List the components of a balanced diet for a human.

 b Give one example of each of the following:

  i a simple sugar ii a complex carbohydrate found in plants

  iii a complex carbohydrate found in animals

2 a Which elements are present in proteins, but are not found in carbohydrates and fats?

 b Name the small molecules that are joined together to make protein molecules.

 c Give three examples of proteins that are made in the body.

 d State the function of each protein that you named in part **c**.

3 a Name the molecules that are reacted together to make a molecule of fat.

 b Give two uses of fats in the body.

4.2 Chemical tests for biological molecules

- State how the following chemical tests are carried out:
  - iodine test for starch
  - Benedict's test for reducing sugars
  - biuret test for protein
  - ethanol emulsion test for fat
  - DCPIP test for vitamin C

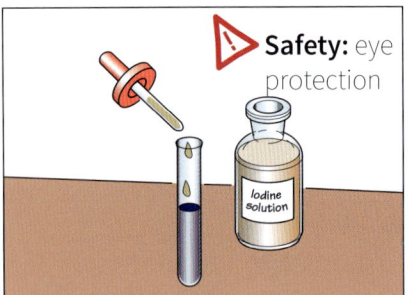

Safety: eye protection

Figure 4.2.1 Test for starch.

EXAM TIP

These tests are often known as 'food tests', but they can be carried out on any plant and animal material or on solutions made up in the laboratory.

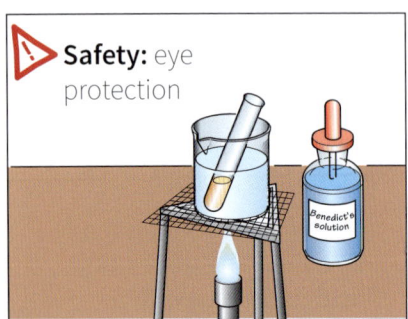

Safety: eye protection

Figure 4.2.2 Test for reducing sugars.

Simple chemical tests can be used to identify starch, sugars, proteins, fat and vitamin C.

First, it is important to carry out these tests on pure forms of these compounds. For example, if you are testing a food for reducing sugars, you should first carry out the chemical test on a glucose solution. Then keep the results to compare with your results from testing other materials, such as foods or animal and plant tissues. You should also do a test with water so that you can see the negative result as well.

You will need to make an **extract** from the material you are testing. This involves grinding up a small amount of the material with some water with a pestle and mortar or putting it into a blender. The chemicals will be in solution in the extract.

PRACTICAL

Safety: Some of the chemicals used in these tests are corrosive, so always wear eye protection.

Testing for starch

- Half-fill a test tube with the food extract you wish to test for starch.
- Add two or three drops of **iodine solution**. Iodine solution usually looks yellow or light brown.
- A positive result for starch is if the iodine solution turns **blue–black**. If the extract remains a yellow or light brown colour it does *not* contain starch.

Testing for reducing sugars

- Put a known volume of the extract you wish to test for reducing sugars in a test tube.
- Place a beaker on a heat-proof mat.
- Carefully half-fill the beaker with boiling water from a kettle (or place the beaker on a tripod and gauze and boil the water with a Bunsen burner).
- Add the same volume of **Benedict's solution** to the test tube containing the food extract and put it into the hot water.
- Benedict's solution is bright blue.
- A positive test for reducing sugars is when Benedict's solution turns red or orange (if you look carefully you can see it turn green and then yellow before turning orange). If you leave the test tube to cool you will also see a precipitate.
- You can use Benedict's test to tell how much simple sugar is present. If the colour changes to green, the extract contains only a little of the reducing sugars. If it turns a deep orange colour then it contains a lot of reducing sugars.
- If the colour remains blue then the extract does *not* contain any reducing sugars.

Testing for protein

- Half-fill a test tube with the extract you wish to test for protein.
- Add five to six drops of **biuret solution** (this solution contains copper sulfate solution and sodium hydroxide solution).

Safety: Take care as sodium hydroxide solution is corrosive.

- Biuret solution usually looks blue in colour.
- A positive test for protein is if the biuret solution turns purple, violet or lilac.
- If the colour remains blue, then the extract does *not* contain protein.

Testing for fats

Fats will not dissolve in water but they will dissolve in **ethanol**. If a solution of fat in ethanol is added to water a cloudy white emulsion is formed.

- Chop up or grind a small amount of material you wish to test for fats. (Do not add water to make the extract this time.)
- Put the extract into a clean test tube and add enough ethanol to cover it.
- Put a stopper over the test tube and shake up the contents.
- Add some distilled water to make the test tube half full.
- Shake the contents of the test tube once more.
- A white emulsion that looks cloudy white or a milky colour is a positive test for fats.
- If this does not happen, the extract does *not* contain fat.

Testing for vitamin C

Vitamin C is in juices, such as freshly squeezed lemon juice. A solution of vitamin C can be made from vitamin C tablets. DCPIP is a blue liquid that loses its colour when it comes into contact with vitamin C.

1 Put a known volume of DCPIP solution in a test tube.

2 Fill a syringe or dropping pipette with a solution of vitamin C or with a juice, e.g., orange juice.

3 Add the liquid one drop at a time.

If the colour of DCPIP disappears the test is positive for vitamin C. If the blue colour persists then the test is negative.

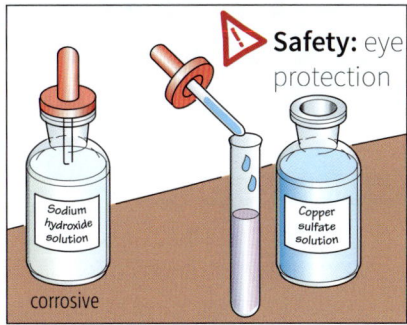

Safety: eye protection

Sodium hydroxide solution

corrosive

Copper sulfate solution

Figure 4.2.3 Test for protein.

SUMMARY QUESTIONS

1 Describe what you would see in each of the following:

   a a positive test for starch

   b a positive test for protein

   c a positive test for a reducing sugar

   d a negative test for starch

   e a positive test for vitamin C

2 Describe how you would carry out each of the following procedures:

   a making an extract of a food to test for reducing sugar

   b demonstrating how to carry out the reducing sugar test using a glucose solution

3 You are given four test tubes, each containing a different concentration of glucose. They are labelled **A**, **B**, **C** and **D** and you do not know the concentration of glucose in each tube.

You have been told to find out which solution is the most concentrated and which is the least concentrated. Describe the practical procedures that you would follow and the observations you would expect.

4 Describe how you would use the DCPIP test to compare the vitamin C content of different fruit juices.

KEY POINTS

1 Iodine solution gives a blue–black colour when added to starch.

2 Benedict's solution turns orange when boiled with reducing sugars.

3 Biuret solution gives a purple or violet colour when added to protein.

4 A white emulsion forms when fat dissolved in ethanol is added to water.

4.3 DNA

Supplement

LEARNING OUTCOMES

- Describe the structure of DNA as two strands coiled together to form a double helix
- State that DNA strands contain chemicals called bases
- Describe how the bases always pair up in the same way

EXAM TIP

A nucleotide is a 'building block' of DNA. The diagrams show how the bases, A, C, T and G, are arranged in DNA. You do not need to remember the structure of a nucleotide shown in Figure 4.3.2.

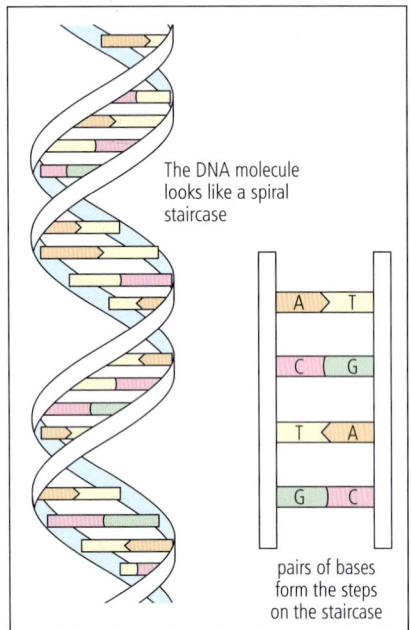

The DNA molecule looks like a spiral staircase

pairs of bases form the steps on the staircase

Figure 4.3.1 A DNA model.

The structure of DNA

Each chromosome is made up of thousands of **genes** arranged like beads in a necklace. It is the genes that carry the genetic information that affects how we grow and what we look like. For instance, there are genes for eye colour, hair colour and height. Some genes code for the production of enzymes that control all the chemical reaction that take place in cells (see page 198).

If we could unravel a chromosome, it would form an extremely long thread. The thread would be made up of a chemical called **DNA (deoxyribonucleic acid)**. A gene is made up of a short length of DNA so the long thread that makes up a chromosome contains hundreds of genes.

DNA belongs to a complex group of biological molecules known as **nucleic acids**. Each DNA molecule is made up of thousands of units each called a **nucleotide**. A single nucleotide is made up of three molecules:

- a phosphate
- a sugar
- a base

The sugar and phosphate molecules join up and form the backbone of the DNA strand. The bases are attached to the sugar molecules. If you look at Figure 4.3.1 you will see that DNA is made up of *two* strands of nucleotides. It is rather like a ladder. The whole molecule is twisted into a **double helix** – a bit like a spiral staircase with the bases as the steps.

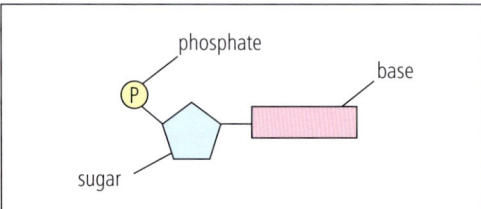

Figure 4.3.2 A single nucleotide. The sugars and phosphates make up the 'uprights' of the ladder and the bases make up the 'rungs'.

Base pairing

So how is the DNA molecule held together?

If you look at Figure 4.3.3 you can see that the bases join together. Each pair of bases is held next to each other between the two strands by weak bonds. There are four different bases in DNA:

- **thymine (T)**
- **adenine (A)**
- **cytosine (C)**
- **guanine (G)**

(You don't have to remember each of the names, just their letters.)

The bases always pair up in the same way:

- adenine (A) pairs with thymine (T)
- cytosine (C) pairs with guanine (G)

Although the bonds holding the two chains of nucleotides together are weak, there are many of them. So altogether they keep the DNA double helix in shape.

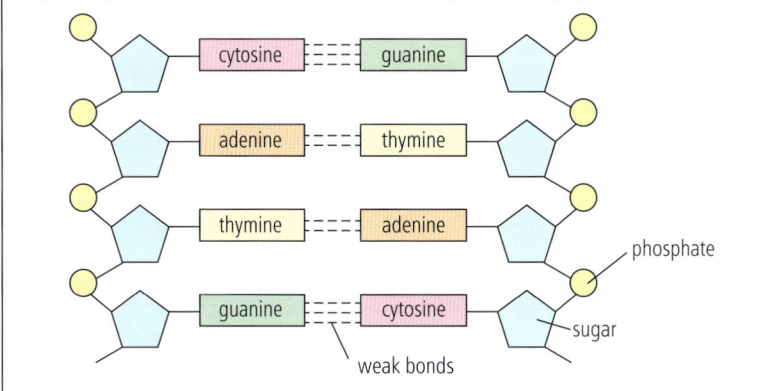

Figure 4.3.3 How the bases pair up in DNA.

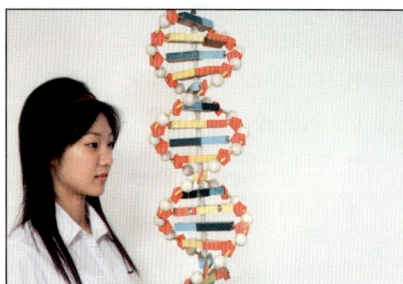

Figure 4.3.4 A student examines a model of DNA that shows the pairs of bases. In this model G (guanine) is yellow and C (cytosine) is red.

Figure 4.3.5 James Watson and Francis Crick worked out the structure of DNA by building models. They are posing with their final model that they made in Cambridge in the early 1950s.

SUMMARY QUESTIONS

1 The diagram below shows a partially completed section of the DNA molecule. Copy and complete the diagram by writing in the letters of the missing bases.

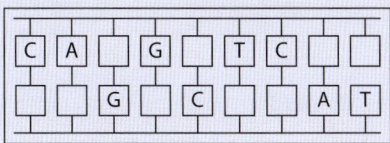

2 The table below shows the percentages of the four bases in DNA from four different organisms.

organism	percentage of each base			
	A	C	G	T
human	31	19	19	31
locust	29	21	21	29
yeast	32	18	18	32
bacterium	15	35	35	15

a Explain the pattern shown in the table.

b In an organism 26% of the bases in DNA are found to be A. What percentage would be C? Show your working.

c Explain how it is possible to have two very different organisms such as humans and yeast with very similar percentages of DNA bases.

KEY POINTS

1 The DNA molecule consists of two strands coiled together to form a double helix.

2 Each strand contains chemicals called bases (A, C, G and T), which bond together.

3 A always bonds with T; C always bonds with G.

4 A chromosome is made up of a long super-coiled strand of DNA.

5 A gene is a length of DNA that codes for the production of a particular protein.

Practice questions

1 What are the chemical elements in proteins?

  A carbon, hydrogen and oxygen

  B carbon, hydrogen, nitrogen and oxygen

  C carbon, hydrogen, nitrogen, oxygen and sulfur

  D carbon, nitrogen, oxygen and sulfur

(Paper 1) [1]

2 The small molecules that are built up into proteins are:

  A amino acids   **C** glucose

  B fatty acids   **D** glycerol

(Paper 1) [1]

3 Which row gives the positive results for the four chemical tests?

	Benedict's test	biuret test	iodine test	emulsion test
A	green	violet	yellow	no suspension
B	blue	blue	yellow	no suspension
C	yellow	red	blue	cloudy suspension
D	orange	violet	blue-black	cloudy suspension

(Paper 1) [1]

4 Glucose is used in the formation of which large molecules?

  1 starch

  2 glycogen

  3 cellulose

  A 1 only   **B** 1 and 3 only

  C 2 and 3 only   **D** 1, 2 and 3

(Paper 1)

5 Enzymes and antibodies are proteins. Special regions of these proteins have specific shapes that bind to other molecules. Proteins can form specific shapes because they have:

  A molecules of different lengths

  B molecules with different numbers of amino acids

  C molecules with different sequences of amino acids

  D molecules with other molecules or ions attached to them

(Paper 2) [1]

Use the diagram of DNA for questions 6–8.

DNA molecules are composed of sub-unit molecules known as nucleotides. The diagram shows a small part of a DNA molecule.

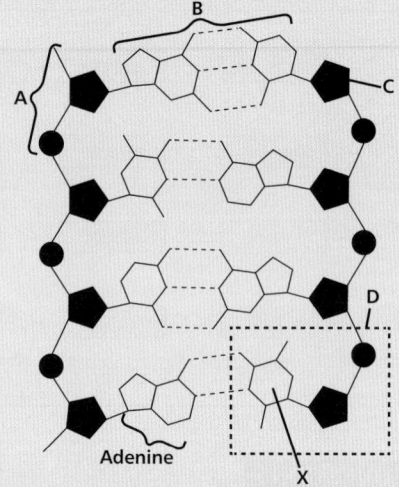

6 Which of the labelled parts of DNA, A, B, C or D, indicates a base pair?

(Paper 2) [1]

7 Which of the labelled parts of DNA, A, B, C or D, indicates a nucleotide?

(Paper 2) [1]

8 What is the base labelled X?

  A adenine (A)   **B** cytosine (C)

  C guanine (G)   **D** thymine (T)

(Paper 2) [1]

9 The diagram below shows how a starch molecule increases in length.

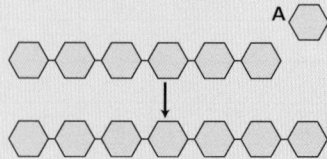

(a) Name molecule **A**. [1]

(b) Name two places where starch is stored in plants. [2]

(c) Describe how you would find out whether a plant tissue contained starch. Include the practical details of the test in your answer. [4]

(Paper 3)

10 The diagram shows a protein molecule:

(a) Name the sub-unit molecules that are assembled into the protein. [1]

(b) Name two proteins found in the human body. [2]

(c) State two places in the human body where many proteins are made. [2]

(d) Some seeds are a good source of protein. Describe the chemical test to determine whether different types of seed are good sources of protein. Include the practical details of the test in your answer. [4]

(Paper 3)

11 The diagram shows a molecule of a fat.

(a) Name the components of the molecule labelled **X** and **Y**. [2]

(b) List two places in the human body where fats are stored. [2]

(c) State three roles of fats in the human body. [3]

(d) Oils have the same molecular structure as fats. Many seeds are rich in oils. Describe a chemical test to find out whether seeds from different flowering plant species contain oil or not. Include the practical details of the test in your answer. [4]

(Paper 3)

12 A student used the DCPIP test to find out how much vitamin C is present in different fruit juices. The concentration of the DCPIP solution was $10\,g\,dm^{-3}$. The student followed these instructions.

1 Put $2\,cm^3$ of DCPIP solution into a test tube.

2 Use a graduated pipette or burette to add a $10\,g\,dm^{-3}$ vitamin C solution drop by drop to the DCPIP solution. Shake the tube gently after adding each drop. Continue to add the vitamin C solution until the colour of the DCPIP solution disappears.

3 Record the volume of vitamin C solution that was added.

4 Repeat the procedure two more times and calculate a mean volume.

5 Repeat steps 1 to 4 with the juices to be tested.

The results are shown in the table.

test substance	volume added to DCPIP solution / cm^3			
	1	2	3	mean
$10\,g\,dm^{-3}$ vitamin C solution	1.8	1.9	2.1	1.9
orange juice	2.4	2.6	2.3	2.4
grapefruit juice	3.2	3.1	2.9	3.1
apple juice	9.6	9.1	9.4	9.4

(a) Explain why the student took three readings for each test substance. [3]

(b) Use the results to calculate the concentration of vitamin C in the three fruit juices. Show your working. [3]

(Paper 6)

13 (a) DNA molecules have a double helix structure. Explain what this means. [2]

(b) State the precise site of DNA in an animal cell. [2]

The sequence of bases in a small sample of DNA is:

ATAGATCCCGAA

(c) Write out the sequence of bases in the opposite strand in this DNA molecule. [1]

(d) The DNA sequence determines a small part of the structure of a protein. Use the sequence above to explain how base sequences determine the sequences of amino acids in a protein. [1]

(e) In 2007, scientists compared the degree of similarity between some of the DNA taken from the nuclei of white blood cells of a domesticated cat and several other mammalian species. The results are below.

percentage similarity between domesticated cat and:					
wild cat	human	dog	cow	rat	mouse
99	90	82	80	69	67

What do you conclude from these results? [3]

(Paper 4)

5.1 Structure and action of enzymes

A **catalyst** speeds up a chemical reaction and remains chemically unchanged at the end of the reaction. Enzymes are proteins, produced by organisms, that speed up chemical reactions. They are known as **biological catalysts**. Enzymes are important because they control the rates of the reactions that support all life.

How enzymes work

Many chemical reactions take place in organisms. These reactions happen too slowly to keep organisms alive unless they are speeded up by enzymes. There are many different types of enzyme as each one catalyses a different reaction. Most enzymes work inside cells, but many of those that we will discuss here work outside cells, for example in the gut (Topics 7.4, 7.6 and 7.7).

The reactions that enzymes catalyse can be divided into three types.

1 Breaking large molecules into small ones

This is important in nutrition when large food molecules are broken down into small ones so that they can be absorbed and then used. Bacteria and fungi release enzymes to break down their food and we release enzymes into the gut for the same reason.

2 Building up large molecules from small ones

Small molecules, such as glucose, are joined together to make large molecules. These enzymes work inside cells to speed up the formation of storage molecules, such as starch, and structural molecules such as cellulose for cell walls of plants.

3 Converting one small molecule into another

Many of the chemical reactions that occur inside cells involve small changes to molecules, such as adding or removing atoms or groups of atoms. For example, there are enzymes that remove hydrogen from compounds during respiration.

Figure 5.1.2 shows the way in which an enzyme catalyses the breakdown of a molecule.

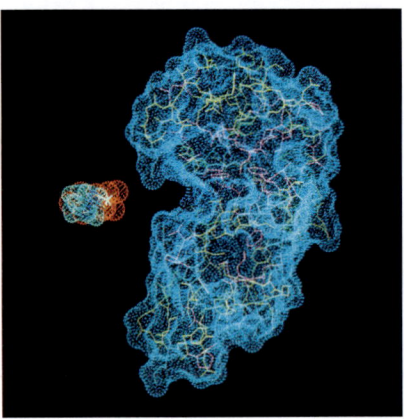

Figure 5.1.1 This computer-generated image of an enzyme (on the right) shows its 3D shape.

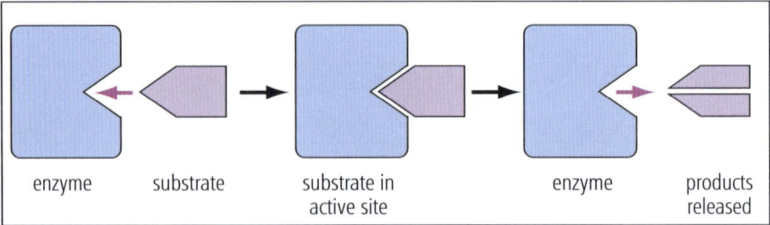

enzyme   substrate   substrate in active site   enzyme   products released

Figure 5.1.2

Figure 5.1.3 shows how an enzyme is involved in building a molecule from two smaller molecules.

Enzymes are made of protein

All enzymes have five important properties.

1 They are all proteins.
2 Each enzyme catalyses one particular metabolic reaction.
3 They can be used again and again.
4 They are influenced by temperature.
5 They are influenced by pH.

Enzymes are made of protein molecules. These molecules can be folded into many different shapes. Each type of enzyme molecule has a shape that makes it suitable for catalysing one type of reaction. This explains why there are many different enzymes – one enzyme for each particular reaction.

Enzymes catalyse reactions in which **substrates** are converted into **products**. Look at Figure 5.1.2. Notice that the shape of part of the enzyme matches the shape of the substrate molecule. As you can see in Figure 5.1.3, the part of the enzyme where the substrate(s) fit and where the reaction takes place is the **active site**. The active site and the substrate have shapes that are **complementary** so they fit together. Once they fit together the reaction can take place. Other substrates have the wrong shape to fit into the enzyme so will not be involved in the reaction catalysed by this enzyme. When the reaction is over the product or products leave the enzyme and another substrate molecule enters.

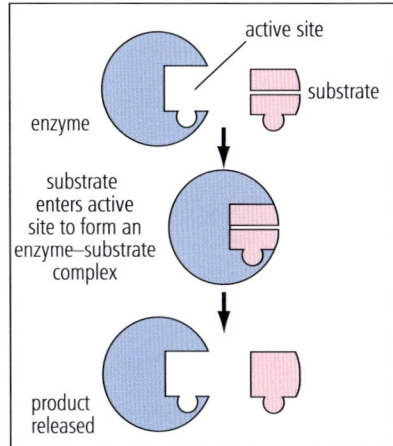

Figure 5.1.3 This shows how an enzyme can join two molecules together.

Figure 5.1.4 Only substrate A will fit the enzyme's active site.

Supplement

An enzyme can only control one particular reaction so we say that an enzyme is **specific** to that reaction. The enzyme and substrate combine like a key entering a lock to form an **enzyme–substrate complex**.

SUMMARY QUESTIONS

1 Copy and complete the sentences using these words:

speed   catalysts   proteins   reactions   substrate

Enzymes are biological _____ that _____ up the rate of chemical _____. Enzymes are all _____ and each enzyme acts on one type of _____.

S 2 Use the terms complementary and active site to explain what is meant by the 'lock and key' model of enzyme action.

3 Use the 'lock and key' model to explain:

a why each enzyme will act only on one substrate

b why an enzyme can be used to catalyse a large quantity of substrate

c why destroying the active site by heating the enzyme to a high temperature stops the enzyme from working.

KEY POINTS

1 Enzymes are biological catalysts that increase the rate of chemical reactions.

2 Enzymes catalyse reactions in which substrate molecules are converted to product molecules, either by building up or breaking down.

3 The active site is the part of the enzyme where the reaction occurs. The active site has a complementary shape to the substrate molecule(s).

4 The shape of the active site makes each enzyme specific to one reaction. **S**

5.2 Factors affecting enzyme action: temperature

LEARNING OUTCOMES

- Investigate the effects of temperature on enzyme activity

- Describe the effects of temperature on enzyme activity

S - Explain the effects of temperature on enzyme activity

The activity of an enzyme is determined by measuring the rate of the reaction that the enzyme catalyses. This may be done either by measuring how much product is formed or by measuring how much substrate is used over a period of time. The rate is like the speed of the reaction measured as the quantity of product or substrate per unit of time, e.g., per minute.

Effect of temperature on enzymes

The activity of enzymes is influenced by temperature. This graph shows the effect of increasing temperature on the rate of an enzyme-catalysed reaction.

Look at the graph and observe that the rate of reaction:

- is slow at low temperatures, e.g., at 10 °C
- increases as the temperature increases to 40 °C
- reaches a maximum at 40 °C
- decreases at temperatures greater than 40 °C
- is zero at 60 °C.

The temperature at which the maximum rate of reaction occurs is called the **optimum temperature**. This is the best temperature for the enzyme. Here are some examples of optimum temperatures:

- fungal and plant enzymes: approximately 20 °C (see page 250)
- human enzymes: 37 °C (body temperature – Topics 7.4 and 7.6)
- some of the enzymes produced by bacteria for use in industry: 90 °C.

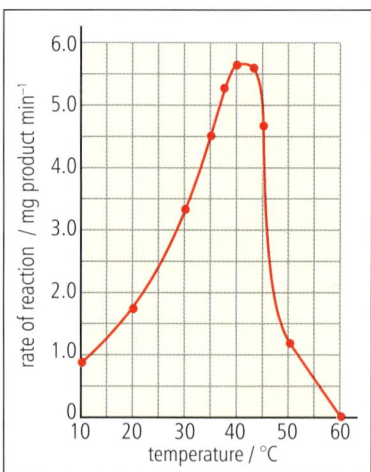

Figure 5.2.1 Effect of increasing temperature on the rate of an enzyme-catalysed reaction

EXAM TIP

When you are describing a graph like this one, always use some figures taken from the graph in your answer. Here you can say the highest rate is 5.6 mg product per minute at 40 °C.

Supplement

As we saw in Topic 5.1, an enzyme molecule is folded into a shape that accepts the substrate molecules. The shape of an enzyme's active site is maintained by bonds between different parts of the molecule. Remember that the shape of the active site determines whether the substrate will fit into the enzyme for the reaction to occur.

At first, increasing the temperature of an enzyme-controlled reaction will increase the rate. This is because enzyme and substrate molecules have greater **kinetic energy**. They move around more quickly and there are more chances of them colliding, the substrate fitting into the active site and a reaction taking place.

However, at higher temperatures the bonds holding the enzyme molecule together start to break down. This changes the shape of the active site, so the substrate no longer fits.

We say that the enzyme has been **denatured** and it can no longer catalyse the reaction.

PRACTICAL

Investigating the effect of temperature on enzyme action

Amylase is an enzyme that breaks down starch to maltose. Starch changes iodine solution blue–black. When starch is broken down, iodine solution will not change colour. Draw a table for your results like this:

temperature at which amylase and starch was kept / °C	time taken for starch to be fully broken down / min

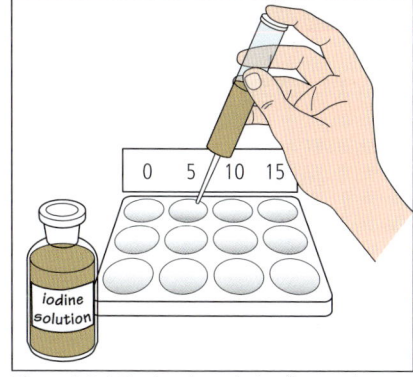

Figure 5.2.2 Testing the solution for starch using iodine solution.

1 Put 5 cm³ of starch solution into each of four test tubes labelled A, B, C, and D.
2 Put test tube A into the refrigerator and note the temperature.
3 Put test tube B into a test tube rack and note the temperature.
4 Put test tube C into a water bath at about 35 °C and note the temperature.
5 Put test tube D into a water bath at about 80 °C and note the temperature.
6 Collect 10 cm³ of amylase solution in a clean boiling tube.
7 Put a drop of iodine solution into each of the wells on a spotting tile.
8 Add 2 cm³ of amylase solution to each of the four tubes and note the time.
9 At two-minute intervals use a clean pipette to remove some of the test solution from each test tube and add it to a drop of iodine solution on the spotting tile.
10 When the amylase solution has broken down all the starch the iodine solution will no longer turn blue–black. They will stay light brown. What does this tell you about the effects of temperature on the amylase?

SUMMARY QUESTIONS

1 a Sketch a graph to show the effect of increasing temperature on the rate of a reaction catalysed by a human enzyme, such as salivary amylase.
  b Describe, in words, what is shown by your graph.
2 Suggest what each of the following would do to the rate of reaction catalysed by a human enzyme:
  a a temperature below 10 °C
  b a temperature of 37 °C
  c a temperature of 50 °C
S 3 Use the 'lock and key' model to explain what happens when enzymes are denatured at high temperatures.

EXAM TIP

You may be asked to <u>describe</u> the effect of temperature and pH on enzymes, but only in Paper 4 will you be asked to <u>explain</u> how these factors influence enzymes. When you explain, remember that enzymes are proteins and are denatured by high temperatures and extremes of pH.

KEY POINTS

1 Increasing the temperature of an enzyme-controlled reaction increases the rate of reaction up to a maximum, which occurs at the optimum temperature.

   S This is because greater kinetic energy is causing a greater number of collisions between enzyme and substrate molecules.

2 At higher temperatures the rate of reaction decreases until it stops acting as a catalyst.

   S This is due to a change in the shape of the active site, which means the substrate can no longer fit. The enzyme is now denatured.

5.3 Enzymes and pH

The rate of enzyme-controlled reaction can be affected by the pH of its surroundings. Hydrochloric acid is present in the stomach, so the enzymes that are active in the stomach work best in acidic conditions (i.e., at a low pH).

Alkaline bile is present in the first part of the small intestine, so the enzymes that are active here work best in alkaline conditions (i.e., at a high pH).

Most enzymes work best inside cells where the conditions are neutral. The optimum pH of these enzymes is therefore about pH 7.0.

Enzymes and pH

Enzymes are influenced by the pH of their surroundings.

Many enzymes work best in neutral conditions, but some work best in acidic conditions and some in alkaline conditions.

Look at the graph in Figure 5.3.1 showing the action of enzymes X and Y.

- Enzyme X works best at pH 2.0 – that is its optimum pH.

- The optimum pH for enzyme Y is pH 8.0.

- Up to pH 2.0 the rate of reaction increases for enzyme X and then between pH 2.0 and pH 5.5 it decreases. There is no reaction above pH 5.5.

- Between pH 4.5 and pH 8.0 the rate of reaction increases for enzyme Y and then between pH 8.0 and pH 10.0 it decreases. There is no reaction below pH 4.5 and above pH 10.0.

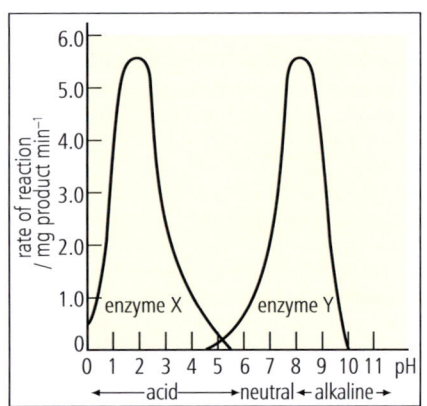

Figure 5.3.1

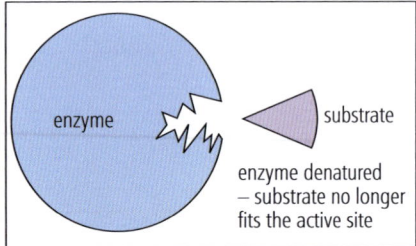

Figure 5.3.2

Supplement

As we have seen, the three-dimensional shape of an enzyme is vital if it is to function. Many of the chemical bonds holding the structure of the enzyme in shape are weak bonds. If these bonds that hold the enzyme molecule in shape are broken by changes in pH, then the shape of the active site can be altered. When the rate of reaction is zero, the shape of the active site has changed so much that the substrate molecules will no longer fit.

At these values of pH, enzymes are denatured. Small changes in pH can affect the rate of reaction without denaturing the enzyme. But at the extremes of its pH range, an enzyme becomes unstable and denatures, as shown in Figure 5.2.2.

PRACTICAL

The effect of pH on the action of a protease

Egg white contains a lot of protein. In this experiment you will investigate the effect of protease from the stomach on the protein in egg white. You will also see how pH affects the way the enzyme works.

1 Set up a water bath at 40 °C and label three test tubes 1–3.

2 Using a syringe put 5 cm³ of egg white into each test tube.

3 Add 2 cm³ of sodium carbonate (an alkali) to tube 1.

4 Add 2 cm³ of water to tube 2.

5 Add 2 cm³ of dilute hydrochloric acid to tube 3.

6 Compare the pH of each test tube by using a clean glass rod for each tube to dip into the solution and then touch the pH test paper.

7 Place all three test tubes in the water bath at 40 °C for five minutes.

8 Add 1 cm³ of protease to each test tube.

9 Compare the appearance of the tubes every minute until there is no further change.

Complete the table below.

tube	egg white + protease plus:	pH	appearance after 5 minutes
1	2 cm³ of sodium carbonate solution		
2	2 cm³ of water		
3	2 cm³ of hydrochloric acid		

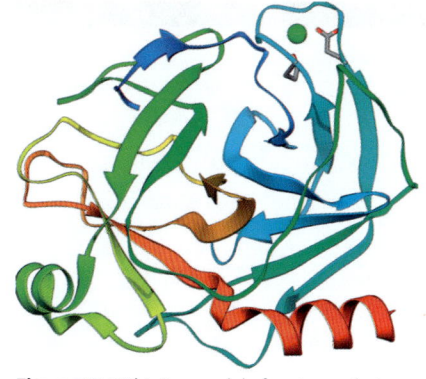

Figure 5.3.3 This is a model of protease that breaks down proteins to amino acids in the small intestine. It works best at a pH of about 8.5 and at 37 °C.

KEY POINTS

1 Most enzymes work best at one value of pH.

2 At either side of their optimum pH value, enzyme activity decreases.

3 At some values of pH the shape of the active site changes so that substrate molecules no longer fit. **S**

SUMMARY QUESTIONS

1 Sketch a graph to show the effect of increasing pH on the rate of an enzyme-catalysed reaction, where the optimum pH is 7.0 and there is no activity below pH 4.0 and none above pH 10.0.

S 2 Use the 'lock and key' model to explain what happens when enzymes are denatured at extremes of pH.

3 An investigation was carried out into the effects of pH on the action of the enzyme amylase on starch. Eight test tubes were set up at a different pH. They were incubated in a water bath at 30 °C for 1 hour. The concentration of reducing sugar (product) was then estimated. The results are shown in the table.

pH	4.0	5.0	6.0	6.5	7.0	8.0	9.0	10.0
relative concentration of reducing sugar produced	1	12	26	32	33	27	13	5

   a Plot a graph to show these results.

S b Explain the effects of pH on the action of amylase in this investigation.

Practice questions

1 Which is the correct statement about enzymes?

  A Enzymes are biological catalysts that function inside and outside cells.

  B Enzymes are catalysts made of protein that only function inside cells.

  C Enzymes are continually made by cells as they are used up during the reactions that they catalyse.

  D Enzymes only catalyse reactions in which substances are broken down into smaller molecules.

  (Paper 1)   [1]

2 The equation shows a reaction catalysed by an enzyme.

$$\text{fat} + \text{water} \xrightarrow{\text{lipase}} \text{fatty acids} + \text{glycerol}$$

  The progress of the reaction can be followed by detecting the decrease in pH. This is because:

  A fat is an acidic substance

  B fatty acids are one of the products and they lower the pH

  C the reaction stops when the pH decreases no further

  D water is a reactant

  (Paper 1)   [1]

3 Starch and human salivary amylase were mixed together at different temperatures. Which temperature will give the fastest rate of starch digestion?

  A 20 °C

  B 60 °C

  C 37 °C

  D 10 °C

  (Paper 1)   [1]

4 The pH of enzyme-controlled reactions may be changed from pH 7 to pH 8. How does this affect the rate of the reactions?

  A always increases the rate

  B always decreases the rate

  C has no effect on the rate

  D may change the rate or have no effect on the rate.

  (Paper 1)   [1]

5 The optimum temperature for an enzyme-catalysed reaction is the temperature at which:

  A least substrate is broken down

  B most product is formed

  C the rate of reaction is at its fastest

  D the reaction is completed in the longest time

  (Paper 1)   [1]

6 In an enzyme-catalysed reaction, the substrate binds to the part of the enzyme molecule known as the:

  A active site

  B complementary site

  C reaction site

  D substrate site

  (Paper 2)   [1]

7 Albumen is a protein found in egg white and in the blood. A solution of protease was kept at 75 °C and then added to a solution of albumen. No reaction occurred.

  Which is the most likely explanation?

  A 75 °C is below the optimum temperature of the protease

  B albumen cannot be broken down by proteases

  C the protease is denatured

  D there was not enough kinetic energy for the reaction

  (Paper 2)   [1]

8 The diagram shows an enzyme and four possible substrates.

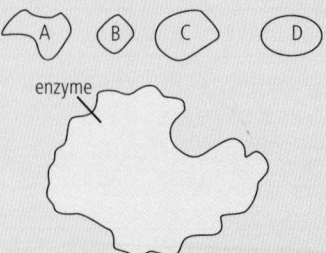

  Which of the substances shown is the substrate for this enzyme?

  (Paper 2)   [1]

9 (a) Explain the following statements.

  (i) Only small quantities of enzymes are required inside cells. [2]

  (ii) Cells make many different types of enzymes, not just one. [2]

 (b) Explain why amylase digests starch, but proteases do not. [1]

 (c) Explain why human enzymes catalyse reactions at 37 °C but not at 73 °C. [2]

(Paper 3)

10 The table shows the effect of temperature on the relative activity of an enzyme kept in a solution at pH 7.

temp / °C	relative activity
5	4
15	8
25	16
35	32
45	30
55	7

 (a) Explain why all the reaction mixtures were kept at the same pH? [2]

 (b) Draw a graph of the results in the table. [6]

 (c) Describe the results shown in the graph. [4]

(Paper 6)

11 The table shows the relative activity of a human enzyme in solutions of different pH kept at 35 °C.

pH	relative activity
3	5
5	14
7	32
9	10
11	4

 (a) Explain why all the reaction mixtures were kept at 35 °C? [3]

 (b) Draw a graph of the results in the table. [6]

 (c) Describe the results shown in the graph. [4]

(Paper 6)

12 The diagram shows an enzyme-catalysed reaction.

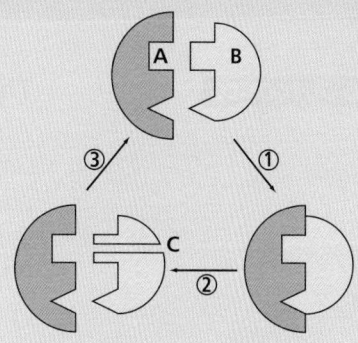

 (a) State the names of A, B and C. [3]

 (b) Use the diagram to describe what happens when an enzyme catalyses a reaction. [4]

 (c) Suggest why the model is known as 'lock and key'. [2]

 (d) Use the model to explain what happens when the enzyme molecule changes shape in high temperatures or extremes of pH. [4]

(Paper 4)

13 Hydrogen peroxide is a very toxic substance that is produced by cells as part of their metabolism. The enzyme catalase breaks down hydrogen peroxide and is found in organisms in all kingdoms.

$$\text{Hydrogen peroxide} \xrightarrow{\text{catalase}} \text{oxygen + water}$$

A student investigated the activity of the enzyme catalase in yeast by measuring the volume of oxygen collected over time. The graph shows the results.

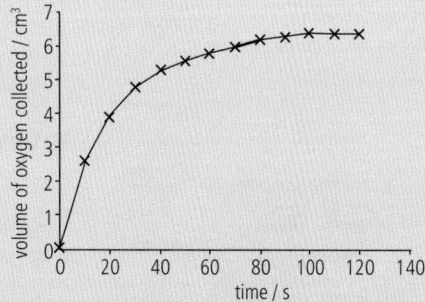

 (a) Use the results shown to describe the effect of catalase on hydrogen peroxide. [4]

 (b) Explain why catalase is the only enzyme to catalyse the breakdown of hydrogen peroxide. [2]

 (c) Describe what happens when an enzyme catalyses a reaction. [4]

(Paper 4)

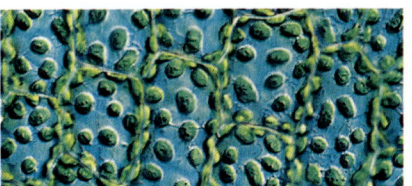

Figure 6.1.1 These plant cells are full of chloroplasts.

EXAM TIP

Make sure the equation is balanced when you write it out. Make sure there are the same number of atoms of carbon, hydrogen and oxygen on both sides of the arrow.

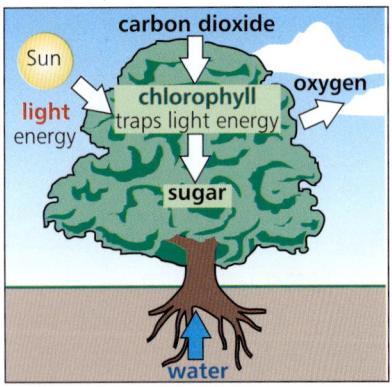

Figure 6.1.2 A simple overview of photosynthesis.

The process of photosynthesis

When you look at some leaf cells using the light microscope you see many round, green structures called **chloroplasts**. Green plants use light energy to convert carbon dioxide and water from their surroundings into simple sugars. This process is called **photosynthesis**.

Photosynthesis is the process by which plants make carbohydrates from raw materials using energy from light.

Photosynthesis can be summarised in this word equation:

$$\text{carbon dioxide} + \text{water} \xrightarrow{\text{light and chlorophyll}} \text{glucose} + \text{oxygen}$$

Supplement

The balanced chemical symbol equation for photosynthesis is:

$$6CO_2 + 6H_2O \xrightarrow{\text{light and chlorophyll}} C_6H_{12}O_6 + 6O_2$$

This equation suggests that photosynthesis is one simple reaction. This is not the case. There are many reactions in photosynthesis, each catalysed by a different enzyme. We can think of the symbol equation above as a summary of the whole process.

Light energy is absorbed by chlorophyll. The energy from light is transferred by chlorophyll as chemical energy to drive the reactions that form carbohydrates from water and carbon dioxide. In the process, water is broken down and oxygen is released.

Photosynthesis provides energy for plants and for all other organisms that feed on plants directly or indirectly (see page 232). The equations for photosynthesis show that oxygen is produced as a by-product. The plant may use oxygen in its own respiration or it may diffuse out into the atmosphere where it is used by other organisms.

Investigating photosynthesis

Green plants make simple sugars from carbon dioxide and water. However, if too much sugar is dissolved in the cell sap of plant cells it would make a very concentrated solution. This would make water move in from other cells by osmosis and cause the cells to swell so much that they would need to make very thick cell walls. To prevent this, glucose molecules are linked together to form larger starch molecules, which are insoluble and have no effect on osmosis. You can use iodine solution to test for the presence of starch in leaves. As you will see in Topic 6.2 we can use this test to find out whether photosynthesis has been happening or not.

PRACTICAL

Testing a leaf for starch

1 Submerge a leaf in boiling water for one minute. This kills the leaf as it destroys membranes, making it easier to extract the chlorophyll.

2 At this point, **turn off** any Bunsen burners used to boil the water.

3 Put the leaf into a test tube of ethanol. The chlorophyll is extracted by dissolving into the ethanol.

4 Stand the test tube in a beaker of hot water for about 10 minutes.

5 Wash the leaf in cold water. This removes the ethanol and rehydrates the leaf, which softens it and makes it easy to spread out.

6 Spread the leaf out flat on a white surface or in a Petri dish and put some drops of iodine solution on it.

7 If the leaf goes blue–black, starch is present. If it stays red/brown there is no starch.

Safety: wear eye protection

Safety: ethanol is flammable

boiling water

ethanol

turn off Bunsen burner

iodine

Safety: be careful not to burn yourself

Figure 6.1.3 Testing a leaf for starch.

SUMMARY QUESTIONS

1 Copy and complete the sentences using these words:

> **oxygen   starch   chlorophyll   carbon dioxide**
> **energy   water   photosynthesis**

Green plants make their own food by _____. Green _____ in the chloroplasts of the leaves traps the Sun's _____. Raw materials for this process are carbon _____ and _____. Simple sugars are made in the leaves and some are changed to _____. The waste product of this process is the gas _____.

2 To find out if photosynthesis has taken place in a leaf we can carry out a test.

Copy out the stages listed on the left. Match each one with the correct reason on the right.

Stage in test	Reason
wash the leaf in cold water	to test for starch
boil the leaf in ethanol	to soften it
add drops of iodine solution to the leaf	to remove the ethanol and rehydrate the leaf
dip the leaf in boiling water	to extract the chlorophyll

3 Write out the word equation for photosynthesis and identify the raw materials and products.

4 Using only information in this topic, explain why plants are vital to our survival.

KEY POINTS

1 Chlorophyll absorbs light energy that is used in photosynthesis.

2 Photosynthesis is the process by which plants make glucose from raw materials using energy from light.

3 Carbon dioxide and water are the raw materials for photosynthesis.

4 The products of photosynthesis are glucose and oxygen.

5 A leaf that has been exposed to light will give a positive test for starch.

Figure 6.2.1 It is important to understand the factors that affect photosynthesis in order to increase crop yields.

The requirements for photosynthesis

Green plants need the following in order to carry out photosynthesis:

- **light**, which provides energy for the process
- **chlorophyll**, a green pigment found in the chloroplasts of the leaves that absorbs the energy from light.
- **carbon dioxide**, which diffuses into the leaves from the air
- **water**, which is absorbed by the plant's roots from the soil.

Water and carbon dioxide are the **raw materials** for photosynthesis.

Investigations demonstrate that chlorophyll, carbon dioxide and light are needed for photosynthesis.

If a plant carries out photosynthesis it will make simple sugars and store them as starch. If it cannot photosynthesise then it will not make starch.

To be certain that the investigation is valid, we must make sure that the leaves have no starch at the start. To do this, a plant is left in the dark for at least 48 hours. This is called **de-starching**. The plant is then given all the things that it needs *except* for the substance that we are testing. (It is difficult to show that water is needed for photosynthesis, as water is required by a plant for so many other things; most of the plant is water so you can hardly 'take it away'.)

PRACTICAL

Showing that chlorophyll is needed for photosynthesis

Take a destarched, variegated plant such as a geranium.

('Variegated' means some parts of the leaves are white because there is no chlorophyll there.)

Place the plant in sunlight for about 6 hours.

Draw one leaf to show the white and green parts.

Now test this variegated leaf for starch using the starch test described in Topic 6.1.

You should find that only the green parts of the leaf go blue–black.

Figure 6.2.2 Variegated geranium leaves.

The green parts contain chlorophyll, which is needed for photosynthesis to make starch.

The white parts contain no chlorophyll, so no photosynthesis occurs here.

Therefore, the white parts of the leaf give a negative result with the starch test.

⚠️ **Safety:** wear eye protection

Showing that carbon dioxide and light are needed for photosynthesis

You could prove that a plant needs **carbon dioxide** to make its own food by providing it with everything it needs for photosynthesis *except* carbon dioxide. Then test to see if it has made starch.

PRACTICAL

Showing that carbon dioxide is needed for photosynthesis

1 Take a destarched plant. Enclose it in a plastic bag with a chemical that absorbs carbon dioxide. (**Soda lime** absorbs carbon dioxide.)

2 Leave the plant in the light for a few hours. Test a leaf for starch as described in Topic 6.1. The leaf should show a negative result for the starch test. Deprived of carbon dioxide the leaf is unable to photosynthesise and make starch.

A **control** experiment should be set up in exactly the same way but without the soda lime. This means the plant in the control experiment does not have carbon dioxide removed. Then we can be sure it was the absence of carbon dioxide that caused the lack of starch, and not keeping the plant inside the plastic bag.

Figure 6.2.3 Showing carbon dioxide is needed for photosynthesis.

You can also show that a plant needs **light** to carry out photosynthesis.

PRACTICAL

Showing that light is needed for photosynthesis

1 Take a destarched plant. Cover part of the leaf with some aluminium foil to prevent light getting through.

2 Leave the plant in the light for a few hours.

3 Test the leaf for starch as described in Topic 6.1. Only the parts of the test leaf that were left uncovered go blue–black. The parts of the leaf that were covered did not receive light and could not carry out photosynthesis and so could not make starch.

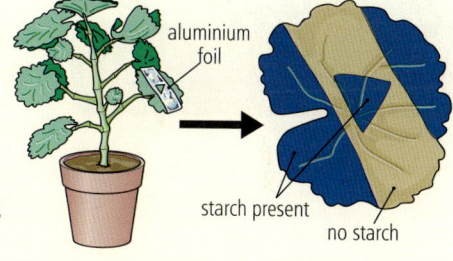

Figure 6.2.4 Showing light is needed for photosynthesis.

SUMMARY QUESTIONS

1 a Describe how you would de-starch a plant.

  b Explain why it is necessary to use a destarched plant to show that plants need light to carry out photosynthesis.

  c What is a variegated leaf?

  d Describe how you could carry out an investigation to show that carbon dioxide is needed for photosynthesis.

2 State whether the following statements are true or false.

  a Plants get all their food from the soil.

  b Plants need chlorophyll to carry out photosynthesis.

  c Water for photosynthesis is absorbed through the leaves.

  d Leaves that are destarched give a blue–black colour when tested with iodine.

3 Explain what is meant by a *control experiment*.

KEY POINTS

1 Plants need light, chlorophyll, carbon dioxide and water in order to carry out photosynthesis.

2 We can test to see that light, chlorophyll and carbon dioxide are needed for photosynthesis by not giving plants each of these and testing leaves for starch.

6.3 Products of photosynthesis

Figure 6.3.1 Photosynthesis makes food in these wheat plants.

The products of photosynthesis are simple sugars, such as glucose, and oxygen.

Most of the sugars that a plant needs for food are made in the leaves.

- Some of the **glucose** is used for respiration in the leaf.
- Some of the glucose is changed into **starch** and stored in the leaves for use in the future, e.g., at night.
- Some of the glucose is used to make **cellulose**, which is needed to make cell walls.
- Glucose is converted to **sucrose** and transported to other parts of the plant in the phloem (Topic 8.5).

Glucose can also be converted to other substances:

- Plants get nitrogen by absorbing nitrate ions from the soil.
- Glucose and nitrate are used to form **amino acids**, which are built up into **proteins**.
- Plants need proteins for growth and cell repair and for making enzymes and hormones.
- **Sugars** are converted to **oils**, which are an efficient way to store energy in seeds.

The importance of photosynthesis

Think about the food you have eaten in the last 24 hours. A lot of what you eat comes from plants like rice, maize, wheat, potatoes, soya beans and nuts.

If you eat meat or fish then that comes from animals that have eaten plants.

Products like cooking oil and margarine are made from plants.

Oxygen is a by-product of photosynthesis

Oxygen is a product of photosynthesis. It is not the main product. Some is used by the plant's respiration, but usually there is more than is needed, so most diffuses out of the leaves into the atmosphere. That is why the oxygen is called a **by-product**.

Plant products

Plants provide us with our food and also food for our livestock. There is more about this on page 246.

Plants also provide a range of raw materials for industry, such as timber and cotton. Many medicines have been discovered in plants. These include digitalis, which is a heart drug, and two anti-cancer drugs extracted from the rosy periwinkle from Madagascar.

Plants are the dominant organisms in almost all environments. As such they provide habitats for animals and microorganisms. The rainforests make up only 6% of the Earth's land surface, but they support more than half the world's species of animals and plants.

The concentrations of gases in the atmosphere are kept constant by photosynthesis. Without green plants, the concentration of carbon dioxide in the air would increase and the concentration of oxygen would decrease.

Figure 6.3.2 How much of this food depends upon plants?

PRACTICAL

Oxygen produced in photosynthesis

1 Set up the apparatus to collect bubbles of gas given off by the Canadian pondweed (*Elodea canadensis*).

A lamp provides the light energy that the plant needs. Carbon dioxide is dissolved in the water. Add some sodium hydrogencarbonate powder to the water and stir to dissolve.

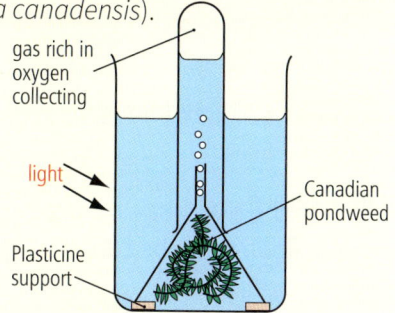

Figure 6.3.3 Showing that photossynthesis produces oxygen.

This is to make sure that there is always enough carbon dioxide disolved in the water.

If there is not enough carbon dioxide then photosynthesis slows down.

2 Place the apparatus in the light and allow a few hours for the gas to collect.

3 Test the gas for oxygen with a glowing splint.

Standing the funnel on plasticine gives a gap to allow carbon dioxide dissolved in the water to reach the pondweed so photosynthesis does not slow down.

A water plant is used because it is easy to collect the gas by downward displacement of water.

SUMMARY QUESTIONS

1 What are the following products made in plants used for?
   a glucose **b** starch **c** cellulose **d** protein **e** oils

2 **a** Describe how you would show that plants produce oxygen as a by-product of photosynthesis. Remember to include all practical details.
   b How would you show that light is necessary for oxygen production?

3 Plants are important because they provide us with:
   a food **b** fuels **c** building materials **d** medicines
   Give three examples of plants that provide us with each of these.

KEY POINTS

1 Products of photosynthesis include glucose, which is used to make starch and cellulose; glucose is also converted to amino acids, which are used to build proteins.

2 Photosynthesis is important as it provides us with food and medicines; it also helps to keep constant concentrations of carbon dioxide and oxygen in the atmosphere.

- Describe how to investigate the effect of varying light intensity, temperature and carbon dioxide concentration on the rate of photosynthesis and describe the results

When plants carry out photosynthesis they produce oxygen. We can determine how fast or slow photosynthesis occurs by measuring the products that are produced. This can be done by measuring how much starch or oxygen is made. Starch production can be measured by finding out the change in dry mass, but it is not an easy method to use. Instead, measuring how much oxygen is produced is much easier as we can use an aquatic plant as described in Topic 6.3. This is done by counting bubbles or by measuring the volume of oxygen produced.

Environmental conditions such as light intensity, temperature and carbon dioxide concentration influence the rate of photosynthesis.

PRACTICAL

Photosynthesis and light intensity

You can use this apparatus to measure the effects of light intensity on the rate of photosynthesis:

1 Cut a piece of pondweed about 5 cm in length.

2 Put a paperclip on the pondweed to stop it floating to the surface.

3 Put a lamp close to the plant and measure the distance between the plant and the lamp.

4 Count the number of bubbles released over 5 minutes. Repeat several times and calculate the average.

5 Repeat this procedure with the lamp at different distances from the plant.

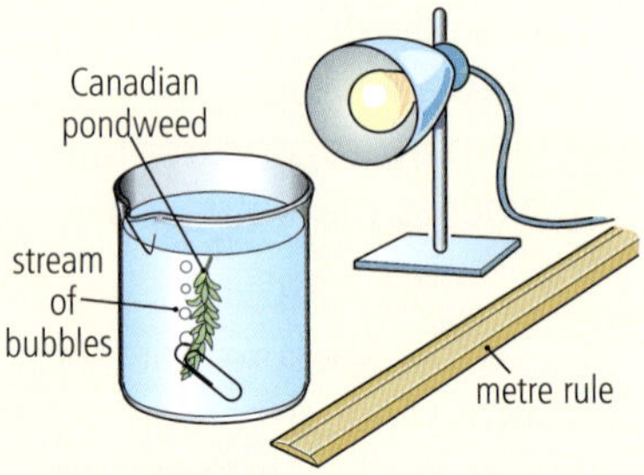

Canadian pondweed

stream of bubbles

metre rule

Figure 6.4.1 Apparatus for showing the effect of light intensity on photosynthesis

The number of bubbles should decrease as the distance between the lamp and the plant increases.

When the lamp is close to the plant the light intensity is high. This gives the plant lots of energy so the rate of photosynthesis is high. As the lamp is moved further away the light intensity decreases. This means there is less energy for the plant and the rate of photosynthesis decreases.

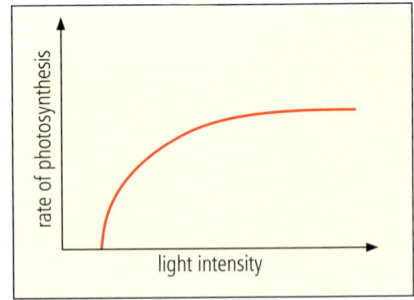

Figure 6.4.2 The effect of increasing light intensity on the rate of photosynthesis.

The graph shows some results of an experiment to measure the effect of increasing light intensity on the rate of photosynthesis.

In the dark there is no photosynthesis. At low light intensities there are very few bubbles of oxygen produced. The rate of photosynthesis increases as the light intensity increases until at a certain light intensity the rate does not increase any further. Increasing the light intensity increases the energy available to the plant for photosynthesis.

When investigating the effect of different light intensities it is important to keep the other environmental conditions the same. Temperature can be kept constant by putting a thermometer into the beaker of water and adding hot or cold water if the temperature changes. The lamp may heat up the water, so the temperature may increase. The concentration of carbon dioxide in the water can be kept constant by adding sodium hydrogencarbonate to the water and letting it dissolve. This ensures that the plant will not run out of carbon dioxide.

PRACTICAL

Photosynthesis and temperature

To find the effect of temperature similar apparatus can be used. Pieces of pondweed are placed into test tubes, which are put inside beakers of water at different temperatures. The minimum is five different temperatures across a range of 10 °C to about 40 °C. The lamp is kept at one distance all the time so that the light intensity is constant. Sodium hydrogencarbonate is added to the water in the test tubes to make sure carbon dioxide concentration is constant. The number of bubbles of oxygen produced at different temperatures can be counted in the same way.

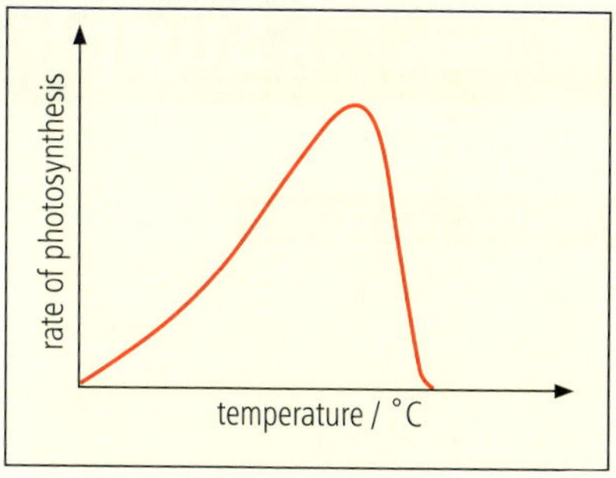

Figure 6.4.3 shows the results you might expect. The maximum rate of photosynthesis occurs at the **optimum temperature**. The actual temperature depends on the plant that is used. Plants that grow in cooler temperate latitudes have optimum temperatures that are lower than those that grow in the hotter tropics.

Figure 6.4.3 The effect of increasing temperature on the rate of photosynthesis. The rate decreases at high temperatures because enzymes in the chloroplasts are denatured.

PRACTICAL

The effect of carbon dioxide on the rate of photosynthesis

The same apparatus can be modified for this investigation.

Adding different quantities of sodium hydrogencarbonate to the water increases the concentration of carbon dioxide. At least five different test tubes can be used, each with a different mass of sodium hydrogencarbonate added.

The light intensity and the temperature must be kept constant.

You can see that the effect of increasing carbon dioxide on the rate of photosynthesis is the same as the effect of increasing the light intensity. If there is no carbon dioxide there can be no photosynthesis. As the carbon dioxide concentration increases there is an increase in the rate of photosynthesis. Above a certain concentration the rate remains constant.

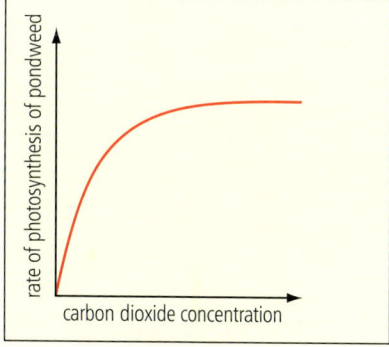

Figure 6.4.4 The effect of carbon dioxide on the rate of photosynthesis.

SUMMARY QUESTIONS

1 List the main environmental factors that influence the rate of photosynthesis.

2 A student used the apparatus shown in Figure 6.4.1. He wrote down the distances and the numbers of bubbles per minute: 20 cm and 25, 30 cm and 24, 40 cm and 18, 60 cm and 10, 100 cm and 6.
   a Explain how the student kept other conditions constant.
   b Make a table of the results and plot them as a line graph.
   c Explain why your graph does not look like Figure 6.4.2.

3 Describe the effect on the rate of photosynthesis of
   a increasing carbon dioxide concentration, and
   b increasing the temperature.

KEY POINTS

1 The release of oxygen from aquatic plants is used to measure the rate of photosynthesis.

2 Light intensity, temperature and the concentration of carbon dioxide are three environmental factors that influence the rate of photosynthesis.

LEARNING OUTCOMES

- Explain the term *limiting factor*
- Identify and explain the limiting factors of photosynthesis in different environmental conditions

We have seen that light intensity, temperature and carbon dioxide concentration influence the rate of photosynthesis in plants.

- Light intensity determines the energy available to photosynthesis.
- Temperature influences the activity of enzymes in the chloroplasts.
- Carbon dioxide is a raw material for photosynthesis, so if the concentration increases there is more of it available for chloroplast enzymes to use to make carbohydrates.

These three factors are **limiting factors**.

Limiting factors

A limiting factor is something present in the environment in such short supply that it restricts life processes.

Light intensity influences the rate of photosynthesis. If the light intensity is low, it does not matter if the plant has lots of carbon dioxide and water and a warm temperature because there is a shortage of energy for photosynthesis and the rate cannot be very high. In this case light intensity is the limiting factor for the rate of photosynthesis. In Figure 6.5.1 you can see that as the light intensity increases the rate of photosynthesis increases. The rate increases because light intensity is the limiting factor. You can also see that the rate becomes constant however much the light intensity is increased. Light intensity is now no longer the limiting factor, it must be something else such as temperature or carbon dioxide concentration.

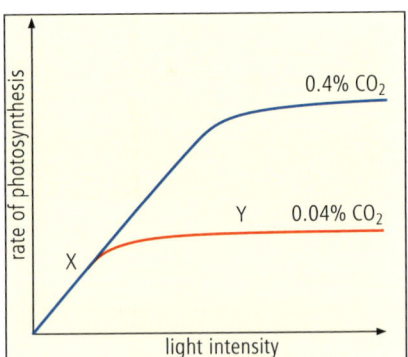

Figure 6.5.1 The effect of increasing light intensity on the rate of photosynthesis at two concentrations of carbon dioxide.

Chemical reactions catalysed by enzymes increase with **temperature**. You can see from the graph in Figure 6.4.3 on page 61 that the rate of photosynthesis increases until a certain temperature when the rate reaches a maximum and then decreases. The rate increases because the kinetic energy of the substrates and enzymes in the chloroplast increase and they collide more often (Topic 5.2). The rate of photosynthesis decreases at higher temperatures because the enzymes in chloroplasts are denatured (Topic 5.3).

EXAM TIP

When you use a graph to answer a question, always move a ruler along the graph from left to right and follow the trends or patterns. Here you can rule lines on the graph to find the conditions when the rate starts to remain constant.

Look at the graph in Figure 6.5.1. Light intensity is limiting the rate of photosynthesis at point X. Another factor is limiting the rate of photosynthesis at point Y. This could be carbon dioxide concentration or temperature.

However, you can see that the rate of photosynthesis increases if carbon dioxide concentration is increased from 0.04% to 0.4% as there is more of this raw material available. This means that the limiting factor at Y is carbon dioxide concentration.

Glasshouse production

Growers try to improve the yield of their crops by giving them the best possible environmental conditions for photosynthesis.

Conditions inside a glasshouse allow plants to:

- grow earlier in the year
- grow in places where they would not normally grow well.

The following conditions inside glasshouses are controlled.

- **Temperature**

  Sunlight heats up the inside of the glasshouse.

  The glass stops a lot of this heat from escaping.

  Electric heaters are used in cold weather.

  Ventilator flaps are opened to cool the glasshouse on hot days.

- **Light**

  The glass lets in sunlight. Artificial lighting can be used to grow plants when light intensity gets too low. Blinds keep out very strong light and shading lowers the temperature in tropical countries.

- **Carbon dioxide**

  Growers can pump carbon dioxide into glasshouses to increase carbon dioxide concentration. They can also burn butane or natural gas, which provides carbon dioxide and also heat to raise the temperature of glasshouses in cold weather.

- **Water**

  Many glasshouses have automatic watering systems using sprinklers and humidifiers, which ensure plants always get enough water.

Glasshouse production provides a good example of how the applications of limiting factors can be used by growers to maximize crop yield.

Figure 6.5.2 Glasshouses can control limiting factors, such as temperature, light intensity and humidity.

Figure 6.5.3 Water misting of a glasshouse crop.

KEY POINTS

1 A limiting factor is a factor in the environment that is in such short supply that it restricts a life process, such as photosynthesis.

2 Light intensity, temperature and carbon dioxide concentration are limiting factors of photosynthesis.

3 Carbon dioxide concentration, light intensity, humidity and temperature are controlled in modern glasshouse systems to give optimum conditions. This ensures that the rate of photosynthesis is kept high so plants produce maximum yield.

SUMMARY QUESTIONS

1 Explain what is meant by the term *limiting factor*. Use the process of photosynthesis to help explain your answer.

2 State which factors would be limiting photosynthesis in the following. Give your reasons in each case.

  a Plants growing on a high mountainside early in the morning.

  b Plants growing in a dense tropical forest at midday.

  c A field of maize early on a very bright morning.

  d The same field late in the afternoon.

  e A house plant in a shaded position in a warm room.

3 Explain how each of the following limiting factors can be controlled in glasshouses to give optimum conditions for growth:

  a carbon dioxide concentration

  b light intensity   c temperature.

6.6 Leaves

Figure 6.6.1 A leaf has an ideal shape for photosynthesis.

The shape of a leaf makes it ideally adapted to carry out its functions of absorbing light for photosynthesis and allowing gases to diffuse into and out of the leaf tissues.

Leaves have:

- **a large surface area** – to absorb light rays
- **a thin shape** – so gases can diffuse in and out easily
- **many chloroplasts** – to absorb light for the reactions that take place in photosynthesis
- **veins** – to support the leaf surface and to carry water and ions to the leaf cells, and to take sucrose and amino acids away from the leaf to all other parts of the plant.

Inside a leaf

To find out how a leaf works we can look at the structure of a typical leaf.

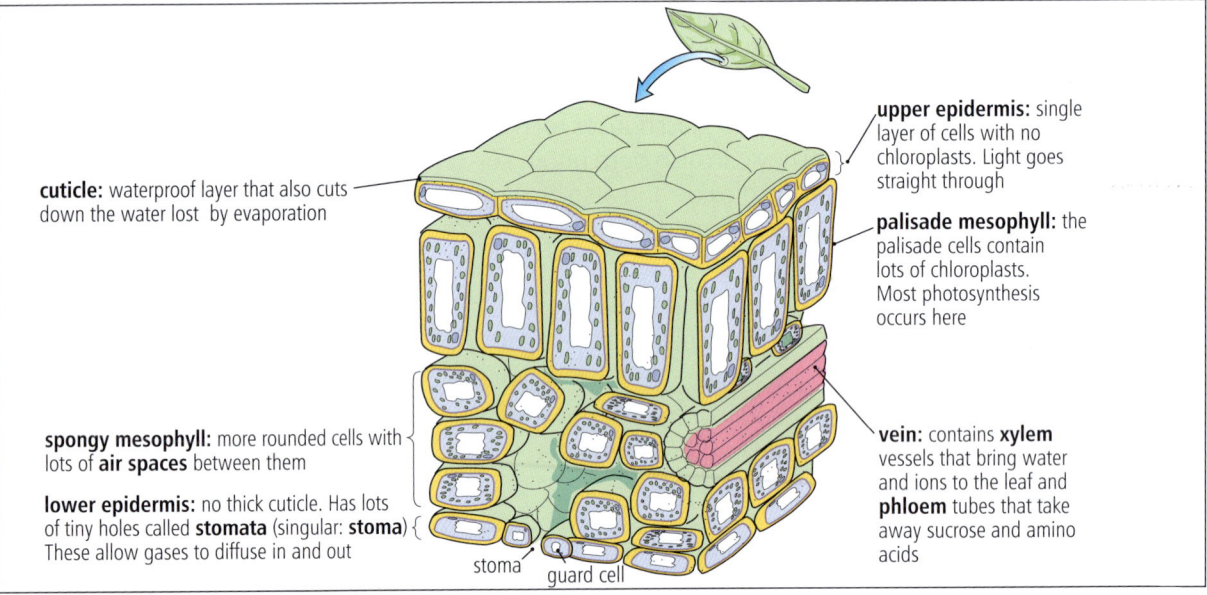

cuticle: waterproof layer that also cuts down the water lost by evaporation

upper epidermis: single layer of cells with no chloroplasts. Light goes straight through

palisade mesophyll: the palisade cells contain lots of chloroplasts. Most photosynthesis occurs here

spongy mesophyll: more rounded cells with lots of **air spaces** between them

lower epidermis: no thick cuticle. Has lots of tiny holes called **stomata** (singular: **stoma**) These allow gases to diffuse in and out

vein: contains **xylem** vessels that bring water and ions to the leaf and **phloem** tubes that take away sucrose and amino acids

stoma   guard cell

Figure 6.6.2 Leaf structure.

Some of the internal features are adaptations for photosynthesis.

- Palisade mesophyll cells are packed tightly together near the upper surface of the leaf to maximise absorption of light where its intensity is highest.
- There are many chloroplasts in the palisade mesophyll cells to absorb as much light as possible.
- Stomata (usually in the lower epidermis) open to allow carbon dioxide to diffuse into the leaf. Carbon dioxide is a raw material for photosynthesis.
- Leaves are thin so that carbon dioxide does not have to diffuse far from the atmosphere to the cells of the palisade and spongy mesophyll.

- There are large intercellular air spaces within the spongy mesophyll layer. This makes it easy for carbon dioxide to diffuse to all the mesophyll cells. Diffusion through air is much faster than diffusion from cell to cell.
- Xylem in veins bring water and ions to the mesophyll cells. Water is a raw material for photosynthesis and magnesium ions are needed by the cells to make chlorophyll.
- The sugar produced in photosynthesis is converted to sucrose and transported away from the leaf in the phloem in veins.

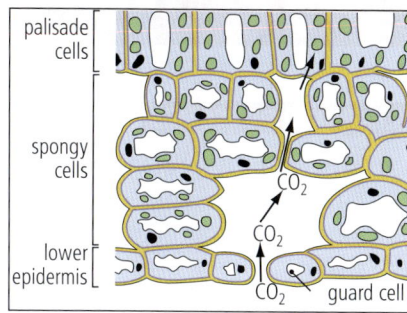

Figure 6.6.3 Carbon dioxide diffuses into the leaf for photosynthesis.

Stomata

Stomata are small pores (holes) in the epidermis that allow gases to diffuse into and out of the leaf. Stomata are usually in the lower epidermis, but some plants like water lilies have them in the upper epidermis.

In sunlight:

- carbon dioxide diffuses in for photosynthesis
- oxygen made in photosynthesis diffuses out
- water vapour diffuses out.

Opening and closing

Stomata are opened and closed by **guard cells**.

Stomata usually open during the day. Water passes into the guard cells by **osmosis**. This makes them bend so the stoma opens.

Carbon dioxide diffuses into the leaf for photosynthesis and oxygen diffuses out. Water vapour also diffuses out.

At night the stomata close. Water passes out of the guard cells by osmosis and they straighten and move closer together so closing the stomata pores. The stomata also close in hot, dry weather to help prevent the plant wilting.

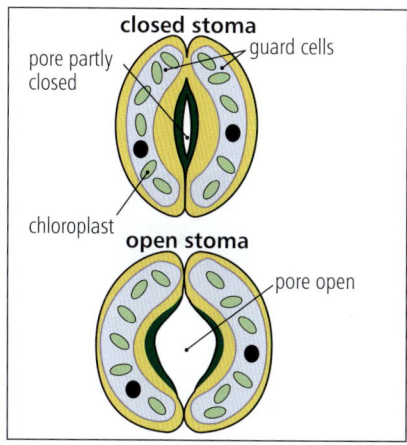

Figure 6.6.4 Closed stoma and open stoma.

SUMMARY QUESTIONS

1 Match each of the leaf parts on the left with their correct function on the right:

stomata	carry water up the stem
palisade mesophyll	allow gases to pass into and out of leaf
spongy mesophyll	contain chloroplasts for photosynthesis
cuticle	contain air spaces for gas exchange
veins	prevents too much water being lost

2 **a** Which gas would pass into a leaf during daylight?

b Which gas passes out of a leaf during daylight?

3 Leaves have a large surface area to absorb light.

Lay a leaf onto graph paper, draw around it and work out its area by counting the squares.

Find out if leaves from the top and bottom of a plant have the same leaf area. Record your results and explain them.

KEY POINTS

1 The diagram of the internal structure of a leaf shows different tissues, which each have important functions to carry out.

2 Stomata control the diffusion of gases into and out of a leaf.

6.7 Mineral requirements

Plants need more than light, carbon dioxide and water for healthy growth. They also need **mineral ions** (also called **plant nutrients**). Nutrients are absorbed from the soil in small quantities as ions by active transport in the roots (Topic 3.4).

Plant nutrients

Nutrients are needed for healthy growth of plants. They are used for a variety of purposes in plants. If these nutrients are lacking in the soil then plants do not grow well and show certain symptoms known as deficiency symptoms. Plants need **nitrate** ions to make amino acids, which are used to make proteins. As proteins are required for growth, plants deficient in nitrate show poor growth.

Magnesium ions are absorbed by plants and used to make chlorophyll. Leaves of plants deficient in chlorophyll look yellow, a condition known as chlorosis. Plants need phosphate for making compounds such as DNA and for respiration.

If the soil does not contain enough nutrients, farmers and growers may add more as **fertiliser** to replace the missing nutrients.

Figure 6.7.1 This plant has magnesium deficiency.

Supplement

PRACTICAL

Some maize plants were grown in water culture to investigate the need for nitrate ions and magnesium ions in plant growth.

Four test tubes were set up as shown in Figure 6.7.2. Each solution lacked a certain nutrient except for test tube 1, which had both nitrate ions and magnesium ions. Each test tube was covered in black paper. The apparatus was then kept in good light for 6 weeks, after which the plants were examined for:

- colour and size of leaves
- length of stems and roots.

1 How would you measure these changes?

2 Describe the results in:
   a   test tube 2 (no magnesium ions)
   b   test tube 3 (no nitrate ions)
   c   test tube 4 (distilled water).

3 Why do you think the test tubes were covered in black paper? What other plants might have grown in the water?

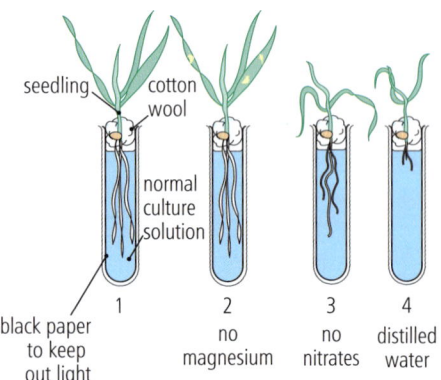

seedling   cotton wool

normal culture solution

black paper to keep out light

1   2   3   4
         no   no   distilled
         magnesium nitrates water

Figure 6.7.2 The effect of nutrients on growth.

Gas exchange in plants

All living organisms, including plants, carry out respiration all the time. Plants are no different from other organisms. To carry out respiration they need a supply of oxygen and they produce carbon dioxide as a waste. At night, plants exchange these gases with their surroundings. During the day, plants carry out photosynthesis, as well as respiration. In bright light, when there is a high rate of photosynthesis, some of the oxygen produced in their chloroplasts is used by mitochondria for aerobic respiration. The rest of the oxygen is not required so diffuses out of the plant. The carbon dioxide produced by mitochondria in respiration is used by chloroplasts for photosynthesis. However, this is not enough so carbon dioxide diffuses in from the surroundings.

Gas exchange in aquatic plants is investigated with hydrogencarbonate indicator solution. This solution contains hydrogencarbonate to provide carbon dioxide and two pH indicators. Carbon dioxide is an acidic gas. When it dissolves in water it forms carbonic acid, which is a weak acid. Hydrogencarbonate indicator solution is prepared by bubbling atmospheric air through it. This gives the indicator a red colour.

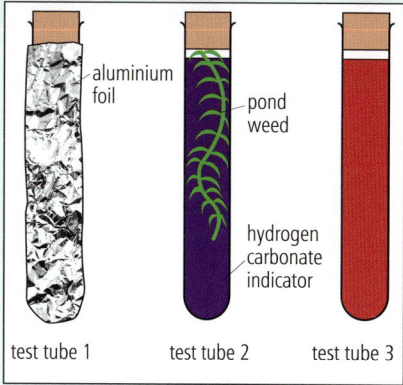

Figure 6.7.3 Experiment to show photosynthesis and respiration in plants

PRACTICAL

Photosynthesis and respiration in plants

1 Set up three test tubes as shown in Figure 6.7.3.
2 Add hydrogencarbonate indicator solution to each test tube.
3 Cut two pieces of pondweed to 10 cm in length and place one in each of test tubes 1 and 2.
4 Put test tube 1 in darkness by covering it in foil.
5 Leave the test tubes near a light for 2–3 hours.

If carbon dioxide is added to the water by the plant, the hydrogencarbonate indicator solution will turn from red to yellow. If carbon dioxide is taken up from the water by the plant, the hydrogen carbonate indicator solution will turn from red to purple.

- In test tube 1, the indicator turned yellow because carbon dioxide was released by the pondweed during respiration.
- In test tube 2, the indicator turned purple because carbon dioxide was taken up by the pondweed for photosynthesis.
- The purpose of test tube 3 was to act as a control.

The indicator solution only responds to changes in carbon dioxide. The colour changes are nothing to do with the oxygen released by the plant in the light or absorbed by the plant in the dark. When oxygen dissolves in water it does not change the pH. If you wanted to monitor the changes in oxygen concentration you would have to use an oxygen probe and a data logger.

If pond weed is kept in a test tube of hydrogencarbonate indicator solution at low light intensity, the colour of the solution usually remains red as the rates of photosynthesis and respiration are the same.

SUMMARY QUESTIONS

1 Explain why plants need mineral ions.
2 Describe the likely appearance of plant growth without nitrate ions.
3 Describe and explain the gas exchange that occurs between a plant and its surroundings **a** at night and **b** in bright sunshine.

KEY POINTS

1 Plant roots absorb nutrients including nitrate ions needed for healthy growth.
2 Nitrate ions are needed to make amino acids; magnesium ions are used by the plants to make chlorophyll.
3 If mineral ions are lacking, a plant develops deficiency symptoms.
4 In light plants use up carbon dioxide in photosynthesis and release oxygen. At night plants produce only carbon dioxide by respiration.
5 Hydrogencarbonate indicator solution is used to detect changes in carbon dioxide concentrations as a result of gas exchange by aquatic plants.

Practice questions

1 Photosynthesis occurs in palisade mesophyll cells. Which structure inside these cells carries out photosynthesis?

  A chloroplast

  B cytoplasm

  C nucleus

  D vacuole

(Paper 1)   [1]

2 The raw materials for photosynthesis are:

  A carbon dioxide and nitrate ions

  B light energy and water

  C light energy, carbon dioxide and water

  D water and carbon dioxide

(Paper 1)   [1]

3 Magnesium ions are used by plants for making:

  A cellulose

  B chlorophyll

  C fats and oils

  D proteins

(Paper 1)   [1]

4 Which represents the movement of gases through the open stomata of a leaf during the day?

	gas into leaf	gas out of leaf
A	oxygen	carbon dioxide
B	water vapour	carbon dioxide
C	carbon dioxide	oxygen
D	oxygen	water vapour

(Paper 1)   [1]

5 Which feature is an adaptation to ensure a good supply of carbon dioxide to the cells inside a leaf?

  A large internal air spaces

  B many chloroplasts in palisade mesophyll cells

  C thin cuticle on the lower epidermis

  D xylem and phloem in veins

(Paper 1)   [1]

6 Hydrogencarbonate indicator solution is red when atmospheric air is bubbled through it. A student placed a piece of pondweed inside test tube Y that contained this indicator solution. Tube Y was covered completely in foil. This was repeated for test tube Z, but the tube was left uncovered. Both tubes were left in bright sunshine. After 60 minutes the colour of the indicator in tube Y was yellow and in tube Z was purple. What is the correct reason for this?

  A pondweed in Y had absorbed carbon dioxide and the pondweed in Z had produced carbon dioxide

  B pondweed in Y had produced carbon dioxide and the pondweed in Z had produced oxygen

  C pondweed in Y had produced carbon dioxide, but the pondweed in Z had absorbed carbon dioxide

  D pondweed in Y had produced oxygen, but the pondweed in Z had absorbed oxygen

(Paper 1)   [1]

7 Which does *not* occur during photosynthesis?

  A absorption of light by chlorophyll

  B conversion of glucose to sucrose

  C reduction of carbon dioxide to a carbohydrate

  D splitting of water molecules using light energy

(Paper 2)   [1]

8 Which row of the table would provide the conditions for the highest rate of photosynthesis of a tropical plant?

	light intensity	carbon dioxide concentration	temperature / °C
A	high	low	30
B	low	high	20
C	low	low	30
D	high	high	35

(Paper 2)   [1]

9 The diagram on the next page shows a transverse section through a leaf as you would see it in a microscope.

  (a) Name the tissues labelled **E**, **F**, **G** and **H**.   [4]

  (b) Describe the functions of the parts of the leaf labelled **I** and **J**.   [3]

(c) Explain how cells in the tissues labelled **F** and **G** are adapted for photosynthesis. [3]

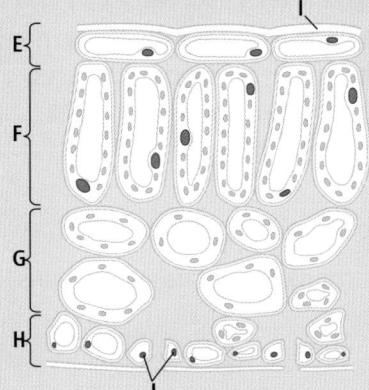

E {
F {
G {
H {

I
J

(Paper 3)

10 Two plants, **K** and **L**, were placed in the dark for several days. They were then placed into the apparatus shown below in the light to find out if carbon dioxide is required for photosynthesis. **L** was provided with carbon dioxide.

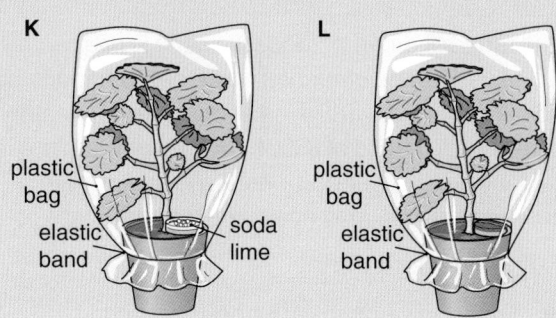

K   L
plastic bag   soda   plastic bag
elastic band   lime   elastic band

(a) Explain why the plants were put in the dark for several days. [2]

(b) Explain why the plants were well watered and kept in bright light. [2]

(c) Describe how you would show that plant **K** did not carry out any photosynthesis while inside the bag, but plant **L** did. [4]

(d) Explain why plant **L** was included. [2]

(Paper 3)

11 A student investigated the effect of varying the light intensity on the rate of photosynthesis using the apparatus shown below.

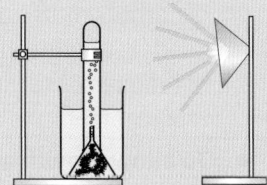

(a) State how the student would:
  (i) vary the light intensity; [2]
  (ii) determine the rate of photosynthesis; [2]
  (iii) maintain a constant temperature and constant carbon dioxide concentration throughout the investigation. [3]

(b) The graph below shows the student's results.

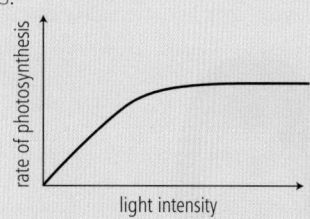

rate of photosynthesis
light intensity

Using the term *limiting factors*, explain the results shown in the graph. [4]

(Paper 4)

12 (a) Write the balanced chemical equation for photosynthesis [3]

Two groups of seedlings were grown in solutions that were deficient in two ions that plants need. Group **M** was grown without nitrate ions; group **N** was grown without magnesium ions. Group **O** received all the ions that plants require.

The results are shown in the table.

group of seedlings	colour of leaves	growth
M	yellow (especially the lower leaves)	reduced growth but not as much as seedlings in group **N**
N	lower leaves – yellow upper leaves – pale green	severely reduced growth
O	green	normal growth

(b) Explain the effects of the deficiencies of nitrate ions and magnesium ions as shown in the table. [4]

(c) Why was it necessary to include the seedlings in group **O**? [2]

(Paper 4)

7.1 A balanced diet

Figure 7.1.1 What other food items would you give this child apart from French fries?

The quantity and quality of the different food types that we eat is known as our diet. These foods are made up of **nutrients**. Nutrients can be divided into **macronutrients**, which are food substances that you eat in relatively large amounts, e.g., fats, proteins and carbohydrates, and **micronutrients**, which you eat in small quantities, e.g., vitamins and minerals.

A **balanced diet** relates to your needs and must include:

- enough energy – provided by carbohydrates, fats and proteins.
- essential amino acids from proteins
- essential fatty acids from fats
- micronutrients – vitamins and minerals
- water to replace the water lost in your urine, sweat, breath and faeces
- fibre for moving food from the mouth to the anus along the gut.
- If eaten in the correct proportions, these components make up a balanced diet which is a healthy diet.

Portion sizes

For healthy adults, the range of different foods we need is pretty much the same for all of us. But the *amount* that we need changes from person to person. Portion sizes are given as averages for healthy adults based on an energy allowance of 8300 kJ. However if you are tall or very active you may need larger portions. If you are a small person or trying to lose weight, you may need smaller portions.

Suggested portions

- Fruit and vegetables: 5 portions per day
- Starchy carbohydrates: 3–4 portions per day
- Beans, pulses, fish, eggs, meat and other proteins: 2–3 portions per day
- Dairy and alternatives: 2–3 portions per day

Nutrient functions

Each of these carries out one or more of three basic functions.

- **To provide energy** – this is mainly the role of carbohydrates and fats. Proteins are used for energy only if they are in excess of requirements, development, repair and replacement.
- **To allow growth and repair** of body cells and tissues. Proteins in the diet provide a source of amino acids for cells to make their own proteins (Topic 4.1).

Figure 7.1.2 A chart from the Caribbean Food and Nutrition Institute showing the different food groups

- **To regulate the body's metabolism** – vitamins and minerals are needed in very small quantities in the diet to help regulate our metabolism.

Even if you are lying in bed and completely inactive you are still using energy to keep your heart beating, your lungs working and your body temperature constant. The chemical reactions in your body, such as those involved in growth and repair, are occurring and require energy.

The energy required for these body functions is the **basal metabolic rate (BMR)**. The BMR varies from person to person, but an adult requires about 7000 kJ per day. So even if you lie down and do very little you still need this energy.

Children have a greater energy requirement than adults because they have a higher basal metabolic rate (BMR). A young child may weigh less than an adult but they have a higher BMR per kilogram per day because they are still growing. Elderly people generally have lower energy needs and have a lower BMR. However they still need to eat a balanced diet if they are to stay healthy. The energy needs of occupations that involved physical activity are greater than less active jobs that involve sitting at a desk. A woman in the last three months of pregnancy may require about 800 kJ more energy than the 9000 kJ recommended for a non-pregnant woman.

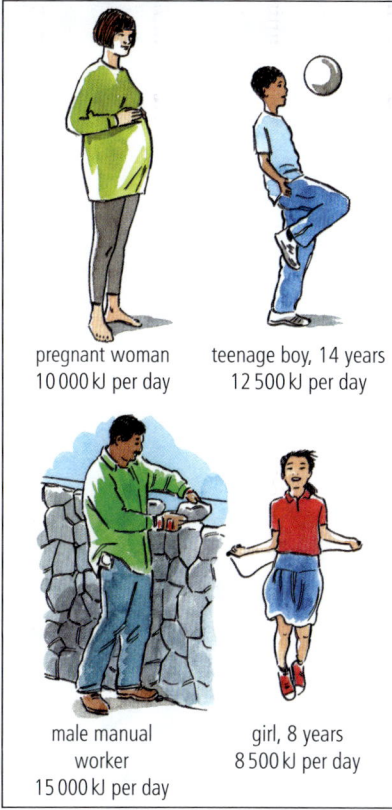

| pregnant woman 10 000 kJ per day | teenage boy, 14 years 12 500 kJ per day |
| male manual worker 15 000 kJ per day | girl, 8 years 8 500 kJ per day |

Figure 7.1.3 Typical energy needs.

SUMMARY QUESTIONS

1 a What is meant by the term *balanced diet*?
   b What must be in included in a balanced diet?
   c Explain the importance of 'portion size' in a balanced diet.

2 The following are the number of calories found in 1 g:
   Carbohydrate : 4   Protein : 4   Fat : 9
   a Why are carbohydrates the main energy source rather than proteins?
   b What is the vital role of proteins in the body?
   c Why are fats a good energy store?

3 a Write out a menu for **i** a well balanced meal and **ii** a poorly balanced meal.
   b In each case explain the reasons for your choices.

KEY POINTS

1 A balanced diet contains the correct nutrients present in the appropriate amounts.

2 The nutrients that make up a balanced diet are carbohydrates, proteins, fats, vitamins, minerals salts, water and fibre.

3 Carbohydrates are the main energy source, proteins are needed for growth and repair of cells, fats for a good energy store, vitamins and minerals are needed in small amounts for the body's metabolism, water allows chemical reactions to take place and is needed for transport and fibre allows food to be pushed along the gut.

7.2 Sources of nutrients

Eating healthily

Diets in different parts of the world vary. However, we should all eat from each of the food groups shown in the food guide pyramid.

Carbohydrates

Carbohydrates include sugars and starches, which provide us with a ready source of energy that is easily respired. Simple sugars are absorbed almost immediately by the stomach into the blood to give an immediate source of energy. This energy is released as a result of respiration. Good sources of carbohydrates are rice, potatoes, bread, yams, sugar and honey.

Figure 7.2.1 Sacks of pulses for sale in a market in Seville, Spain. Pulses are a protein-rich food suitable for those who do not eat foods of animal origin.

Proteins

In order for your body to grow and develop, it needs protein to make new cells. You digest protein into amino acids and then use these to make your own proteins. There are 20 different types of amino acid and your body must have all of these to make its own protein (Topic 4.1).

Cell membranes and cytoplasm contain a great deal of protein. Your body may need to replace old or damaged cells so you need enough protein in your food for this as well. If protein is not used for growth and repair it may be respired to provide energy. Good sources are meat, fish, milk and nuts.

Fats

Fats are important as a long-term energy store. The fat is stored under the skin and around the heart and kidneys. One gram of fat releases more than twice as much energy as can be released from one gram of carbohydrate or protein. When we are short of energy our body uses the fat.

Fats are also good thermal insulators, since they cut down wasted energy to the environment. Fats also give buoyancy to marine animals, for example whales have a thick layer of blubber. Good sources are butter, cheese, meat, fish and nuts.

Water

Water makes up two-thirds of your body mass. You take in water when you drink or eat.

You could go without food for a number of weeks, but you would die in a few days without water. Water is needed for chemical reactions to take place in solution. The blood transports substances dissolved in water. Waste chemicals are passed out of

EXAM TIP

You can now add more information to the table you started in Topic 4.1. For each feature you include (e.g., energy content in kJ) make sure you write something for each nutrient even if the answer is 0 (e.g., fibre, vitamins, minerals and water do not provide any energy).

our bodies in solution in the urine, and water in our sweat cools us down. It is important that your intake of water each day equals your loss of water in urine, faeces, sweat and breath.

Vitamins and minerals

We need small, regular amounts of vitamins and minerals if we are to be healthy. If these are lacking in the diet we can develop symptoms of deficiency diseases:

Table 7.2.1 The source, use and deficiency symptoms of some vitamins and minerals

name	rich food source	use in body	deficiency symptoms (deficiency disease)
vitamin C	oranges, lemons, other citrus fruits	tissue repair, resistance to disease	bleeding gums (scurvy)
vitamin D	fish oil, milk, butter (also made by skin in the Sun)	strengthens bones and teeth	soft bones, legs bow outwards (rickets)
iron	liver, meat, cocoa, eggs	used in formation of haemoglobin in red blood cells for transport of oxygen	tiredness, lack of energy (anaemia)
calcium	milk, fish, green vegetables	strengthens bones and teeth	weak, brittle bones and teeth (rickets), muscle weakness and cramps

Fibre

Dietary fibre or roughage comes from plants. It is made up mainly of cellulose from plant cell walls. Although it cannot be digested, it is an important part of the diet. High-fibre foods include bran cereals, cabbage, sweetcorn and celery.

Fibre adds bulk to our food. Since it is not digested, it passes down the entire gut from mouth to anus and does not provide any energy.

The muscles of the gut wall need something to push against. Fibre helps the movement of food in the alimentary canal by peristalsis so preventing **constipation**. Fibre absorbs poisonous waste from bacteria in our gut. Many doctors believe that a high-fibre diet lowers the concentration of **cholesterol** in the blood. Fibre reduces the risk of heart disease and bowel cancer.

Figure 7.2.2 Tangerines are a good source of vitamin C.

Figure 7.2.3 What can you recognise among these foods rich in fibre and other nutrients considered essential for good health, such as vitamins and minerals?

KEY POINTS

1 Carbohydrates are a quick source of energy, while fats are an energy store.

2 Protein is needed for growth and repair of tissues.

3 Lack of vitamin D can cause rickets (soft bones); and lack of vitamin C causes scurvy.

4 Roughage helps the passage of food through the gut.

SUMMARY QUESTIONS

1 List the components of a balanced diet and give a good source of each one.

2 Explain why the diet must contain each component.

3 Explain the term *deficiency disease*.

7.3 Nutrient deficiency

LEARNING OUTCOMES

- Describe the cause and effects of scurvy
- Discuss the causes and effects of vitamin D and iron deficiencies

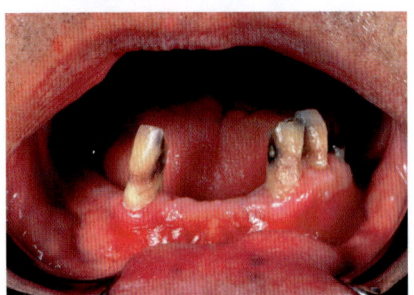

Figure 7.3.1 The mouth of a person who has scurvy. The gums are inflamed and the teeth have fallen out as a result of not enough vitamin C in the diet.

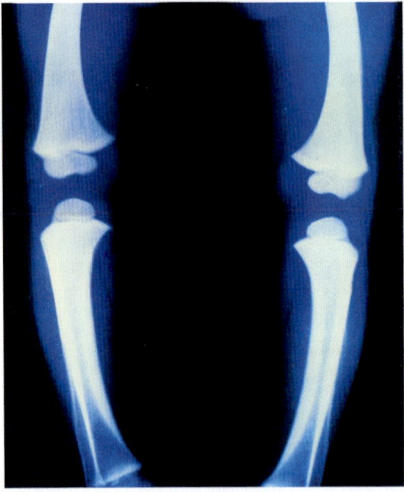

Figure 7.3.2 An X-ray showing the weakened bones and bowed legs of a child with rickets.

As we saw in 7.2, vitamins and minerals are nutrients (or micronutrients) that are needed in only small amounts in the body, but if lacking can result in deficiency diseases.

Vitamins

These are a group of organic compounds that do not have the same chemical structure but which do share certain characteristics.

They are all:

- essential in small quantities to ensure a healthy body
- able to work with enzymes to control the chemical reactions of the body.

They cannot:

- be digested or broken down by enzymes
- be used to build up body structures.

Green plants are able to synthesise vitamins in their leaves, but animals obtain them in their diet, either by eating plants or by eating other animals. If a particular vitamin in missing from the diet, then a deficiency disease may develop. This sort of disease can be cured at an early stage by simply adding the vitamin in question to the diet.

Vitamin C is found in oranges, lemons and other citrus fruits. As we have seen, a deficiency of vitamin C leads to the disease called scurvy. This can cause bleeding in parts of the body, particularly the gums. Vitamin C helps bond cells together and aids in the use of calcium by our bones and teeth. We use vitamin C to form an important protein used to make skin, tendons, ligaments and blood vessels. It also helps in the healing of wounds and the formation of scar tissue. Vitamin C cannot be stored in the body so daily intake is needed.

Vitamin D is needed for the small intestine to absorb calcium. It is also needed to regulate the deposition of calcium in bone cells. Deficiency symptoms include the lack of calcium in bones causing **rickets** in children. Bones fail to grow properly and become soft, so when children start walking the bones bend with the weight of the body. We can prevent rickets by eating foods rich in vitamin D like fish-liver oil, butter, eggs and milk. In adults, deficiency gives rise to the condition known as **osteomalacia**, which leads to a softening of the bones and an increased chance of fractures.

Vitamin D can be manufactured in your body when the skin is exposed to sunlight. But people living in regions with relatively low numbers of sunny days per year are not exposed to enough of the UV rays in sunlight to be able to make vitamin D in their skin at certain times. So in order to obtain enough vitamin D, these people need to include it in their diet. People in these regions with darker skin, such as those of African, African-Caribbean and South Asian origin, are at greater risk because their skin reduces UV light absorption.

Iron is needed to make the blood protein **haemoglobin**, which is found in red blood cells. Haemoglobin is a protein combined with iron. It is needed to carry oxygen around the body.

Foods rich in iron include liver, meat, cocoa. If there is a lack of iron in the diet, a disease called **anaemia** can occur. An adult needs about 16 mg of iron each day. At puberty, girls need to make sure they have enough iron in their diet as they start to **menstruate** (have periods).

The lack of iron leads to a reduction in the number of red blood cells. The main symptoms of anaemia are tiredness and lack of energy, shortness of breath, heart palpitations and a pale complexion.

Calcium is deposited in bones and teeth as calcium phosphate, giving them their hardness. Calcium is present in the blood plasma and plays an important part in blood clotting (see page 111). It is also involved in complex chemical reactions that enable muscle contraction and the transmission of nerve impulses. Rich sources of calcium include milk, cheese, fish and green vegetables, but it is present in most of the foods that we eat.

Vitamin D helps in the absorption of calcium from the intestines, and the deposition of calcium salts in the bones. Perhaps not surprisingly, calcium deficiency can result in rickets, as we have seen with a deficiency of vitamin D.

EXAM TIP

Summarise the information about calcium, iron, vitamin C and vitamin D in a table. Use columns for listing the sources, roles in the body, deficiency diseases, symptoms, and ways to prevent deficiency diseases.

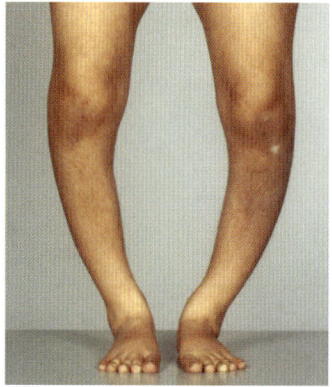

Figure 7.3.3 Rickets causes bowing of the legs, and widening of elbows and knee joints.

SUMMARY QUESTIONS

1. a Why are vitamins and mineral sometimes called 'micronutrients'?
   b Why can vitamins or minerals not be digested or broken down by enzymes?
   c Why may deficiency diseases result from a lack of certain vitamins and minerals in the diet?

2. a Why is vitamin D sometimes called the 'sunshine vitamin'?
   b Describe the role of vitamin D in the body.
   c Which two deficiency diseases can result from a lack of vitamin D in the diet?

3. a What is the deficiency disease caused by a lack of iron in the diet?
   b Why does a pregnant woman need lots of iron and calcium in her diet?
   c Why might a 13-year-old girl starting her periods become a bit anaemic?

KEY POINTS

1 Scurvy is caused by a deficiency of vitamin C in the diet and can cause bleeding in parts of the body, particularly the gums.

2 Lack of iron in the diet can cause anaemia, which affects the ability of the blood to transport of oxygen.

3 Vitamin D deficiency can cause a softening of the bones, causing rickets and osteomalacia.

7.4 Digestion

- Describe physical (mechanical) and chemical digestion
- Describe ingestion, absorption, assimilation and egestion
- State the function of amylase, protease and lipase, listing the substrates and the end-products

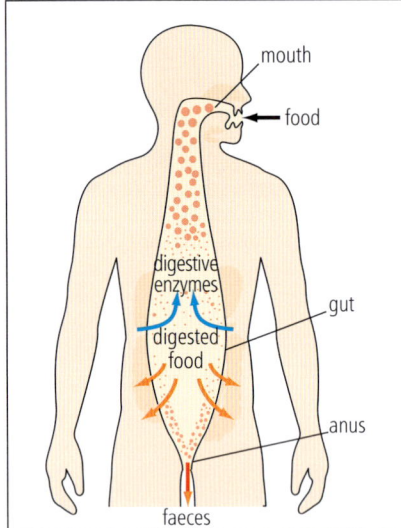

Figure 7.4.1 Digestion in the alimentary canal.

EXAM TIP

Learn definitions of 'digestion', 'physical digestion' and 'chemical digestion' and be prepared to distinguish (give the differences) between them.

During the passage of food along the alimentary canal, a number of different processes can be identified.

- **Ingestion** is the taking of substances (food and drink) into the body through the mouth.
- **Digestion** is the breaking down of large insoluble molecules into small soluble molecules so that they can pass through the gut wall into the blood.
- **Absorption** is the movement of small food molecules and ions through the wall of the intestine into the blood.
- **Assimilation** is the movement of digestive food molecules into the cells of the body where they are used, becoming part of the cells.
- **Egestion** is the passing out of food that has not been digested or absorbed, as faeces through the anus.

The large food molecules are starch, proteins and fats. Digestion occurs physically (mechanically) and chemically. This happens in the alimentary canal, or gut, which together with the liver and pancreas form your digestive system (Topic 2.4). When the food has been digested, it is **absorbed** through the wall of the small intestine into the blood. Before the food molecules can go through the gut wall they must be dissolved. Large food molecules are insoluble; they will not dissolve, therefore cannot get through the wall of the small intestine. On the other hand, small food molecules are soluble. They will dissolve so they can get through the wall of the small intestine and enter the blood and the lymph, which is another body fluid that absorbs and transports fat.

Physical digestion

Physical digestion is the breakdown of large pieces of food into smaller pieces of food without changing the food molecules. This starts in the mouth where chewing breaks down food into smaller pieces that can be swallowed. Muscular contractions of the stomach continue this process. In the small intestine, large globules of fat are broken into smaller globules by emulsification by bile. There is more about physical digestion in the next few topics.

Chemical digestion

Before the body can use food that has been eaten, the food must be broken down into small molecules.

Chemical digestion is the breakdown of the large insoluble food molecules into smaller soluble molecules by the action of **enzymes**. This occurs in the mouth, stomach and small intestine. Physical digestion gives a larger surface area for the enzymes to work on.

The three types of enzymes in the alimentary canal are: proteases, carbohydrases and lipases. Each type of food molecule needs a specific enzyme because they have different shapes. Topic 5.1 to remind yourself about the reason for having different enzymes to break down proteins, carbohydrates and fats.

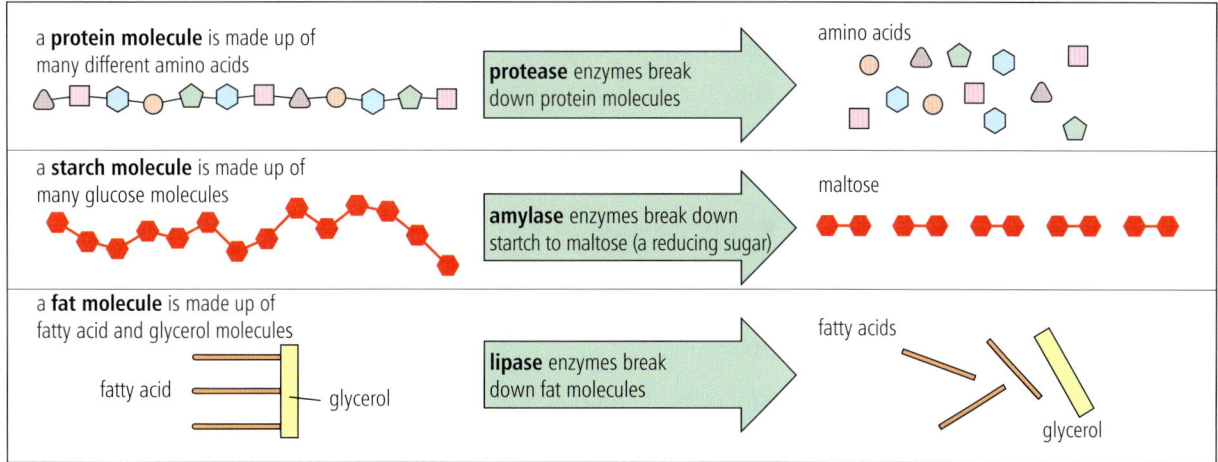

a **protein molecule** is made up of many different amino acids

protease enzymes break down protein molecules

amino acids

a **starch molecule** is made up of many glucose molecules

amylase enzymes break down starch to maltose (a reducing sugar)

maltose

a **fat molecule** is made up of fatty acid and glycerol molecules

fatty acid — glycerol

lipase enzymes break down fat molecules

fatty acids

glycerol

Figure 7.4.2 Amylase is the carbohydrase enzyme that breaks down starch into maltose. Another carbohydrase breaks down maltose to glucose.

The carbohydrase enzyme that breaks down starch is called **amylase**.

SUMMARY QUESTIONS

1 Name the enzymes and end-products involved in the chemical digestion of:
   a starch **b** protein **c** fat

2 **a** Explain why digestion is necessary.
   b Explain the difference between physical digestion and chemical digestion.
   c Outline what happens to food after it has been digested.

3 The graph shows the activity of amylase and pepsin, the protease that acts in the stomach.

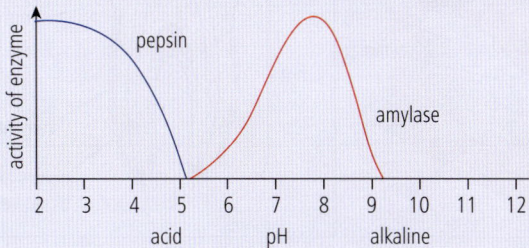

   a **i** At what pH does pepsin work best?
      ii What are formed when pepsin breaks down proteins?
   b **i** At what pH does amylase work best?
      ii What is formed when the amylase breaks down starch?
      iii Why is it useful that saliva contains an alkali?

KEY POINTS

1 Physical digestion is the physical breakdown of food particles to increase their surface area. Chemical digestion is the breakdown of large, insoluble food molecules into smaller, water-soluble molecules by the action of enzymes.

2 Amylase (a carbohydrase enzyme) breaks down starch molecules to glucose molecules.

   Protease breaks down protein molecules to amino acids.

   Lipase breaks down fats to fatty acids and glycerol.

7.5 Teeth

LEARNING OUTCOMES

- Describe the structure of human teeth and their function in physical digestion
- Identify the four types of teeth in diagrams and images

The teeth carry out physical digestion when chewing food. This breaks down food into smaller pieces so making it easier to swallow. The surface area of food is increased so making it easier for enzymes to catalyse the reactions needed for chemical digestion.

Adult humans have 32 teeth, but not all of the teeth are the same. They have different shapes for performing different functions:

- incisors are chisel-shaped for biting and cutting
- canines are pointed for piercing and tearing
- premolars have uneven 'cusps' for grinding and chewing
- molars are like premolars and are for chewing up food.

Counting the number of each type of tooth, you should find that there are:

8 incisors, 4 canines, 8 premolars, and 12 molars, which makes 32 teeth in all.

During our lives we have two sets of teeth. The first set, or **milk teeth**, are small teeth and there are only 8 molars as our jaws are small. Between the ages of 6 and 12 these teeth gradually fall out, to be replaced by our **permanent teeth**. The last of our permanent teeth will come through when we are at least 18, if at all! These are our back molars or **wisdom teeth**.

The teeth are embedded in the bone that forms the upper and lower jaws. The gums that cover the jawbones are composed of soft tissue with many blood vessels and nerves.

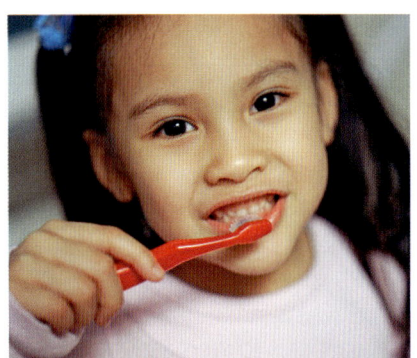

Figure 7.5.1 A child usually has 20 milk teeth.

EXAM TIP

Make sure that you can identify the different types of human teeth from photographs and diagrams and can state their functions in physical digestion.

EXAM TIP

Note that fluoride strengthens the enamel of the teeth, giving them a hard exterior surface.

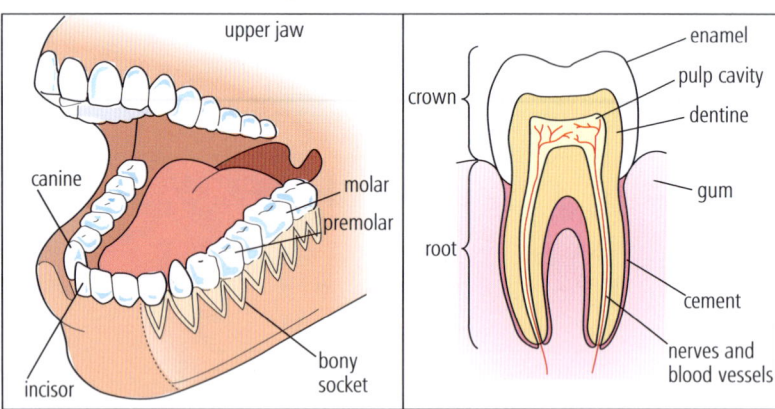

Figure 7.5.2 Adult teeth.   Figure 7.5.3 Vertical section through a molar.

Enamel forms the hard, outer layer of the crown of the tooth, which is the part above the gum. Inside this is softer **dentine**, which is more like bone in structure. A layer of **cement** fixes the root of the tooth into a bony socket in the jaw. The root is the part that is below the gum.

The **pulp cavity** is a space in the tooth containing nerves and blood vessels.

Figure 7.5.4 is an X-ray of the teeth of a young adult. Compare this image of the teeth in the X-ray with the diagram in Figure 7.5.2 and identify the four types of teeth. Look carefully at the structure of the roots of the teeth embedded in the jaw bones. You can see that the four back molars are not fully erupted.

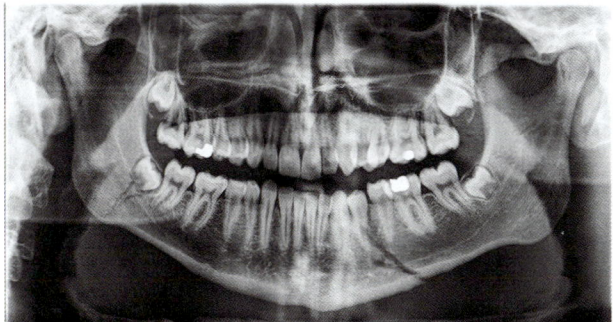

Figure 7.5.4 A panoramic X-ray of the full set of permanent teeth in a young adult.

Figure 7.5.5 is an X-ray of the teeth of a 12 year old. You can see some of the milk teeth and some permanent teeth that have not yet erupted. Compare this image of the teeth with the diagram in Figure 7.5.2 and the X-ray of the teeth of the young adult.

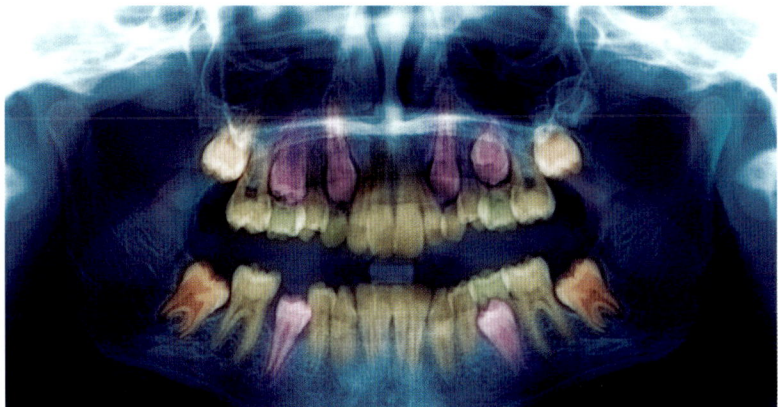

Figure 7.5.5 A panoramic X-ray of teeth of a 12-year old.

SUMMARY QUESTIONS

1 Copy and complete the sentences using these words:

> **dentine   crown   physical   socket
> root   bite   enamel   chew   bone**

The teeth are used to _____ and _____ your food. This is a type of _____ digestion. Each tooth consists of a _____ above the gum and a _____, which is fitted into a bony _____. The _____ is the hard, outer layer of the tooth; inside this is the softer _____, which is similar to _____.

2 a Name the four main types of teeth.

   b For each type in part **a**, describe its shape and its function.

   c What is the difference between the milk teeth and the permanent teeth?

3 Describe how the number and arrangement of teeth in the X-ray in Figure 7.5.5 differs from the X-ray of the teeth of the young adult in Figure 7.5.4.

KEY POINTS

1 Teeth are used in physical digestion to break food into smaller pieces.

2 Incisors are for biting and cutting, canines for piercing and tearing, and premolars and molars are for grinding and chewing.

3 Each tooth is composed of a hard outer layer of enamel that covers the softer dentine. In the centre is the pulp cavity with nerves and blood vessels.

4 Teeth are embedded in the gums and jaw bones. Cement helps to hold the teeth securely in the bone.

7.6 Mouth, oesophagus and stomach

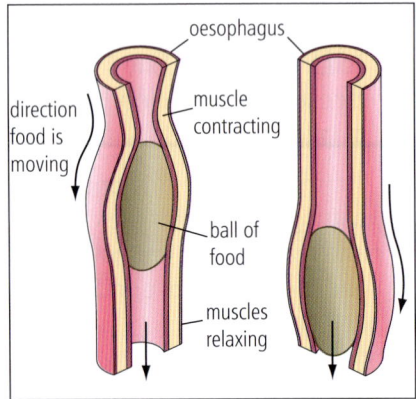

Figure 7.6.2 Peristalsis in the oesophagus is a wave of contraction to pass food to the stomach.

Chewing and swallowing

Food and drink are ingested into the body through the mouth. When food enters the mouth, the incisor teeth and canine teeth bite it into chunks. The premolar and molar teeth grind these chunks of food into much smaller pieces. It is important to chew your food well as this makes digestion easier. The tongue mixes the food with **saliva** and the moistened food is chewed by the teeth. The food is then rolled into a ball or **bolus**.

The **salivary glands** make saliva, which contains:

- **mucus**, a slimy substance that lubricates the passage of the food bolus down the throat
- **amylase**, the enzyme that catalyses the breakdown of starch to maltose.

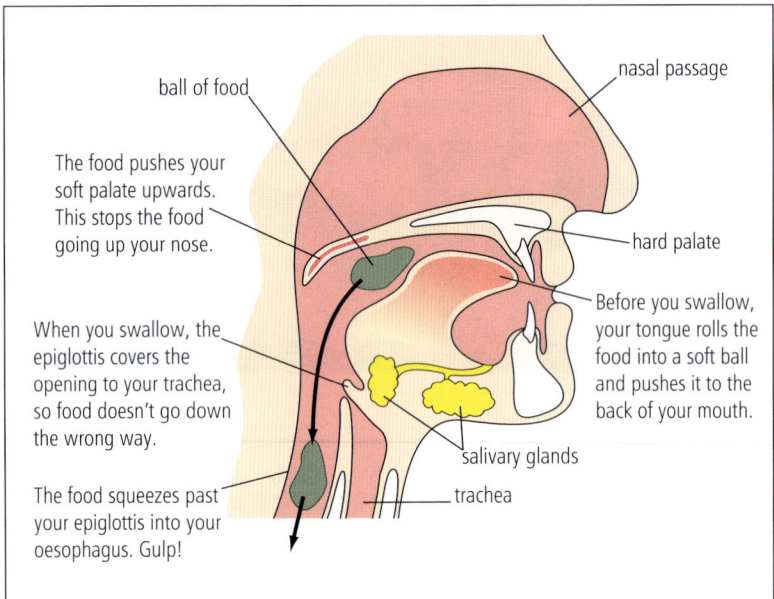

Figure 7.6.1 Events during swallowing.

The oesophagus

Food passes down the **oesophagus** from the mouth to the stomach. The oesophagus (also known as the gullet) has circular and longitudinal muscles in its wall.

The movement of food down the oesophagus occurs by a wave of muscular contraction called **peristalsis**. The circular muscles contract and the longitudinal muscles relax behind the food bolus to push it along. In front of the bolus, the circular muscles relax and the longitudinal muscles contract to widen the oesophagus to allow the bolus of food to move along. Movement of food in the small and large intestines also occurs by peristalsis, though to a lesser extent.

The stomach

The stomach is a muscular bag that usually holds about 1 litre of food but can stretch to accommodate more. When food reaches the stomach a number of things happen to it.

The walls of the stomach make a digestive juice known as **gastric juice**. This contains a protease, which starts the digestion of proteins to smaller molecules called polypeptides.

Gastric juices also contain **hydrochloric acid**, which carries out two important functions due to its low pH:.

- The low pH denatures the enzymes in any harmful microorganisms in the food, so killing them.

- The acid pH of about 1.5 to 2.0 gives the optimum conditions for the action of the protease enzyme.

The muscular walls of the stomach churn up the food making sure that it is mixed well with the juices. This muscular action is a type of physical digestion. After 2–3 hours of churning, the contents of the stomach are like a runny liquid.

A ring of muscle called the **pyloric sphincter**, opens to let the food pass a little at a time into the **duodenum**, which is the first part of the small intestine.

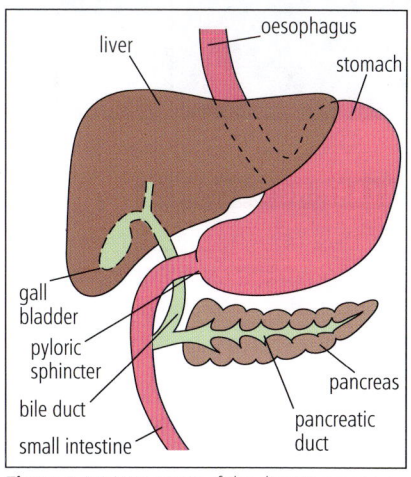

Figure 7.6.3 Upper part of the digestive system.

Supplement

Pepsin is the protease enzyme that starts the digestion of proteins in the stomach.

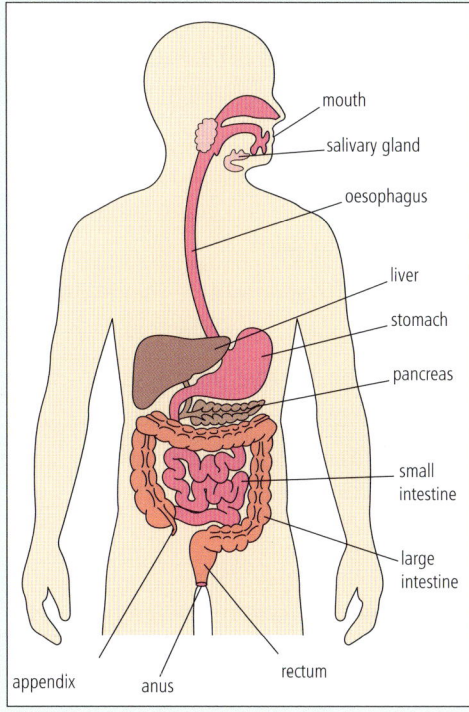

Figure 7.6.4 The digestive system.

KEY POINTS

1   Food is chewed up by the teeth. It is then swallowed by the actions of the tongue, saliva, soft palate and epiglottis.

2   Peristalsis involves a wave of muscular contractions that squeeze the food bolus, carrying it down the oesophagus to the stomach.

3   In the stomach, food is churned up and mixed with gastric juice and hydrochloric acid. This starts the digestion of proteins.

SUMMARY QUESTIONS

1   With reference to the following, describe the events that take place when food is chewed and swallowed:
   a   saliva   **c**   the tongue
   b   mucus   **d**   the epiglottis

2   **a**   Describe one function of the muscle layers in the stomach wall.

   b   Describe the function of the protease produced by the walls of the stomach.

   c   Explain briefly why a different enzyme is required for the digestion of each type of food substance. (Topic 5.1 before you start your answer.)

   d   State the name of the type of protease that acts in the stomach. **S**

7.7 Small intestine and absorption

LEARNING OUTCOMES

- Describe the digestion of food in the small intestine, including the role of bile in emulsifying fats
- Describe how nutrients are absorbed in the small intestine
- Describe how the small intestine is adapted for efficient absorption of food

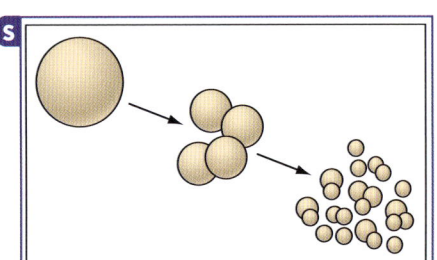

Figure 7.7.1 Large globules of fat are emulsified to smaller globules of fat to increase their surface area.

EXAM TIP

Emulsification increases the surface area of fats for lipase enzymes to act on. Emulsification is not a type of chemical digestion as the fat is not changed chemically.

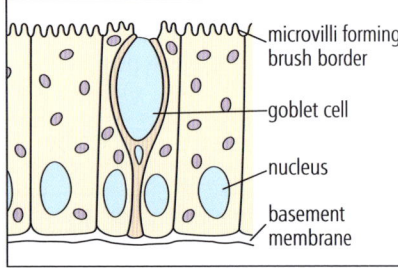

Figure 7.7.2 Detailed structure of the epithelial cells lining the small intestine. Goblet cells release mucus to protect the epithelial cells from enzymes and physical damage.

The small intestine

The small intestine is not really so small as it is about 6 metres long; however its diameter is less than that of the large intestine. The first part, after the stomach, is the **duodenum** that leads into the **ileum**.

The **pancreas** is connected to the duodenum by the **pancreatic duct**. **Pancreatic juice** flows down the duct to meet food arriving from the stomach. Pancreatic juice contains:

- **amylase** – breaks down starch to maltose
- **a protease** – breaks down proteins and polypeptides to peptides. (At Supplementary level you need to know that the name of the protease is **trypsin**)
- **lipase** – breaks down fats to fatty acids and glycerol.

These enzymes do not work well in an acidic environment, so pancreatic juice contains the alkali, sodium hydrogencarbonate. This neutralises the acidic food that enters the duodenum from the stomach.

Supplement

Bile also enters the duodenum along a tube called the **bile duct**. Bile is a yellow-green fluid made in the **liver** and stored in the **gall bladder**.

Bile is alkaline and also neutralises the acid that was added to the food in the stomach. This gives the best pH for enzymes in the small intestine to work.

Bile **emulsifies** fats by breaking down large globules of fats into smaller globules. These minute droplets have a much larger surface area over which the enzyme lipase can act to break down molecules of fat into fatty acids and glycerol.

Cells lining the ileum make enzymes that complete the digestion of food.

- Proteases, which break down peptides to amino acids.
- Sucrase, which breaks down sucrose to glucose and fructose.
- Maltose is broken down by maltase to glucose on the membranes of the epithelium lining the small intestine.

Absorption

Absorption is the movement of digested food molecules through the wall of the intestine into the blood or the lymph. Digested food includes simple sugars, amino acids, fatty acids and glycerol. These molecules pass through the wall of the small intestine either by diffusion or by active transport.

The small intestine is adapted for efficient absorption of food as it has:

- a very large surface area of about 9 square metres
- a thin lining (only one cell thick), so digested food can easily cross the wall into the blood and lymph.

It is possible to fit such a large surface area into such a small space because the small intestine is very long (at least 6 metres in an adult).

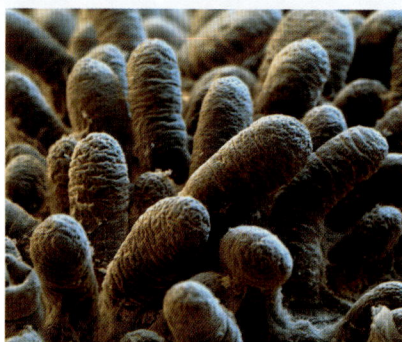

Figure 7.7.3 Villi increase the surface area of the small intestine.

Supplement

The small intestine has a folded inner lining with millions of tiny, finger-like projections called **villi** (singular: **villus**). The epithelial cells lining the villi have microscopic projections called **microvilli**. These vastly increase the absorptive area of the cell membrane of the epithelial cells (Figure 7.7.2).

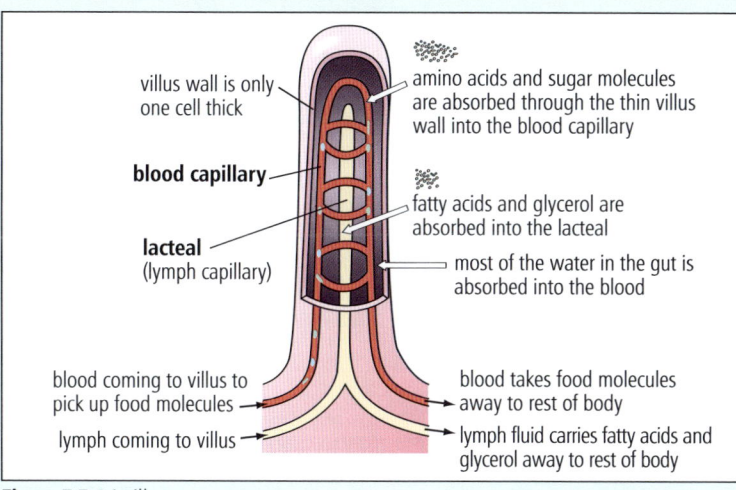

Figure 7.7.4 A villus.

Labels in figure:
- villus wall is only one cell thick
- **blood capillary**
- **lacteal** (lymph capillary)
- amino acids and sugar molecules are absorbed through the thin villus wall into the blood capillary
- fatty acids and glycerol are absorbed into the lacteal
- most of the water in the gut is absorbed into the blood
- blood coming to villus to pick up food molecules
- lymph coming to villus
- blood takes food molecules away to rest of body
- lymph fluid carries fatty acids and glycerol away to rest of body

The digested food reaches the capillaries and **lacteals** (lymph capillaries) in the villi. Absorbed food molecules are transported quickly to the liver by the **hepatic portal vein**. Fatty acids and glycerol are transported in the lymph. Movement of the gut empties the lacteal and the lymph moves slowly through lymphatic vessels, which are thin-walled like veins, eventually to enter into the blood near the heart. As a result, fat does not enter the bloodstream too quickly.

SUMMARY QUESTIONS

1 Describe the function of the following in digestion:
   a oesophagus   b stomach   c liver
   d gall bladder   e pancreas   f duodenum.

S 2 a State: **i** the components of bile, **ii** the organ where bile is produced, **iii** the organ where bile is stored.
   b Describe two ways in which bile aids the digestion of fats.

3 a How are conditions in the stomach different from the conditions in the small intestine?
   b State the composition of: **i** gastric juice, **ii** pancreatic juice.

EXAM TIP

By the time the remains of what you ate gets to your **colon**, there is not much that is useful left. It is now mainly fibre, dead cells, bacteria and some water.

The solid waste (**faeces**) is stored in the **rectum**. Eventually the faeces are passed out through the **anus**. This is called **egestion**.

KEY POINTS

S 1 Bile emulsifies fats in the duodenum. The smaller fat globules have a larger surface area upon which lipase can act.

2 Enzymes produced by the pancreas and the small intestine complete chemical digestion.

3 Absorption is the movement of digested food molecules through the wall of the intestine into the blood and lymph.

4 The small intestine is adapted for absorption by being very long, having a folded inner lining with millions of tiny villi.

Practice questions

1 The breakdown of large, insoluble food molecules into small, water-soluble molecules is:

   A absorption   B digestion
   C egestion   D ingestion

   (Paper 1)   [1]

2 Which is a correct description of the action of a protease?

   A breaks down fat into fatty acids and glycerol

   B breaks down protein into amino acids

   C breaks down starch into glucose

   D converts amino acids into proteins

   (Paper 1)   [1]

3 The drawing shows a vertical section through an incisor tooth.

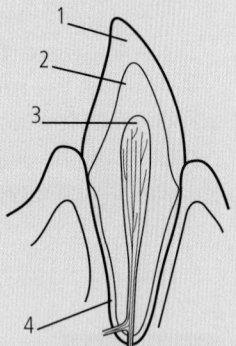

   Which row is the correct identification of the parts shown in the drawing above?

	1	2	3	4
A	cement	pulp	enamel	dentine
B	dentine	enamel	pulp	cement
C	enamel	dentine	pulp	cement
D	pulp	cement	dentine	enamel

   (Paper 1)   [1]

4 Which is the cause of scurvy?

   A lack of calcium in the diet

   B lack of iron in the diet

   C lack of vitamin C in the diet

   D lack of vitamin D in the diet

   (Paper 1)   [1]

5 Which row shows the correct description of the digestion of protein?

	organ of the digestive system	enzyme	pH
A	stomach	pepsin	alkaline
B	stomach	trypsin	acid
C	small intestine	pepsin	acid
D	small intestine	trypsin	alkaline

   (Paper 2)   [1]

6 The drawing shows part of the digestive system.

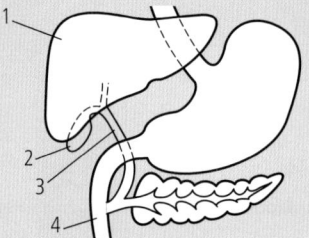

   What is the function of the liquid produced by part 1, stored in part 2 and released into part 3?

   A to digest proteins to amino acids

   B to increase the surface area of fat droplets

   C to acidify the contents of part 4

   D to digest fats to fatty acids and glycerol

   (Paper 2)   [1]

7 The following events occur during the passage of starch through the alimentary canal.

   1 Amylase is secreted in saliva from salivary glands.

   2 Maltase is secreted onto the surface of epithelial cells in the small intestine.

   3 Maltase catalyses the breakdown of maltose to glucose.

   4 Glucose enters the blood plasma.

   5 Glucose is absorbed by active transport into the epithelial cell.

   6 Amylase catalyses the breakdown of starch to maltose.

   Which is the correct sequence for these events?

   A 1,6,2,3,5,4   B 6,1,3,2,4,5
   C 2,3,1,6,5,4   D 3,2,6,1,4,5

   (Paper 2)   [1]

8 Which feature of a villus increases its surface area for absorption?

    A capillaries   **B** goblet cells

    C lacteal   **D** microvilli

(Paper 2)   [1]

9 Copy the diagram from question 6.

  (a) On the diagram label the following: oesophagus, stomach, pancreas and duodenum.   [4]

  (b) Draw arrows on the diagram to show the direction taken by food as it passes through the parts of the alimentary canal on the diagram.   [2]

  (c) On the diagram use label lines and the following letters to show:

    E the site of secretion of hydrochloric acid

    F an organ that secretes molecules of a protease

    G a place with an alkaline pH

    H the organ that stores bile   [4]

(Paper 3)

10 (a) Humans are recommended to eat a balanced diet.

     Explain the term *balanced diet*.   [4]

  (b) Explain the importance of physical digestion in the human digestive system.   [3]

  (c) Describe what is meant by *chemical digestion*.   [2]

  (d) Explain the importance of enzymes in the human digestive system.   [3]

(Paper 3)

11 The liver performs important functions in the digestion and assimilation of food. One of these is the production of an alkaline digestive juice that breaks down large globules of fat.

  (a) (i) Name the digestive juice secreted by the liver.   [1]

    (ii) Explain how this digestive juice reaches the food in the small intestine. [3]

    (iii) Explain why this digestive juice is alkaline.   [2]

    (iv) Outline how this digestive juice breaks down large globules of fat. [4]

  (b) Explain what is meant by *assimilation*. [2]

(Paper 4)

12 The diagram shows a villus from the alimentary canal.

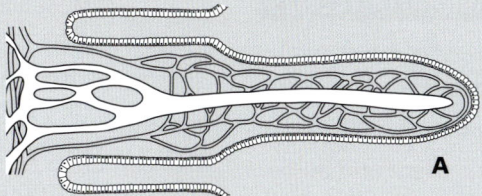

  (a) Copy the diagram and label the capillaries and the lacteal.   [2]

  (b) State the part of the alimentary canal where villi are located.   [1]

  (c) Name three substances that pass from **A** on the diagram into the villus.   [3]

  (d) With reference to the diagram, explain the significance of villi in the alimentary canal. [6]

  (e) Most of the water that enters the alimentary canal is absorbed in the small intestine. Explain how water is absorbed in the small intestine.   [3]

(Paper 4)

13 A meal contains bread and cheese. Bread contains starch, and cheese contains protein.

  (a) Describe how starch is digested in the alimentary canal. In your answer state where the changes that you describe take place. [5]

The diagram shows a small protein molecule and the places along the molecule where proteases **1** and **2** act on the molecule to help break it down.

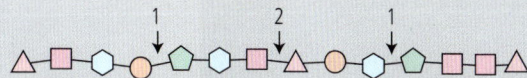

  (b) How many fragments will be produced after the molecule has been digested by the two proteases?   [1]

  (c) Suggest why protease 2 cannot act on the protein in the same places as protease 1. [4]

  (d) State the organs in the alimentary canal where proteases digest protein.   [2]

  (e) The fragments produced by digestion by proteases 1 and 2 are not absorbed into the blood. Explain why this is so.   [2]

  (f) Describe what happens to these fragments so that the protein is completely broken down.   [2]

(Paper 4)

LEARNING OUTCOMES

- Describe the functions of xylem and phloem
- Identify the positions of xylem and phloem in diagrams and images of transverse sections of roots, stems and leaves

Most living organisms need a transport system. Only very small organisms, such as bacteria, protoctists and some worms, do not have transport systems. Their bodies are so small, or so flat, that they can rely on diffusion alone to obtain oxygen and remove carbon dioxide.

Vertebrates and flowering plants are large organisms and have transport systems to carry substances, such as food and water, throughout their bodies. These transport systems move fluids through tubes so that all of the fluid moves in the same direction within each tube. This type of transport is called **mass flow**.

Vertebrates have a circulatory system consisting of a pump (the heart) and blood vessels. Flowering plants have two separate transport systems: **xylem** and **phloem**.

Xylem and phloem

Xylem and phloem are plant tissues composed of cells that are specialised for transport (Topic 2.3). These tissues are found throughout the plant body in roots, stems and leaves.

Xylem tissue transports water and mineral ions. The roots absorb the water and mineral ions from the soil. These enter the xylem in the root and travel upwards in the stem to the leaves, flowers and fruits.

Transport in the xylem, unlike in the phloem, is in one direction only – from roots via stem to leaves.

Figure 8.1.1 *Eucalyptus regnans* is the tallest flowering plant, growing up to 90 metres.

Figure 8.1.2 The world's tallest tree is a giant redwood in California. It is 114 m tall.

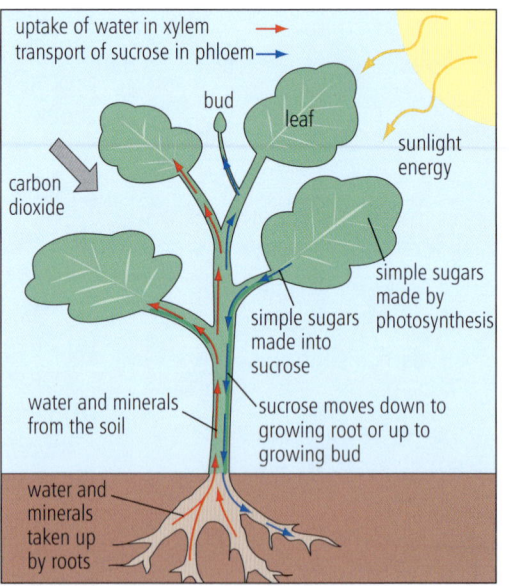

Figure 8.1.3 Transport in the xylem and phloem.

Phloem tissue transports sucrose, amino acids and hormones throughout the plant. Sucrose is a soluble, complex sugar that is made especially for transporting energy. It is made in leaves with the sugars from photosynthesis and from starch in storage organs, such as swollen roots and stems.

Substances are transported in the phloem in two directions: downwards from leaves to roots and upwards from leaves to flowers, fruits and buds; also from storage organs to new stems and leaves.

Hormones control cell division for growth of the stem, roots and leaves. They also control the growth of flowers and fruits (see page 168).

Inside the root and stem

To find the transport tissues of roots and stems, we cut across them to give transverse sections.

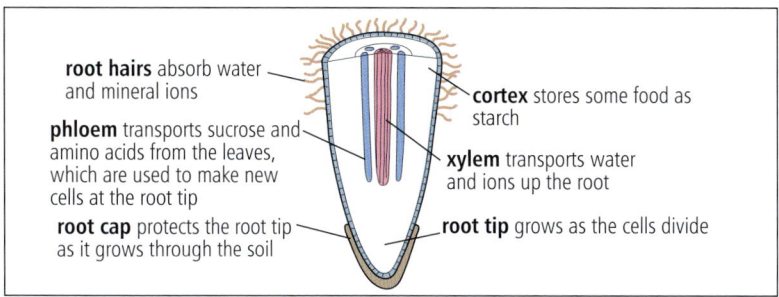

root hairs absorb water and mineral ions

cortex stores some food as starch

phloem transports sucrose and amino acids from the leaves, which are used to make new cells at the root tip

xylem transports water and ions up the root

root cap protects the root tip as it grows through the soil

root tip grows as the cells divide

Figure 8.1.4 The internal structure of a root.

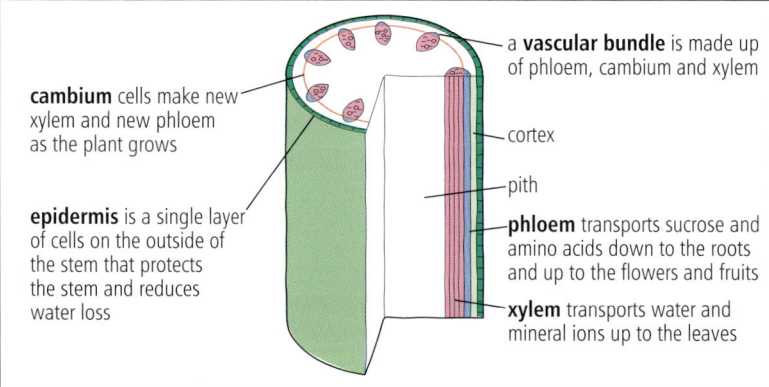

a **vascular bundle** is made up of phloem, cambium and xylem

cambium cells make new xylem and new phloem as the plant grows

cortex

pith

epidermis is a single layer of cells on the outside of the stem that protects the stem and reduces water loss

phloem transports sucrose and amino acids down to the roots and up to the flowers and fruits

xylem transports water and mineral ions up to the leaves

Figure 8.1.6 The internal structure of a stem.

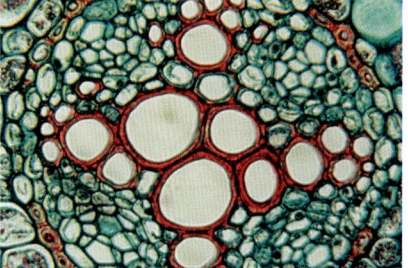

Figure 8.1.5 The tissue in the centre of this root is xylem, which is stained red. The green areas around the xylem are phloem tissue.

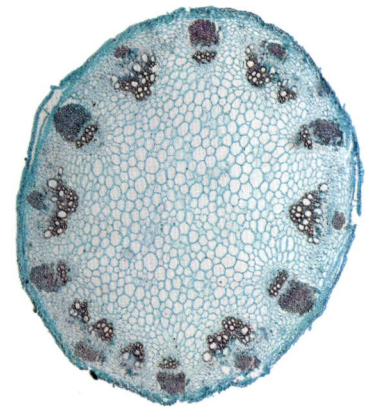

Figure 8.1.7 A transverse section through a young stem. Note the vascular bundles.

See Topic 6.6 for the positions of xylem and phloem in the leaves.

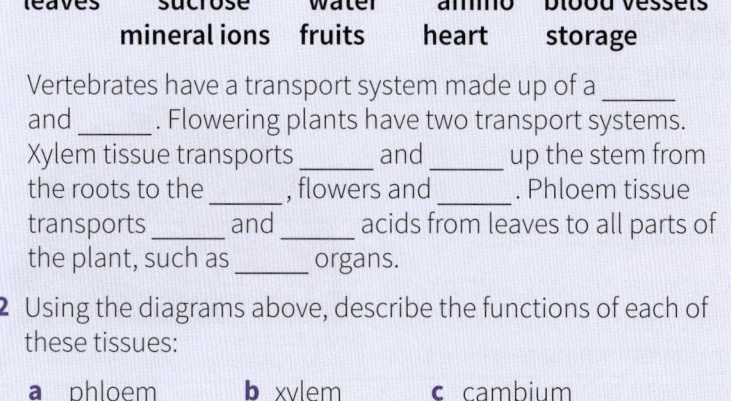

SUMMARY QUESTIONS

1 Copy and complete the sentences using these words:

**leaves   sucrose   water   amino   blood vessels
mineral ions   fruits   heart   storage**

Vertebrates have a transport system made up of a _____ and _____. Flowering plants have two transport systems. Xylem tissue transports _____ and _____ up the stem from the roots to the _____, flowers and _____. Phloem tissue transports _____ and _____ acids from leaves to all parts of the plant, such as _____ organs.

2 Using the diagrams above, describe the functions of each of these tissues:

a phloem   b xylem   c cambium

d epidermis   e root cap   f root cortex

KEY POINTS

1 Xylem vessels transport water and mineral ions from the roots up the stem to the leaves.

2 Phloem vessels transport sucrose, made in the leaves during photosynthesis and in storage organs, to other parts of the plant.

3 Xylem and phloem are distributed in a central core inside a root. Inside a stem they are organised into vascular bundles.

8.2 Water uptake

LEARNING OUTCOMES

- Identify root hairs as seen using the microscope and relate their structure to the uptake of water and mineral ions
- Outline the pathway taken by water through root, stem and leaf
- Investigate water transport in the xylem

Water uptake by roots

The main functions of roots are to anchor the plant in the soil and to take up water and mineral ions. Just behind the root tip are small **root hairs**. Water passes into the root hairs by **osmosis**.

Soil water is a dilute solution of various solutes including mineral ions, such as nitrate and potassium. Root hairs have thin, permeable cell walls and provide a large surface area to absorb water. The cell sap in the root hair cells is a more concentrated solution than soil water. This is because root hair cells also absorb mineral ions and have other solutes such as sugars. The cell membrane is partially permeable so water diffuses from the soil into the root hair cells by osmosis.

Water then passes across the cortex of the root. Most of the water moves through the cell walls and in the spaces between the cells. Some moves from cell to cell. Water eventually reaches the xylem in the centre of the root. From here water moves up the xylem through the stem and to the leaves where it enters the spongy mesophyll cells. Much of the water enters the cell walls, evaporates to form water vapour and then diffuses through stomata to the atmosphere.

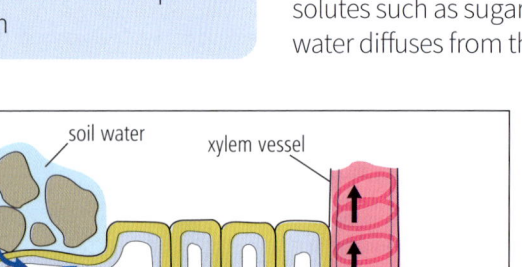

Figure 8.2.1 Absorption of water by a root.

Supplement

We describe the movement of water into a root hair, across the cortex and into a xylem vessel in terms of **water potential** (Topic 3.2). Water passes *down* a **water potential gradient**, from a high water potential in the solution in the soil to a lower water potential in the root hair cell. Water in turn, passes from the root hair cell to a lower water potential in the cells of the cortex and eventually to an even lower water potential in the xylem vessel.

PRACTICAL

Looking at root hairs

Look closely at a section of a young root with a hand lens and then with a microscope.

Look at the root hairs.

Root hairs are long and thin. You can also look at root hairs on a germinating seedling such as a bean or a radish.

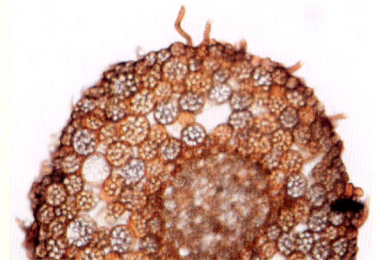

Figure 8.2.2 Part of a section of a young root seen with a microscope.

Figure 8.2.3 Root hairs of a germinating bean seed.

S They have a large surface area through which water is absorbed by osmosis and mineral ions are absorbed by active transport.

Investigation into water transport in the xylem

As we have seen, water and mineral ions travel from the roots up to the leaves via the stem, in a system of xylem vessels. Xylem vessels are like pipes as they are empty except for the water that fills them. But how can we prove that water transport takes place inside the xylem vessels?

First of all stand some celery in water containing red dye for a few hours.

Carefully cut off a length of about 2 cm.

Look at the cut end to see where the dye is located. The dye has been carried up the celery in the xylem vessels.

Carefully cut out a 2 cm length of xylem.

Place it on a slide with a cover-slip.

Look at it using the microscope.

Carefully draw your observations of the xylem vessels viewed lengthways (a longitudinal section).

Figure 8.2.4 Be careful when cutting a length of celery.

Figure 8.2.5 This celery was left standing in a red dye for several hours. The red strands are the xylem vessels filled with dye.

1 Copy and complete the sentences using these words:

> **stomata osmosis large ions
> transpiration hair area**

Water passes into a root _____ cell by _____. Mineral _____ are also taken up. A root hair cell has a _____ surface _____ for taking up water and mineral ions. Water is lost from the leaves by _____. This is controlled by the opening and closing of the _____ .

2 Look at the diagram of the root hair in the soil.

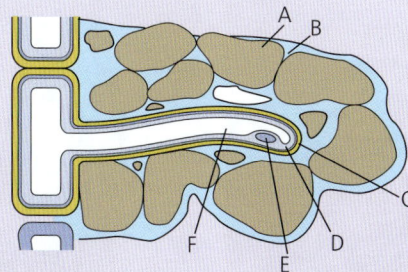

a Name the parts labelled A to F.

b Explain how water gets into a root hair cell from the soil.

c Give two other functions of root hair cells.

s **3** Explain, using the term *water potential*, how water passes into root hair cells, across the cells of the cortex, into the xylem vessels and up to the leaves, where it is lost to the atmosphere.

1 Water enters a root hair cell by osmosis. It then passes across the cells of the root cortex by osmosis before passing into the xylem and then up the stem to the leaves.

2 Root hairs are well adapted for the absorption of water since they have thin cell walls and have a large surface area.

3 Loss of water from the leaves 'pulls' more water up the stem from the roots in the transpiration stream.

8.3 Transpiration

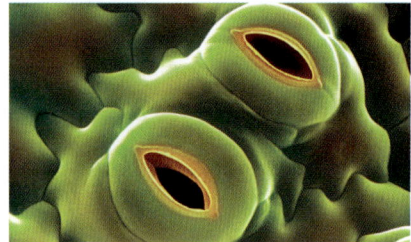

Figure 8.3.1 Two open stomata.

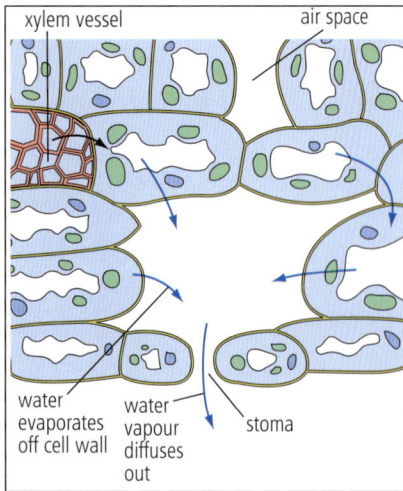

Figure 8.3.2 Loss of water vapour from spongy mesophyll cells of a leaf.

Plants absorb large quantities of water from the soil. Much of this water passes into roots, moves up through stems and enters leaves where it is lost to the atmosphere as water vapour. If you look at the diagram of a section of a leaf you will see that there are many air spaces inside (Figure 8.3.2). These are surrounded by mesophyll cells, which have damp cell walls. Water evaporates from this huge, moist internal surface so the air spaces become saturated with water vapour.

There is usually more water vapour in the air spaces than there is in the outside air, so water vapour diffuses through the stomata into the air (Figure 8.3.1). The water lost from the mesophyll cells is replaced by more water from the xylem.

The loss of water vapour from plant leaves is known as **transpiration**.

Supplement

The size of the internal surface area and the number and size of stomata affect the rate of water loss by transpiration. The large air spaces that extend throughout the spongy mesophyll provide a very large surface area for the evaporation of water. The leaves of some species are more compact and have a smaller internal surface area, so lose less water vapour. These leaves often have few stomata to reduce the loss of water.

More transpiration takes place during the day than at night because the stomata are open during the day and closed at night. The stomata open during the day so that carbon dioxide can diffuse into the leaf for use in photosynthesis by chloroplasts in mesophyll cells.

The stomata close at night to reduce the volume of water lost by transpiration. They may also close in hot, dry conditions during the day as water lost in transpiration is not being replaced by water from the soil.

Figure 8.3.3 Turgid plant with plenty of water.

Figure 8.3.4 Wilted plant with not enough water.

Wilting

The stomata close up to reduce transpiration. If the plant still does not get enough water it will start to **wilt** (Figure 8.3.4). Its cells have lost so much water that they are no longer turgid or full of water.

Turgid cells are firm and give the plant support. If the cells become flaccid then the plant becomes soft. The stem is no longer upright and the leaves droop.

Wilting is not a bad thing. The leaves move downwards so are out of the direct rays of the Sun so do not get as hot. When the temperature decreases and the plant can absorb more water than is lost by transpiration the leaves will recover.

Structure and function of xylem vessels

There is a continuous flow of water from the roots to the leaves. This movement of water up the xylem is called the **transpiration stream**. The xylem is composed of long, continuous columns of cells called **xylem vessels**.

Xylem vessels form when a column of simple cells lose their end walls. The walls of these tubes become strengthened by the addition of a woody material called **lignin**. This increases the mechanical strength of the vessels and enables them to withstand the strong pressures that occur during water transport.

Lignin is impermeable, so materials cannot pass into xylem cells and their contents cannot survive. This means that the xylem vessels become hollow and there are no cell contents to restrict the flow of water.

Young xylem vessels may contain rings or spirals of lignin, but as more lignin is added the vessels become much stronger.

Most of the water that enters a leaf is lost by transpiration. A little is used in photosynthesis and some enters cell vacuoles to keep cells turgid. The evaporation of water from the mesophyll and the diffusion of water vapour through the stomata create a **transpiration pull** that moves water up the xylem vessels in the stem from the roots to the leaves.

As water is used up or lost from the leaves, more is sucked up from the xylem vessels, rather like water being sucked up a straw.

This flow of water up the xylem relies on two properties of water:

- **Cohesion** – the water molecules tend to attract each other, sticking together and moving upwards together.
- **Adhesion** – the water molecules tend to stick to the inside of the xylem vessel so reducing the chance that the columns of water will be broken.

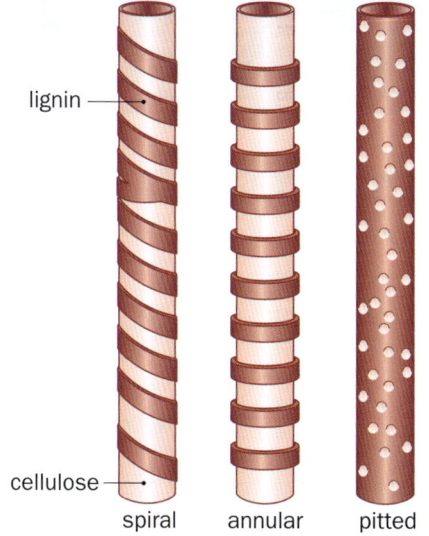

lignin

cellulose

spiral   annular   pitted

Figure 8.3.5 Xylem vessels showing different types of thickening with lignin. The walls of pitted vessels are covered in lignin apart from small areas of cellulose to allow water to flow out to surrounding cells.

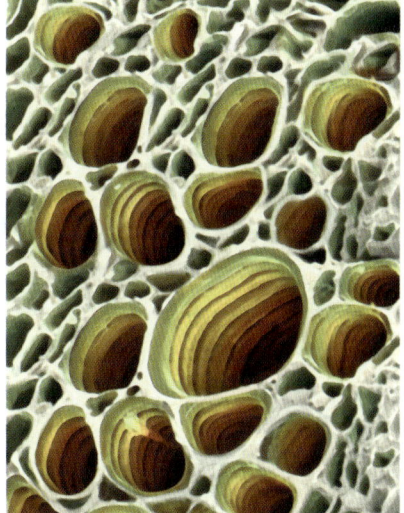

Figure 8.3.6 Rings of lignin are visible in this image of xylem vessels.

KEY POINTS

1 Transpiration is the evaporation of water at the surfaces of the mesophyll cells followed by the diffusion of water vapour through stomata to the atmosphere.

2 The structure of xylem is related to its functions of water transport and mechanical support.

3 If water loss is greater than water uptake, plants wilt because of lack of turgor in cells in stems and leaves.

SUMMARY QUESTIONS

1 State the function of xylem tissue.

2 Make simple diagrams to show the distribution of xylem tissue in transverse sections of a leaf, a stem and a root.

S 3 Explain how the structure of xylem vessels is related to their function in transport.

4 Explain how a molecule of water moves from a xylem vessel in a leaf to the air outside a stoma.

5 a Explain how wilting takes place.
   b Explain how wilting helps plants survive in their environment.

8.4 Measuring transpiration

It is not easy to measure the rate of transpiration, but you can use a **potometer** to measure the rate of water uptake.

The volume of water lost is slightly less than the volume of water taken in by the roots. This is because some of the water is used up in photosynthesis and in keeping cells turgid. You could use a simple piece of apparatus to investigate the rate of water uptake.

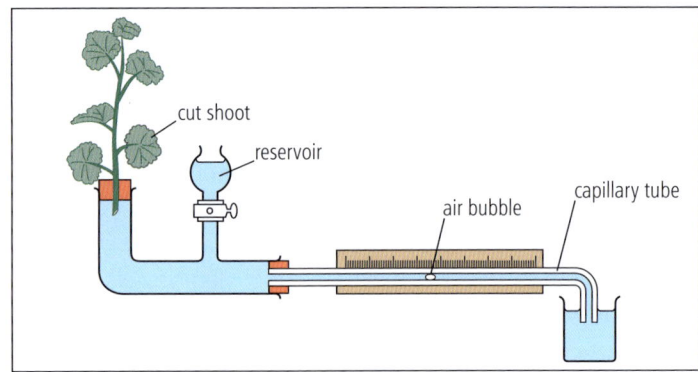

Figure 8.4.1 A potometer.

PRACTICAL

Measuring the rate of water uptake

Fill a capillary tube by submerging it in water and set up apparatus as shown in Figure 8.4.2.

Cut the end of a shoot under the water and add the shoot to the apparatus.

Clamp the apparatus in the position shown.

See how far the bubble travels along the tube in five minutes.

- Investigate the effects of wind speed by placing a fan at different distances away from the apparatus and recording the results.
- Investigate the effect of temperature by placing the apparatus near a heater or in a cold place, such as a fridge.

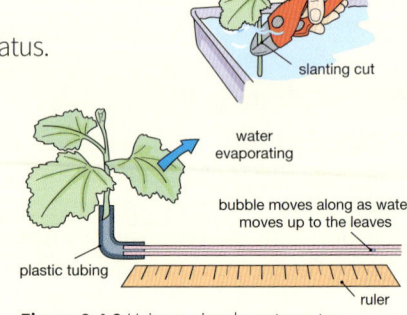

Figure 8.4.2 Using a simple potometer.

Factors affecting transpiration

The effects of increasing wind speed and temperature are shown in Figure 8.4.3.

- The rate of transpiration increases as the air **temperature** increases.
- The rate of transpiration increases as the **wind speed** increases.

Light intensity also affects the rate of transpiration. As the light intensity increases, the rate of transpiration increases up to a maximum that is determined by the other conditions such as the temperature, wind speed and humidity in the air.

Supplement

Explanations

Light causes stomata to open. As the light intensity increases, stomata open wider allowing more water vapour to diffuse out of the leaves. As light intensity decreases (for example, when it is cloudy or towards evening) stomata close so that less water is lost by transpiration.

Temperature influences the rate of evaporation of water from the surfaces of cells inside the leaf. As temperature increases, water molecules on the cell surfaces have more kinetic energy and enter the air inside the leaf as water vapour.

An increase in **wind speed** means that water molecules are blown away from the leaf surface. This results in the net diffusion of more water molecules out of the stomata so increasing the rate of transpiration.

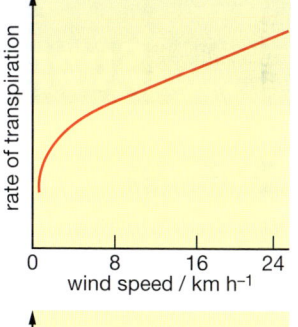

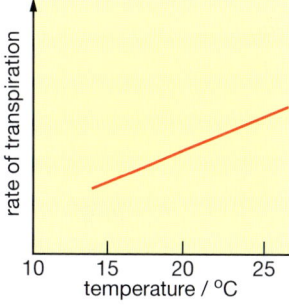

Figure 8.4.3 Two factors affecting the rate of transpiration.

In very **humid** conditions, there may be nearly as much water vapour in the atmosphere as in the air inside the leaves. This means that there is no concentration gradient for water vapour so the rate of transpiration is low. In dry conditions the concentration gradient is very steep so much more water vapour diffuses out through stomata.

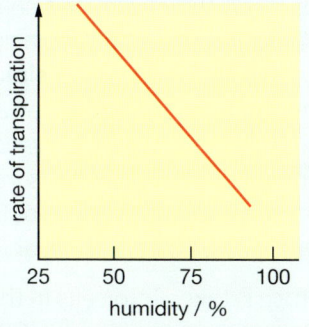

Figure 8.4.4 The effect of humidity on the rate of transpiration.

SUMMARY QUESTIONS

1 **a** Describe how you would set up a potometer to measure the rate of water uptake by a cut shoot?
  b Explain why cut shoots take up water when put into a potometer.

2 Describe how you would change conditions around a cut shoot in a potometer to measure the effect of wind speed and temperature on the rate of transpiration.

3 State the effect of **a** increasing wind speed and **b** increasing temperature on the rate of transpiration.

S 4 Explain how each of these conditions affects the rate of transpiration:
  a an increase in light intensity
  b an increase in temperature.

KEY POINTS

1 A potometer can be used to measure the rate of water uptake, but it cannot directly measure the rate of transpiration.

2 Increased light intensity opens the stomata and increases transpiration rate. Increasing temperature and wind speed increases the rate of transpiration.

S 3 Increasing humidity decreases the rate of transpiration.

4 Temperature affects the rate of evaporation of water inside leaves. Humidity and wind speed affect the rate of diffusion of water vapour out of leaves.

- Describe translocation in plants
- State that sucrose and amino acids move in phloem from a source to a sink
- Explain that some parts of a plant may act as a source and a sink at different times during the life of a plant

Glucose is made in the leaves by photosynthesis. Leaf cells convert glucose into sucrose for transport. Leaves use glucose and nitrate ions to make amino acids. Some of these amino acids are used in leaf cells to make proteins, the rest is exported from the leaves for protein synthesis elsewhere in the plant. Sucrose and amino acids are soluble in water and are transported to other parts of the plant in solution in the phloem. This transport is called **translocation**, which means 'from place to place'. The transport of sucrose and amino acids takes place from organs where they are produced to organs where they are used in synthesis or where sugar is respired to provide energy for growth and other functions.

In translocation, the part of the plant where sucrose and amino acids begin their journey is called a **source**. This is where leaves release sucrose and amino acids into the phloem. The part where they end their journey is a **sink**. This is where sucrose and amino acids are used. Mature leaves are sources while they receive enough light to photosynthesise. Phloem transports sucrose and amino acids from the leaves into the stem and then downwards to the roots and growing root tips or upwards to shoot tips, young leaves, buds, flowers, fruits and seeds.

Sucrose is taken into cells in the sinks where it is broken down by an enzyme to form glucose. Glucose is:

- broken down in respiration to provide energy
- built up into starch for storage in roots and in seeds
- built up into cellulose for new cell walls at growing root tips and shoot tips

Sucrose is also stored in the stems of some plants (e.g., sugar cane) and in some fruits to make them sweet to attract animals. Sucrose is also used to make nectar to attract pollinators (see page 176).

Storage organs, such as potato tubers, are sinks when the potato plants are growing. This is the time when they are filling with starch. The potato plant dies down at the end of the growing season leaving tubers in the soil to survive the harsh conditions of winter. At the beginning of the next growing season the tuber is a source of sucrose and amino acids that travel in the phloem into the new shoots and roots (see pages 174 and 175).

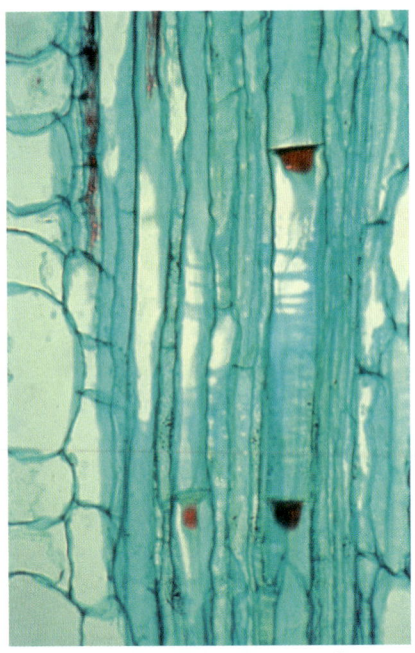

Figure 8.5.1 Phloem tubes are living cells.

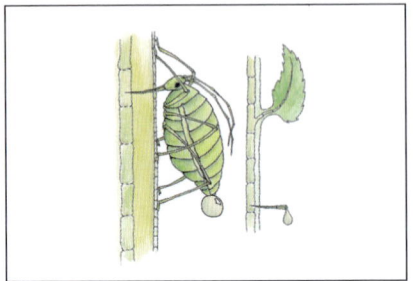

Figure 8.5.2 Aphids are crop pests. They feed on phloem sap by using their piercing mouthparts.

Translocation is an active process that occurs in phloem. Phloem tubes are living cells that contain some cytoplasm. Movement in the phloem requires active transport of sucrose at the source. The sucrose lowers the water potential in the phloem tubes so that water enters by osmosis. A head of pressure builds up in the phloem tubes that forces the phloem sap to the sinks. Translocation is most active on sunny, warm days when plants have high rates of photosynthesis.

Aphids and phloem transport

Aphids are small insects that feed on the sap in the phloem of plants. They have piercing mouthparts called **stylets**, which can be inserted through the surface of the stem into a single phloem tube. The pressure of the sap inside the phloem tube causes it to pass along the stylet and so provide the aphid with food.

Aphids can be used to study the transport of sugars in the phloem. If the stylet of a feeding aphid is cut, the sugary sap will pass out of the stylet for some time.

Sucrose that has been labelled with a radioactive isotope can be injected into a leaf and samples collected from aphid stylets that have been inserted into the stem above and below the leaf. The contents of the sap can be analysed for the isotope and the passage of the sucrose traced.

Maple syrup and phloem

Maple syrup is prepared from phloem sap of maple trees. Farmers drill holes into the tree and insert tubes through which the sap flows and collects in buckets.

Figure 8.5.3 Black bean aphids. When there is plenty of food and conditions are favourable, aphids reproduce very rapidly. Farmers control them by spraying insecticide.

SUMMARY QUESTIONS

1 **a** Define the term *translocation*.
  b What substances are translocated in the phloem?
  c What are the materials transported in the phloem used for?
2 Make a table to compare transport in the phloem with transport in the xylem. Include such features as substances transported, direction of movement in the stem and the types of cells in the two tissues. Use three columns headed: features, phloem and xylem.
3 Explain how a potato tuber can be a source and a sink at different times of the year.
4 Aphids feed on plants.
  a How are aphids used to investigate movement in phloem?
  b Farmers and growers control aphids by spraying insecticides. Explain why this is necessary.

KEY POINTS

1 Translocation is the movement of sucrose and amino acids from the leaves to regions of respiration, growth and storage.

2 Regions of production are called sources and regions of storage, respiration and growth are called sinks.

3 Some plant organs, such as potato tubers can be a source and a sink at different times of the year.

Practice questions

1 In which conditions would a plant lose most water by transpiration?

A cool and still air   C hot and still air

B cool and   D hot and
   fast-moving air   fast-moving air

(Paper 1)   [1]

2 Which is the correct pathway taken by a molecule of water as it moves through a plant?

A root cortex, xylem in root, xylem in stem, leaf mesophyll, atmosphere

B root hair cell, root cortex, xylem, leaf mesophyll cells, air spaces in mesophyll, stoma

C stoma, air spaces in leaf, xylem, root cortex, root hair cell

D xylem, root cortex, leaf mesophyll cell, stoma, atmosphere

(Paper 1)   [1]

3 Which absorbs water from the soil?

A cuticle   C root hairs

B mesophyll   D xylem vessel

(Paper 1)   [1]

4 Which statement best describes the pathway and the mechanism by which most water is lost from leaves?

A Diffusion of water vapour from intercellular air spaces through stomata into the atmosphere.

B Evaporation of water and diffusion of water from mesophyll cells through stomata into the atmosphere.

C Evaporation of water from mesophyll cells and diffusion of water vapour through the cuticle into the atmosphere.

D Evaporation of water from mesophyll cells and diffusion of water vapour from air spaces through stomata into the atmosphere.

(Paper 1)   [1]

5 Which row best describes the substances transported in the phloem and xylem?

	phloem	xylem
A	amino acids and ions	sucrose and water
B	amino acids and sucrose	ions and water
C	sucrose and water	amino acids and ions
D	glucose and sucrose	ions and water

(Paper 2)   [1]

6 A sweet potato plant grows in a field in bright sunlight. Which row shows the correct movement of carbohydrate in the phloem?

	type of carbohydrate	source of carbohydrate	destination
A	glucose	leaves	tubers
B	glucose	tubers	leaves
C	sucrose	leaves	tubers
D	sucrose	tubers	leaves

(Paper 2)   [1]

7 A potted plant has wilted after being left without water for 6 days. Which combination of environmental conditions is likely to have caused it to wilt?

	light intensity	humidity	air temperature
A	high	high	low
B	low	low	high
C	high	low	high
D	low	high	low

(Paper 2)   [1]

8 Which statement explains the movement of water from the stem into leaf cells?

Water moves from:

A mesophyll down a water potential gradient

B phloem down a water potential gradient

C xylem down a water potential gradient

D xylem from low water potential to high water potential

(Paper 2)   [1]

9 The diagrams show transverse sections of **A** root, **B** stem and **C** leaf.

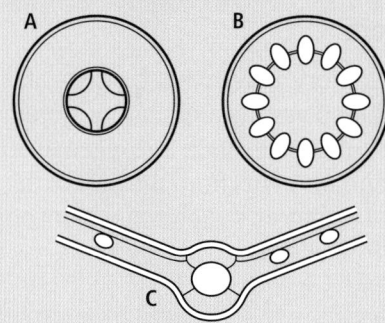

(a) Copy the diagrams and shade the areas that represent the positions of xylem and phloem in **A**, **B** and **C**. [3]

(b) Describe how you would discover the pathway taken by water as it travels through the parts of the plant that are above ground. [4]

(Paper 3)

10 A potometer (Topic 8.3) was used to measure water uptake by a leafy shoot.

	conditions	time taken for the water to move 100 mm / min
A	cool, still air in daylight	2
B	cool, moving air in daylight	4
C	warm, still air, in daylight	3
D	warm, moving air, in daylight	1
E	warm, still air, at night	60

(a) Calculate the rate of water movement in mm min^{-1} for each of the conditions, **A** to **E**, and present your answers in a table. [3]

(b) Draw a bar chart to show the effect of the five conditions on the rate of water movement. [5]

(c) From the results, state three environmental conditions that affect the rate of water movement through the plant. [3]

(d) State two ways in which the air around the shoot would be affected if it was covered with a transparent plastic bag. [2]

(Paper 6)

11 A leafy shoot was cut from a plant and placed into a potometer which was placed on a balance. The table shows the rate of water loss by transpiration and the rate of water absorption by the leafy shoot during a hot, dry day.

time / h	rate of water loss / g h^{-1}	rate of water absorption / g h^{-1}
midnight	4	7
0300	6	9
0600	12	12
0900	18	14
1200	24	17
1500	22	19
1800	18	20
2100	8	14
midnight	6	10

(a) Plot a graph to show the data in the table. [6]

(b) Calculate the percentage change in the rate of water loss between 0600 hours and 1200 hours. Show your working. [2]

(c) Describe the changes in water loss and water absorption over the 24-hour period. [4]

(d) Explain how water moves from the leaves of the shoot to the atmosphere. [4]

(Paper 6)

12 The image shows some xylem vessels.

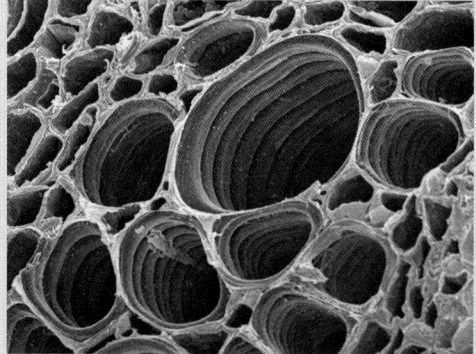

(a) With reference to the image, explain how xylem vessels are adapted to the transport of water through plant organs. [2]

(b) Water moves from the xylem in the root to the atmosphere surrounding the leaves.

Describe the mechanisms involved in moving water through the plant and into the atmosphere. [6]

(Paper 4)

9.1 Circulation

LEARNING OUTCOMES

- Describe the one-way flow of blood around the body
- Describe the heart as a pump for the flow of blood and the role of valves to permit only one-way flow
- **s** Describe the single circulation of a fish
- Describe the double circulation of a mammal and explain its advantages

The circulatory system

Most animals have a circulatory system that consists of blood, some sort of pump and blood vessels. Circulatory systems in animals transport useful substances such as oxygen absorbed from a gas exchange surface and nutrients from the part of the gut that is adapted for absorption. Circulatory systems also transport waste, such as carbon dioxide, from all over the body to the gas exchange surface.

Blood circulation

The **heart** is a pump that circulates blood through blood vessels.

Blood flows in **arteries** away from the heart to the different organs of the body. Blood flows back to the heart in **veins**. The smallest blood vessels that connect arteries to veins are **capillaries**. Cells are very close to capillaries so they receive a good supply of oxygen and nutrients from these capillaries and can have their carbon dioxide and other waste products removed efficiently.

The circulatory system allows a **one-way flow of blood** around the body. The heart pumps blood giving it pressure so that it flows inside arteries and this helps to maintain a one-way flow. This is good for getting blood to the capillaries, but high pressure blood will damage the delicate capillaries so small muscular blood vessels known as arterioles reduce the pressure before the blood enters capillaries (you can find out more about arterioles on page 105). When blood leaves capillaries the blood pressure is even lower.

There are **semi-lunar valves** in veins to make sure blood does not flow backwards away from the heart (Figure 9.1.1). If this happens, veins swell and the blood is not circulated properly. Each valve has three of these pockets. The valves open when the pressure of the blood pushes against them, but they close when blood flows back to fill the pockets.

There are semi-lunar valves in the heart at the places where blood leaves the heart chambers to enter arteries. There are also valves between the chambers of the heart (see page 100).

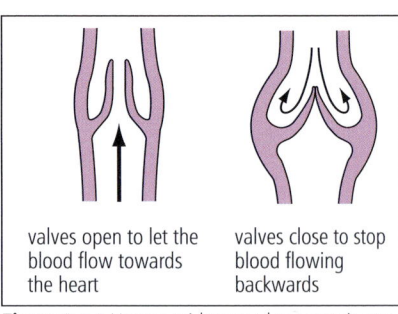

valves open to let the blood flow towards the heart

valves close to stop blood flowing backwards

Figure 9.1.1 How semi-lunar valves permit one-way flow of blood.

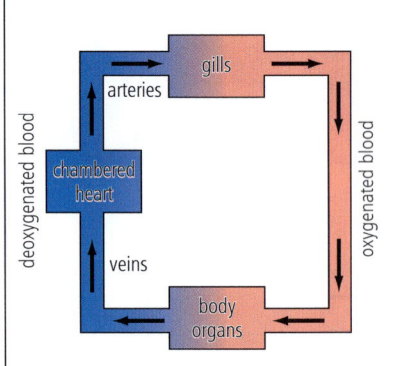

deoxygenated blood

gills

arteries

chambered heart

veins

body organs

oxygenated blood

Figure 9.1.2 Fish have a single circulation in which blood flows through the heart once in a circuit around the body.

Supplement

A single circulation

Fish have a single circulation because blood flows through the heart once during a complete circuit of the body (Figure 9.1.2). Blood from the organs flows into the heart from veins. This blood has little oxygen it and is called **deoxygenated blood**. The heart pumps blood into an artery that takes it to the gills to be oxygenated. The blood flows on from the gills in arteries to the body organs.

Supplement

A double circulation

Mammals have a double circulation. Blood flows through the heart twice during one complete circuit around the body. You can confirm this by following the pathway that blood takes around Figure 9.1.3.

The mammalian heart is divided into two halves: right and left. The septum is a thick wall of muscle that separates the two halves of the heart that stops blood in the right side mixing with blood in the left side.

Blood circulation has two functions:

- The right side of the heart pumps deoxygenated blood to the lungs and back to the heart again. The pressure required to force blood to the lungs is not very high since there is little resistance to flow in the lungs as they are a spongy tissue filled with air. Gas exchange occurs as blood flows through capillaries in the lungs. Blood absorbs oxygen and loses carbon dioxide. The **oxygenated blood** returns to the heart in a vein. This blood is bright red in colour and is shown in red on diagrams of the circulation.

- The left side of the heart pumps oxygenated blood to the rest of the body and back to the heart again. The pressure of blood leaving the left side is much greater than on the right side as there is much more resistance to flow than there is through the lungs. Gas exchange occurs as blood flows through capillaries in organs such as muscles, the gut, liver and kidneys. Oxygen leaves the blood and carbon dioxide enters. The deoxygenated blood that leaves capillaries and flows through veins is dark red in colour, but is always shown as blue.

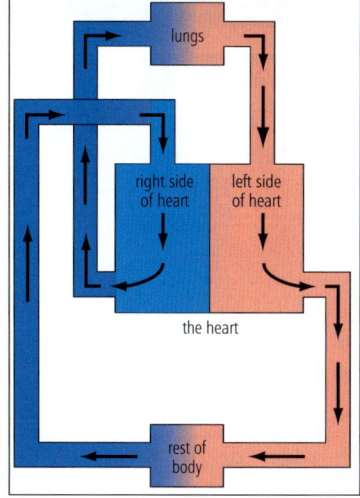

Figure 9.1.3 Mammals have a double circulation in which blood flows through the heart twice in a complete circuit of the body.

SUMMARY QUESTIONS

1 Copy and complete the sentences using these words (some may be used twice):

**oxygenated   body tissues   left   lungs   carbon dioxide
veins   deoxygenated   right   oxygen**

The right side of the mammalian heart pumps blood to the _____ . Here it absorbs _____ and is now called _____ blood.

The gas _____ is removed from the blood in the lungs. The blood returns to the _____ side of the heart and from here it is pumped to the _____ where it gives up its _____ and is now called _____ blood. It passes back to the _____ side of the heart in _____ .

S 2 Explain what is meant by the terms *single circulation* and *double circulation*.

3 Explain the following statements about the mammalian heart.
   a Blood on the right side of the heart is at low pressure.
   b Blood on the left side of the heart is at high pressure.
   c Veins contain semi-lunar valves.

KEY POINTS

1 The heart and valves ensure a one-way flow of blood around the circulatory system.

2 Fish have a single circulation, so blood passes through the heart once during one circuit of the body.

3 Blood flows twice through the heart during one circuit of the body in a mammal.

4 The heart pumps blood at low pressure to the lungs and at high pressure to the rest of the body in a double circulation. **S**

5 Blood flows in arteries away from the heart. Blood flows in veins towards the heart.

6 Advantages of a double circulation: Oxygenated blood and deoxygenated blood are kept separate. Blood is pumped to lungs at low pressure so they are not damaged. Blood is pumped to the rest of the body at high pressure to give fast supply of oxygen and nutrients to respiring tissues.

LEARNING OUTCOMES

- Describe the four chambers of the heart and locate the major blood vessels and valves associated with the heart
- State that blood is pumped away from the heart into arteries and returns to the heart in veins
- **S** State the sequence of events that take place during one heartbeat
- Explain the importance of the septum in separating oxygenated and deoxygenated blood
- Locate and name the valves in the heart
- Explain why the walls of the ventricles are thicker than those of the atria and why the left ventricle is thicker than the right

EXAM TIP

When you look at a diagram or drawing of the heart you are viewing it from the front of the body – so the left side of the heart is on the right hand side of the drawing.

EXAM TIP

Trace the pathway taken by blood through the heart. The red arrows show oxygenated blood; the blue arrows show deoxygenated blood.

Heart structure

The human heart consists of two pumps lying side by side. The right side of the heart is one pump and the left side of the heart is another pump. The heart consists almost entirely of cardiac muscle tissue (see page 24).

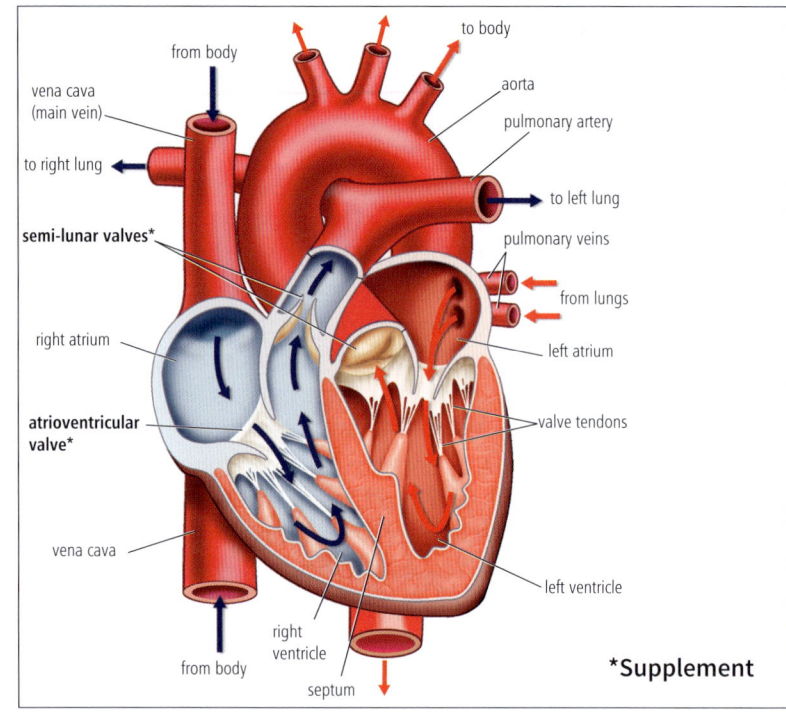

Figure 9.2.1 The arrows show how the blood moves through the heart.

On each side of the heart there are two chambers. The upper chambers are called **atria** (singular: **atrium**). Blood empties into them from veins. When the atria contract they pump blood into **ventricles**, which have much more muscular walls than the atria. When the ventricles contract they pump the blood out into the arteries at higher pressure. The right ventricle pumps blood to the lungs via the pulmonary artery. The left ventricle pumps blood to the rest of the body via the aorta (main artery).

Between each atrium and ventricle is a one-way **valve**. The valves prevent the blood flowing back into the right and left atrium when the right and left ventricles contract. There are one-way valves between the right ventricle and the pulmonary artery and between the left ventricle and the aorta. These prevent blood flowing back into the ventricles when they relax and fill with blood from the atria. The ventricles have more muscular walls than the atria because they have to pump the blood much further than the atria – either to the lungs or to the tissues of the body.

Deoxygenated blood returns to the right atrium in the vena cava (main vein). Oxygenated blood returns to the left atrium in the pulmonary veins.

Supplement

The walls of the atria are thinner than the walls of the ventricles as the atria pump blood the short distance into the ventricles.

The left ventricle has a more muscular wall than the right ventricle because it has to pump blood around the body and has to overcome more resistance to flow than the blood pumped out by the right ventricle that travels only to the lungs.

The septum is a wall of muscular tissue that separates the two ventricles. The septum prevents deoxygenated blood on the right side of the heart mixing with oxygenated blood on the left side. The pressures on the two sides are also different. The left ventricles has a much higher pressure when it contracts compared to the right ventricle.

Heart action

The heart pumps blood when its muscles contract. When the muscles contract, the chamber gets smaller and squeezes the blood out. After each chamber contracts it relaxes so it fills up with blood again.

The two sides of the heart work together. The atria contract and relax at the same time. The ventricles contract and relax at the same time.

During the relaxation phase, blood flows into the atria from the veins.

During the contraction phase:

- The atria contract and force blood into the ventricles. The atrioventricular valves between the atria and ventricles open due to the pressure of blood against them.
- Then the ventricles contract to force blood out into the arteries. The atrioventricular valves close to prevent blood flowing back into the atria.

The semi-lunar valves are at the base of the aorta and pulmonary arteries. They act like the ones in Figure 9.1.1. During the relaxation phase they shut, preventing back flow of blood into the ventricles. During the contraction phase they open allowing blood to leave to the lungs and the body.

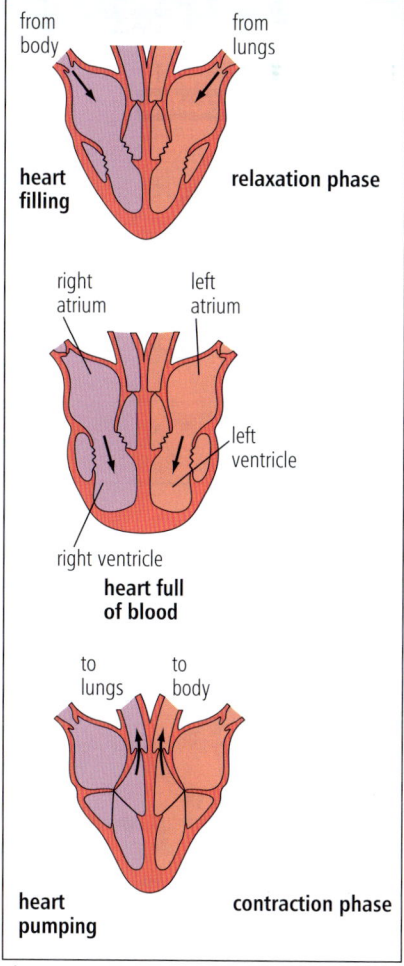

Figure 9.2.2 The cardiac cycle – the changes that occur in the heart during one heartbeat.

SUMMARY QUESTIONS

1. a What are the two upper chambers of the heart called?
   b What are the two lower chambers of the heart called?
   c What is the function of the septum?

2. Explain each statement.
   a The atria have less muscular walls than the ventricles.
   b The left ventricle has a much more muscular wall than the right ventricle.

3. In each of the following cases, which blood vessel transports:
   a blood to the rest of the body
   b blood from the lungs to the left atrium
   c blood from the body to the right atrium

9.3 Heart and exercise

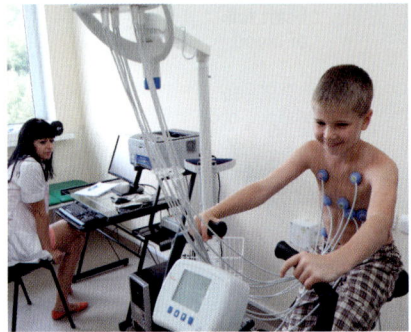

Figure 9.3.1 A boy using an exercise bicycle while having an ECG to monitor his heart function.

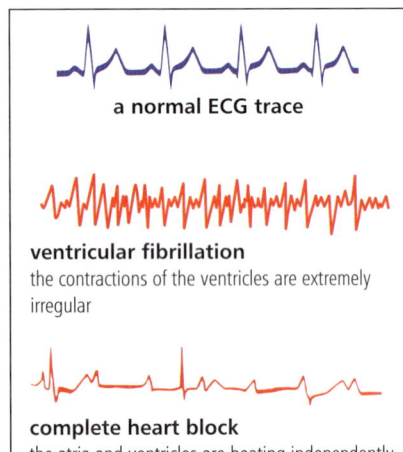

a normal ECG trace

ventricular fibrillation
the contractions of the ventricles are extremely irregular

complete heart block
the atria and ventricles are beating independently

Figure 9.3.3 A normal ECG trace and two ECG traces for heart disorders.

There are different methods by which the activity of the heart may be monitored. These include electrocardiograms (ECGs), pulse rate and listening to the sounds of the heart valves closing.

The electrocardiogram (ECG)

The control of the heartbeat depends upon electrical activity. A variety of heart disorders can produce irregularities in this activity. The electrocardiogram, or ECG, is a useful diagnostic tool that can detect these defects. Electrodes are taped at various positions on the body and the electrical activity of the heart is then displayed on a monitor. Figure 9.3.2 shows a normal ECG, while Figure 9.3.3 shows ECGs for a heart disorder.

The resultant trace can be compared with the normal ECG with its characteristic P, QRS and T waves. The P wave shows the atria contracting, the QRS 'spike' immediately precedes the contraction of the ventricles, and the T wave represents ventricles relaxing.

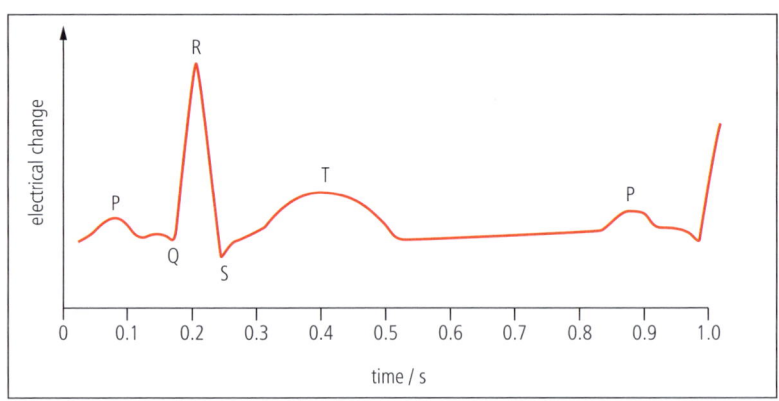

Figure 9.3.2 In this example one heart beats takes 0.8 seconds.

Heart sounds

The heart valves respond to pressure changes during a cardiac cycle. The noise of the blood when the valves open and close make the sound of your heartbeat – 'lub-dub'.

During the contraction phase, the muscular walls of the ventricles contract to force blood out of the pulmonary artery and aorta. The pressure of blood against the atrioventricular valves causes them to shut, preventing blood going back into the atria. This produces the first part of the heart sound – 'lub'.

During the relaxation phase, the ventricles relax. The blood under high pressure in the arteries causes the semi-lunar valves to shut, preventing the blood from going backwards into the ventricles. This produces the second part of the heart sound – 'dub'. Doctors are able to listen to these heart sounds with a stethoscope and identify any irregularity in the heartbeat.

You can detect the flow of blood through arteries as a 'pulse'. Each time the left ventricle beats, a wave passes along the arteries, which you can feel. You can find your pulse by feeling an artery at your wrist. The number of pulses per minute is your **pulse rate** and this is the same as the heart rate.

The **resting pulse rate** gives a good indication of a person's fitness. The fitter you are, the lower your resting pulse rate (see Table 9.3.1).

The same volume of blood is being passed out of the heart per minute but with fewer heartbeats.

The table shows the general relationship between resting pulse rate and levels of fitness.

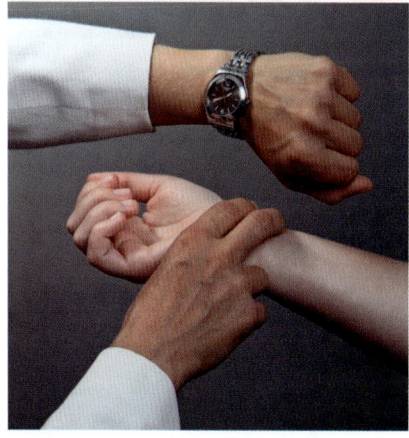

Figure 9.3.4 A person having their pulse taken.

Table 9.3.1 How pulse rate shows your level of fitness

pulse rate	level of fitness
less than 50	outstanding
50–59	excellent
60–69	good
70–79	fair
80 and over	poor

Investigating the effect of exercise on pulse rate

Sit still and get a partner to measure your pulse rate at rest.

Try doing step-ups for 1 minute (light exercise).

As soon as you have finished, sit down and count your pulse rate.

Wait until your pulse returns to the resting rate.

Now do step-ups as quickly as you can for 3 minutes (heavy exercise). Then count your pulse rate.

Remember that to make this a valid investigation you should have the same person doing the exercise, do the same type of exercise each time, rest for 5 minutes between each exercise or until your pulse returns to its resting rate.

You can also repeat the tests and calculate an average to make your results more reliable.

You can explain the results in terms of supplying substances to your muscles and removing their waste products.

Supplement

The heart and exercise

During exercise, your muscles need more energy from respiration in order to contract. So the heart beats faster and arteries supplying muscles **dilate** (widen).

These changes increase the blood flow to muscles and result in:

- an increase in supply of oxygen and glucose
- an increase in removal of carbon dioxide.

KEY POINTS

1 ECG traces, pulse rate and the sounds of valves closing can be used to monitor heart activity.

2 Pulse rate increases with activity and then decreases to resting pulse rate.

3 Physical activity increases the heart rate, increasing blood flow to muscles, supplying more oxygen and glucose and removing carbon dioxide. **S**

SUMMARY QUESTIONS

1 a State what happens to pulse rate when a person exercises.

b Explain how and why pulse rate changes with exercise. **S**

2 a Draw a normal ECG trace.

b Explain what is happening to the heart at P, Q, R, S and T waves.

3 Pulse rate is affected by certain risk factors. Suggest how some of these can be the result of a person's lifestyle. **S**

9.4 Blood vessels

LEARNING OUTCOMES

- Name and identify in diagrams and images the main blood vessels to and from the heart, lungs and kidney

- Describe the structure and function of arteries, veins and capillaries

S • Name the blood vessels to and from the liver

- Explain how the structure of arteries, veins and capillaries are adapted for their functions

Blood flows in arteries to all the organs of the body and flows away from them in veins. This table shows you some of the most important arteries and veins.

Table 9.4.1 Some of our main arteries and veins.

vessel	organs			
	heart	**lungs**	**kidneys**	**liver**
bringing blood to organ	vena cava to right atrium; pulmonary vein to left atrium	pulmonary artery	renal artery (see page 146)	hepatic artery; hepatic portal vein (see page 144)
taking blood away from organ	pulmonary artery from right ventricle; aorta from left ventricle	pulmonary vein	renal vein (see page 146)	hepatic vein (see page 144)

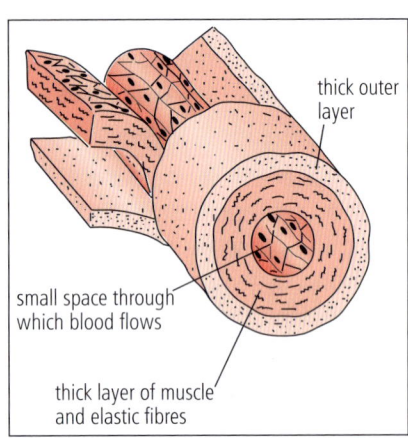

Figure 9.4.2 An artery.

- thick outer layer
- small space through which blood flows
- thick layer of muscle and elastic fibres

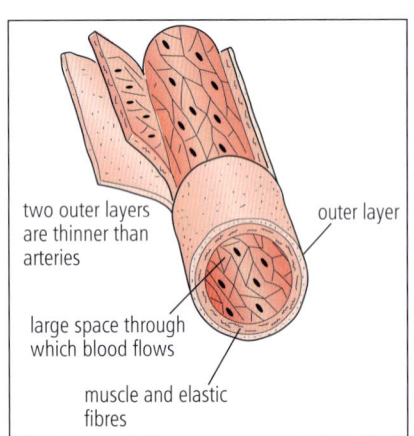

Figure 9.4.3 A vein.

- two outer layers are thinner than arteries
- outer layer
- large space through which blood flows
- muscle and elastic fibres

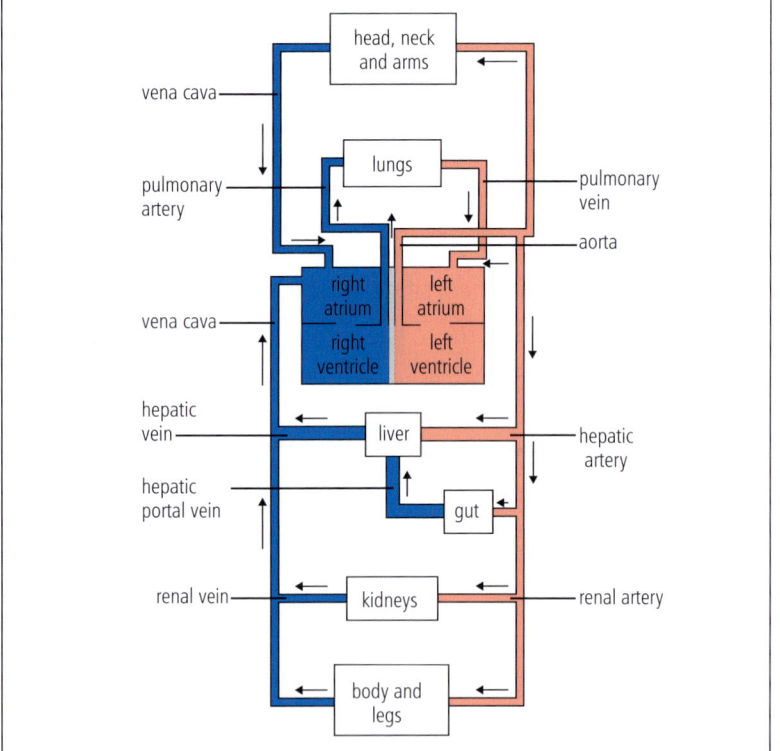

Figure 9.4.1 Main blood vessels of the body.

Structure and function of blood vessels

When your heart muscles contract they force blood into the arteries at a high pressure. Arteries have quite a narrow space in the centre for the blood to flow along. They have thick walls made of muscle and elastic fibres to withstand the pressure of blood.

Veins have a wider space for blood to flow than arteries and are thinner, less muscular and have less elastic walls. The pressure inside veins is much lower than in arteries so they do not need such thick walls.

There are semi-lunar valves at intervals along veins to ensure that blood flows in one direction – back to the heart (see page 98).

The arteries branch many times until the smallest branches form capillaries. Capillaries are very narrow; a red blood cell can only just squeeze through. The wall of a capillary is made from a single layer of very thin cells so it is easy for substances to pass to and from the blood.

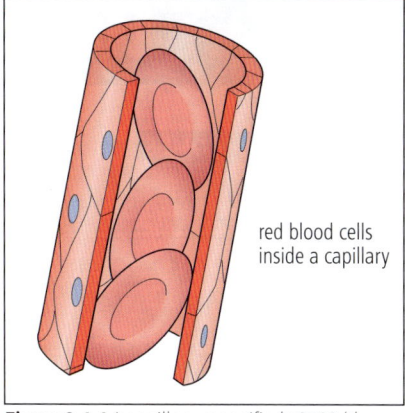

red blood cells inside a capillary

Figure 9.4.4 A capillary magnified ×2500 (the capillary is actually cylindrical, a section is missing so you can view red blood cells).

Supplement

How blood vessels are adapted to their functions

The elastic fibres in the walls of the arteries stretch and then recoil like a piece of elastic when you stretch it and then let it return to its original length. The recoil of elastic fibres helps to push blood along so maintaining its pressure. As a result the pressure at the end of an artery where it enters an organ is only a little less than when it left the heart.

It is estimated that there are over 80 000 km of capillaries in the human body. These tiny vessels provide a huge surface area for exchange between blood and cells. Blood flows through capillaries very slowly (about 1 mm per second) giving time for exchange of substances with cells. As the walls are made from one layer of thin cells the diffusion distance between blood and cells is very short. This makes it easy to supply oxygen and nutrients, such as glucose, and to remove carbon dioxide and other waste products.

As the pressure of blood in the veins is very low, it is squeezed along by pressure from the contraction of body muscles and other organs that surround the veins.

There are other vessels in the circulatory system.

Other vessels

Arterioles are the small subdivisions of arteries that carry blood to capillary networks. Like arteries, they have muscle in their walls. Apart from blood transport, arterioles are important in regulating blood pressure. They receive nerve impulses and respond to various hormones in order to regulate their diameter. This regulates blood flow through the capillaries.

Venules are small blood vessels whose function is to collect blood from the capillary beds. These thin-walled vessels then unite to form veins, which transport deoxygenated blood back to the heart.

A **shunt vessel** is a blood vessel that links an artery directly to a vein, allowing the blood to bypass the capillaries in certain areas. Shunt vessels can control blood flow by constriction and dilation. In endotherms (warm-blooded animals), shunt vessels dilate in response to cold, thereby cutting off the blood flow to the extremities and reducing wasted energy from the body heating the environment.

KEY POINTS

1 Arteries have thicker walls and smaller lumens than veins. There are one-way valves in veins but not in arteries. Capillaries are lined by a single layer of thin cells.

2 Arteries have thick muscular and elastic walls to withstand high blood pressure. Veins have much less muscle and elastic tissue and have thin walls. Blood pressure in veins is low.

3 Capillaries have walls that are one cell thick so that substances can pass easily in and out of the blood in body tissues. **S**

SUMMARY QUESTIONS

1 Make a table to show the main differences between arteries and veins. Include three columns: feature, artery, vein.

2 In what ways are the capillaries well adapted for exchange of substances between the blood and body tissues? **S**

You need a healthy heart to pump blood around your body. Heart muscle needs glucose and oxygen to keep it contracting. These are transported to the heart in the **coronary arteries**. If these arteries get blocked then heart muscle could become starved of oxygen and die. This blockage causes **coronary heart disease**.

Slowing the flow

Healthy arteries have a smooth lining, letting blood flow easily. However, **cholesterol**, which is made in the liver, can stick to their walls. This can narrow the artery and slow down the flow of blood. This condition is called **atherosclerosis**.

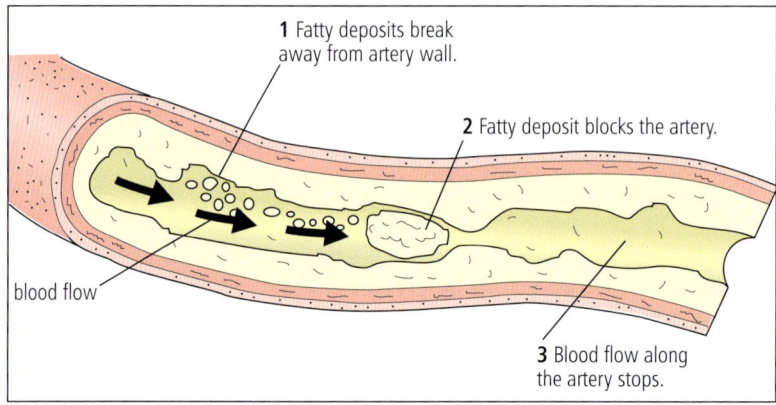

1 Fatty deposits break away from artery wall.

2 Fatty deposit blocks the artery.

blood flow

3 Blood flow along the artery stops.

Figure 9.5.1 Atherosclerosis.

The artery walls can become rough, which can cause the blood to clot and block the vessel. The blockage is called a **thrombosis**.

Narrowing of the coronary artery causes serious problems. If the coronary artery gets partly blocked it can cause chest pains, especially if activity or emotion makes the heart work harder. This is called **angina**. It is caused by not enough oxygen getting to the heart muscle. Angina should act as a warning to the sufferer because it may lead to a heart attack.

A total blockage or thrombosis can cause a heart attack. When this happens, the supply of oxygen is cut off. It causes a severe pain in the chest and the affected part of the heart is damaged. The heart may stop beating altogether – this is called **cardiac arrest**. Death will follow unless the heart starts beating again within minutes.

Around 7.2 million people in the world die from coronary heart disease each year. It is the leading cause of death in many countries. The country with the highest death rate from CHD is Ukraine; Japan has the lowest death rate. There are many reasons for these differences – diet is one of them.

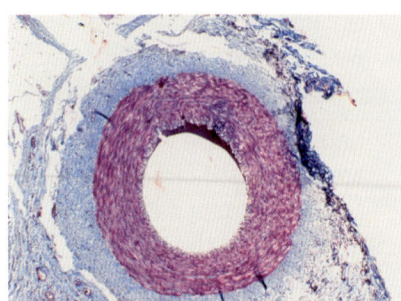

Figure 9.5.2 Cross-section of a healthy artery.

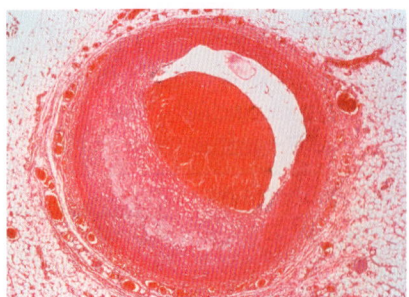

Figure 9.5.3 A coronary artery blocked by fatty deposits. A blood clot has formed to further narrow the artery. Compare with the healthy artery above (Figure 9.5.2).

Risk factors for CHD

- **Genes** inherited from parents – CHD tends to run in families.
- **Age** – the chances of getting CHD increase with age.
- **Sex** – men are more likely to get CHD than women.
- **Smoking** significantly increases the chances of getting CHD.

Ways to reduce the risk of CHD

Diet

Eating a healthy balanced diet reduces the chances of getting CHD. A low-fat, high fibre diet should include plenty of fresh fruit and vegetables (5 portions a day) along with whole grains. There are two types of fat: **saturated** and **unsaturated**. You should avoid foods containing saturated fats, because it will increase the levels of bad cholesterol in your blood. These include meat pies, sausages, fatty cuts of meat, butter and cream. Your diet should include unsaturated fats, which increase the level of good cholesterol and reduce any blockages in your arteries. These foods include oily fish, avocados, vegetable oils, nuts and seeds. Limit the amount of salt as too much will raise your blood pressure (have no more than a teaspoonful a day). Also try to avoid too much sugar, which can increase the risk of diabetes, which is linked to increased chances of developing CHD.

Exercise

Regular exercise, like cycling, raises the heart rate and will make your heart and blood circulatory system more efficient. It will also lower your cholesterol level and keep your blood pressure at a healthy level. Combining a healthy diet and regular exercise is the best way of maintaining a healthy weight, which reduces your chances of developing high blood pressure and CHD. Believe it or not, changes have been found in the coronary arteries of children as young as seven or eight.

Figure 9.5.4 Taking any exercise, like cycling, that raises the heart rate helps to reduce the risk of developing coronary heart disease.

SUMMARY QUESTIONS

1. Describe what is meant by: **a** coronary heart disease **b** thrombosis **c** angina **d** cardiac arrest
2. **a** If the heart is full of blood, why does it need its own blood supply?
   b How is the heart muscle supplied with oxygenated blood?
   c Why does the heart muscle need this oxygenated blood?
3. There are risk factors that can increase the chances of a person getting heart disease.
   a List some of the factors that cannot be controlled.
   b List some of the factors that can be controlled.
   c What advice would you give to a person recovering from a heart attack?

KEY POINTS

1. Coronary heart disease is caused by blockage of the coronary arteries that supply the heart with glucose and oxygen.
2. Risk factors that we cannot control are genes, age and sex. Avoidable risk factors include diet, lack of exercise and smoking.
3. Ways to reduce the risk of CHD by changing diet and taking exercise that raises heart rate.

9.6 Blood

Blood composition

You have about 5 litres of blood in your body. Blood is made of cells and cell fragments suspended in a yellow liquid called **plasma**. The red colour of blood is due to the pigment **haemoglobin** in red blood cells.

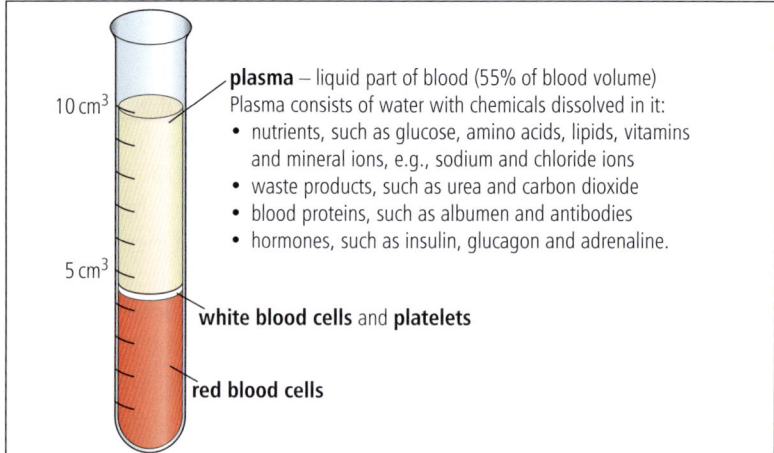

plasma – liquid part of blood (55% of blood volume)
Plasma consists of water with chemicals dissolved in it:
- nutrients, such as glucose, amino acids, lipids, vitamins and mineral ions, e.g., sodium and chloride ions
- waste products, such as urea and carbon dioxide
- blood proteins, such as albumen and antibodies
- hormones, such as insulin, glucagon and adrenaline.

white blood cells and **platelets**

red blood cells

Figure 9.6.1 When blood is spun in a centrifuge (or left to stand for 24 hours), it separates into three layers.

Blood cells

There are three main types of cell in the blood.

Red blood cells have no nuclei and have cytoplasm that is full of many thousands of molecules of haemoglobin for transporting oxygen.

White blood cells have nuclei. They look white when separated by spinning blood in a centrifuge (Figure 9.6.1). Under a microscope they are colourless as they do not have haemoglobin or any other pigment. To see them, stain has to be added as in the blood smear shown in the photograph. One group of white blood cells (phagoocytes) searches out bacteria and takes them into vacuoles where they are digested. This process is known as phagocytosis (Figure 9.7.2). Another group of white blood cells (lymphocytes) makes protein molecules called antibodies. Antibodies protect us against different types of disease-causing organism that invade the body.

Platelets are tiny fragments of cells that cause blood to clot, for example when you cut yourself. Clotting stops blood loss and prevents the entry of pathogens. The **plasma** contains many soluble substances as listed in Figure 9.6.1. Among its functions are the transport of ions, nutrients, such as glucose and amino acids, carbon dioxide and hormones.

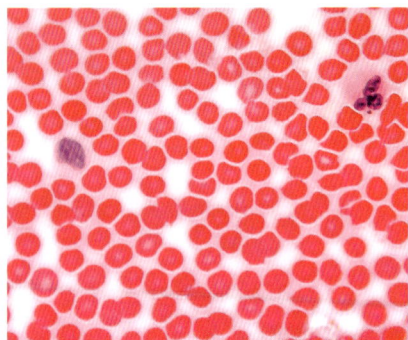

Figure 9.6.2 This blood smear seen with a light microscope shows red blood cells and two white blood cells with their nuclei stained blue.

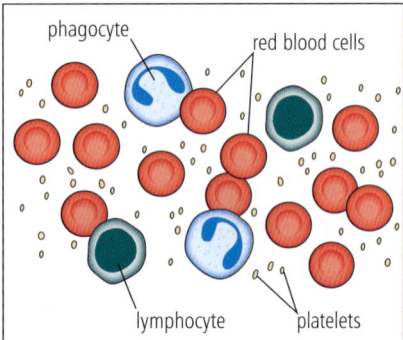

Figure 9.6.3 Red and white blood cells and platelets.

Red blood cells transport oxygen

Red blood cells are disc-shaped with a large surface area compared with their volume and this helps them to absorb oxygen.

Haemoglobin is a special type of protein that contains iron. You must have enough iron in your diet to make enough haemoglobin for your red blood cells (see anaemia on page 75). As the cells have no nuclei this makes more space for haemoglobin.

Haemoglobin combines easily with oxygen in the lungs to form **oxyhaemoglobin**:

haemoglobin + oxygen $\rightarrow$ oxyhaemoglobin

Look at Figure 9.6.6:

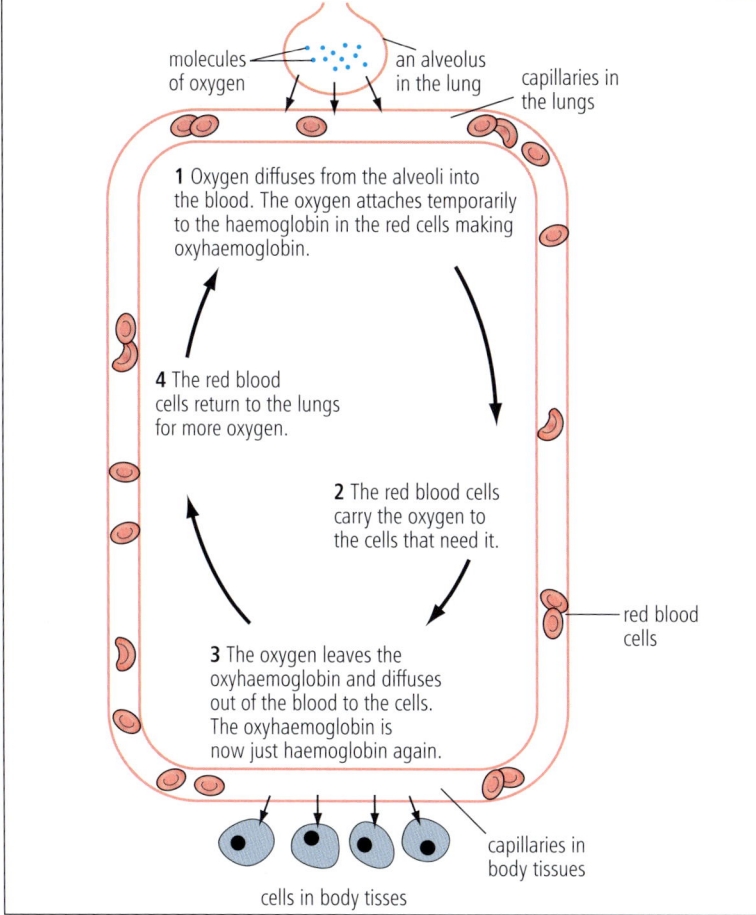

Figure 9.6.6 Transport of oxygen from the lungs to body tissues.

- Notice that oxyhaemoglobin forms while blood flows through capillaries in the alveoli in the lungs.
- Oxyhaemoglobin gives out oxygen as blood flows through capillaries in body tissues.

KEY POINTS

1 The main components of blood are plasma, red blood cells, white blood cells and platelets. Red and white blood cells are visible with the light microscope.

2 Red blood cells contain haemoglobin, which combines with oxygen to form oxyhaemoglobin.

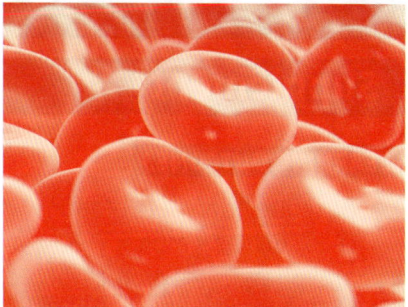

Figure 9.6.4 There are 5 million red blood cells like these in every cubic millimetre (mm³) of blood.

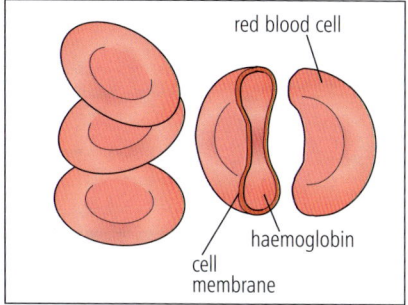

Figure 9.6.5 Red blood cells magnified ×2500.

SUMMARY QUESTIONS

1 a List four chemicals found in plasma.

  b Give three ways in which a red blood cell differs from a white blood cell.

  c Name the two main types of white blood cell.

2 a How is the red blood cell adapted for oxygen transport?

  b i What is haemoglobin?

   ii Where does it combine with oxygen and what molecule is formed?

9.7 Blood in defence

LEARNING OUTCOMES

- Describe how white blood cells such as lymphocytes and phagocytes protect the body from disease
- State the role of lymphocytes in producing antibodies
- Describe the process of blood clotting

White blood cells

White blood cells defend us against disease. There are far fewer of them than red blood cells. The two groups of white blood cells described in Topic 9.6 are **phagocytes** and **lymphocytes**.

Phagocytes

These are white blood cells that ingest pathogens, such as bacteria.

They surround the pathogens, ingest them and take them into food vacuoles. Then they digest them by using enzymes and this kills them. The process is called **phagocytosis**.

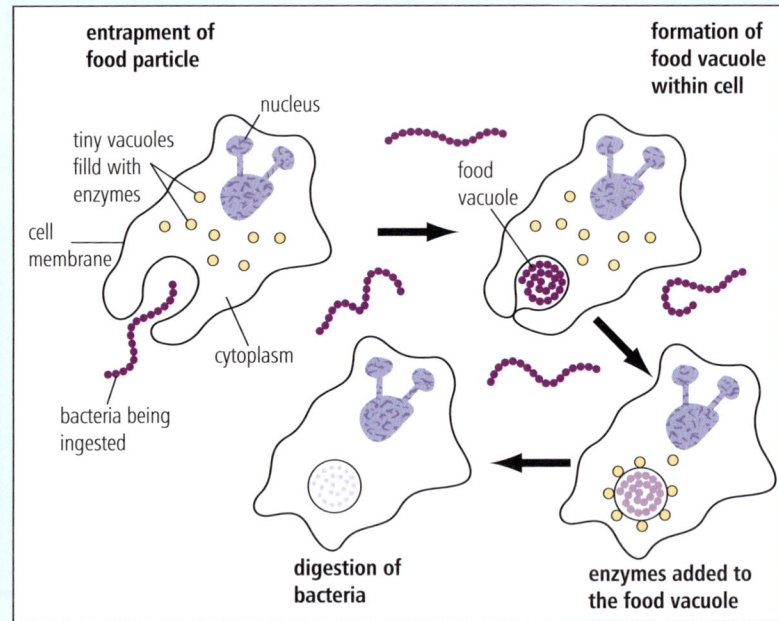

Figure 9.7.3 A white blood cell destroys some bacteria.

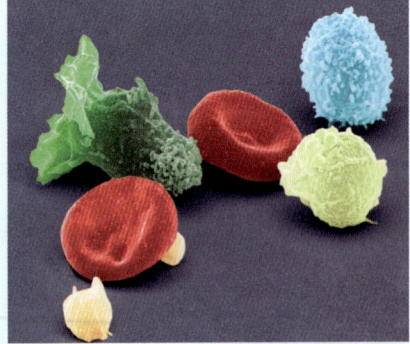

Figure 9.7.1 This image was taken with an electron microscope and then coloured with a computer program. You can see two red blood cells and three white blood cells.

When pathogens invade the body, phagocytes move towards them. They can squeeze through capillary walls.

Lymphocytes

Lymphocytes are a type of white blood cell. When a bacterium or virus enters the body the lymphocytes recognise that it is 'foreign' and should not be there. Lymphocytes then make proteins called **antibodies**.

Like enzymes, antibodies are proteins with many different shapes. Enzymes have active sites that combine with their substrates. Similarly, each type of antibody has a binding site that combines with one type of pathogen.

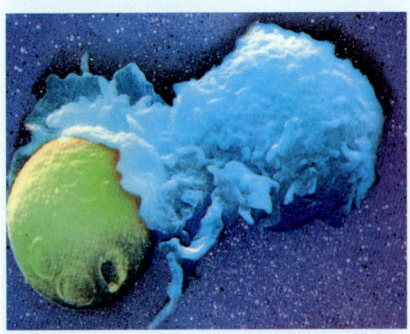

Figure 9.7.2 This image was also taken with an electron microscope and then coloured. It shows a phagocyte ingesting a yeast cell.

Antibodies attack the pathogens in a number of ways:

- they can make them stick together (agglutinate)
- they can dissolve their cell membranes
- they can neutralise the toxins (poisons) that some pathogens, such as some bacteria, produce.

There is a different type of antibody for each type of pathogen.

After you have had a disease such as measles, lymphocytes are ready to produce more of the appropriate antibodies should the pathogen enter the body again. This means you have **immunity** to that particular disease.

Blood clotting

When you cut yourself you bleed. Before long, platelets help the blood to thicken and the bleeding stops. The thickened blood has formed a **clot**. Without clotting, blood would be lost and pathogens would enter.

This is another way that the blood defends against disease.

Platelets are small fragments of cells. They are made in the bone marrow like the blood cells. When a blood vessel is damaged, platelets release substances to change the soluble protein **fibrinogen** in the plasma into the insoluble protein **fibrin**, which forms a meshwork of fibres. Red blood cells get trapped in these threads to make the clot. The clot hardens to make a scab, which keeps the cut clean. With time, the skin heals and the scab falls off.

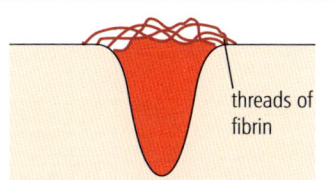

1 The skin is cut and blood starts leaking out of the body. However there is a protein in the blood called **fibrinogen**. When **platelets** come in contact with air they turn the fibrinogen into threads of **fibrin**. The threads make a net over the cut.

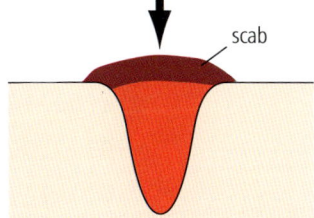

2 Red blood cells get caught in the net and make a blood clot, which seals the cut. The clot dries to make a scab. New skin grows under the scab.

Figure 9.7.5 Blood clotting.

Figure 9.7.4 Like the blood smear on page 108, these two white cells have been stained. The larger white cell in this photograph of a blood smear is a phagocyte. The smaller white cell is a lymphocyte.

KEY POINTS

1 Phagocytes ingest pathogens and digest them using enzymes.

2 Lymphocytes make antibodies that protect against pathogens in the body.

3 After a person has had a disease, the lymphocytes remain to produce more antibodies for that pathogen if the disease is encountered again. This is called immunity.

4 Blood clotting prevents blood loss and the entry of pathogens.

5 Platelets help to convert fibrinogen to fibrin during blood clotting.

SUMMARY QUESTIONS

1 a State where white blood cells are made.

  b Explain how lymphocytes act to defend the body from pathogens.

2 Explain how a person can become immune to an infectious disease like measles.

3 a What would happen if a person's blood did not clot when exposed to the air at a cut?

  b Explain, in detail, the role of platelets in the clotting of blood.

  c Describe the process of phagocytosis.

Practice questions

1 Which chamber of the heart pumps blood into the aorta?

A left atrium

B left ventricle

C right atrium

D right ventricle

(Paper 1) [1]

2 The plan diagram shows the human circulatory system. Which is the pulmonary artery?

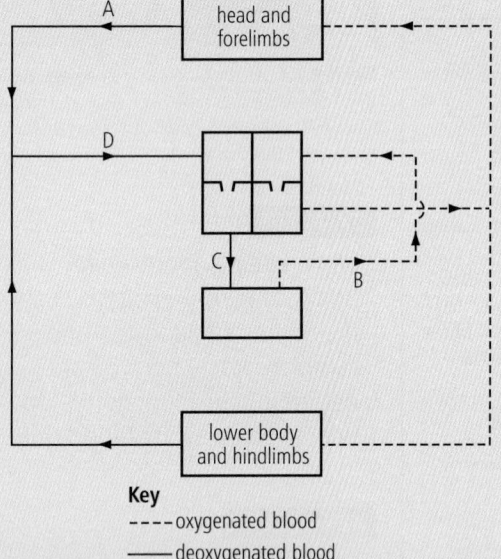

Key
---- oxygenated blood
—— deoxygenated blood

(Paper 1) [1]

3 Which is the correct pathway taken by blood flowing through the human circulatory system?

A body ⟶ vena cava ⟶ left atrium ⟶ left ventricle ⟶ lungs

B left ventricle ⟶ aorta ⟶ body ⟶ vena cava ⟶ right atrium

C lungs ⟶ pulmonary artery ⟶ left atrium ⟶ left ventricle ⟶ aorta

D right ventricle ⟶ aorta ⟶ lungs ⟶ body ⟶ right atrium

(Paper 1) [1]

4 A risk factor influences the chances of developing a disease.

Which is an avoidable risk factor for coronary heart disease?

A age

B gender

C genetic predisposition

D smoking

(Paper 1) [1]

5 Which row shows the correct functions of the components of the blood?

	plasma	platelets	red blood cells
A	antibody formation	oxygen transport	clotting
B	clotting	antibody formation	oxygen transport
C	transport of ions	clotting	oxygen transport
D	oxygen transport	clotting	antibody formation

(Paper 1) [1]

6 In a single circulation of a fish, the blood flows through the heart once during a complete circulation of the body. What type of blood leaves the heart?

A deoxygenated blood at high pressure

B deoxygenated blood at low pressure

C oxygenated blood at high pressure

D oxygenated blood at low pressure

(Paper 2) [1]

7 Arterioles are blood vessels that transport blood from:

A arteries to veins

B arteries to capillaries

C capillaries to veins

D capillaries to venules

(Paper 2) [1]

8 Which describes the position of the valve that regulates the flow of blood from the heart to the lungs?

A between left ventricle and aorta

B between right ventricle and pulmonary artery

C between left atrium and left ventricle

D between right atrium and right ventricle

(Paper 2) [1]

9 This is a drawing made of a blood smear as seen with a light microscope.

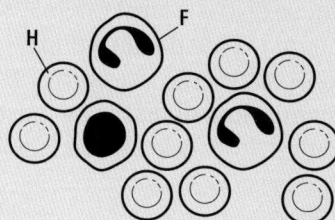

(a) (i) Name the cells **F** and **H**. [2]

(ii) State the functions of the cells **F** and **H**. [2]

(b) The magnification of the drawing is ×1100. Calculate the diameter of cell **H**. Show your working. [2]

(c) Some blood was spun in a centrifuge to separate the components. The diagram shows the results.

(i) Identify the cells that would be found in regions **K** and **L**. Give reasons for your answer. [3]

(ii) Name the fluid at **J** and describe its role in the body. [5]

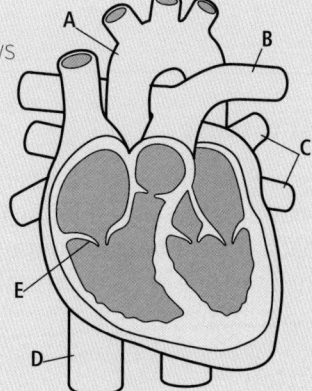

(Paper 3)

10 Taking the pulse is one way to monitor the health of the heart.

(a) What is the pulse? [2]

(b) Describe what happens to the pulse rate during and after exercise. [4]

(c) State two other ways in which the health of the heart can be monitored [2]

(Paper 3)

11 The diagram shows the heart.

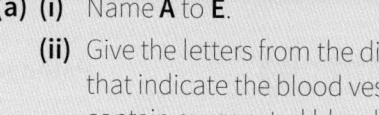

(a) (i) Name **A** to **E**. [5]

(ii) Give the letters from the diagram that indicate the blood vessels that contain oxygenated blood. [1]

(b) Describe what happens to blood as it flows through the capillaries in the lungs. [4]

(c) Explain why blood flowing to the lungs is at a lower pressure than blood flowing to the rest of the body. [3]

(Paper 3)

12 (a) Copy and complete the table to compare arteries, veins and capillaries.

feature	arteries	veins	capillaries
relative thickness of wall		thin	very thin
muscle tissue		some	none
elastic tissue		some	
direction of blood flow	heart to organs		arteries to veins

[5]

(b) Explain how the structure of an artery helps it to carry out its function. [3]

(c) Describe how substances are provided to muscle cells by capillaries. [4]

(d) The liver receives blood from the hepatic artery and the hepatic portal vein. Describe the roles of the liver in assimilating nutrients absorbed by the digestive system. [2]

(e) The liver makes substances that are used in blood clotting.

Describe the process of blood clotting. [3]

(Paper 4)

10.1 Disease

Figure 10.1.1 Even though she is using a handkerchief, droplets of water containing pathogens from her nose and mouth are entering the air and may be inhaled by others.

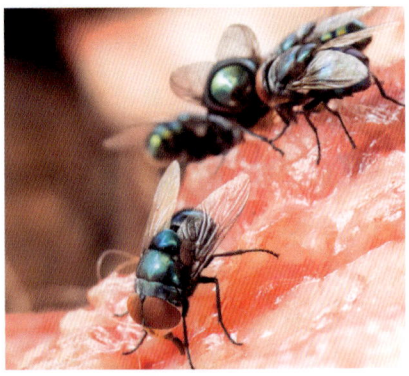

Figure 10.1.2 These houseflies may have picked up bacteria on their bodies while feeding on rubbish and human waste and are now transferring them to uncooked meat.

Pathogens are organisms that cause disease. Most of the organisms that cause disease in humans are bacteria and viruses. Other pathogens are fungi, protoctists and worms. The diseases that they cause are **transmissible diseases** because the pathogens are passed, or transmitted, from one person to another. They are also called infectious diseases.

Table 10.1.1 shows ways in which diseases are transmitted directly and indirectly from person to person.

Table 10.1.1 Methods of transmitting, descriptions and examples of some diseases. D = direct transmission; I = indirect transmission

method of transmission	description	examples of diseases
through the air (I)	pathogens are in tiny droplets of liquid from the airways and lungs of infected people (see photo opposite)	influenza, tuberculosis, common cold
contaminated food and drink (I)	people preparing food do not wash their hands; foods are not cooked properly; human faeces contaminate water supplies; flies transfer pathogens on their bodies	cholera, typhoid
direct contact (D)	uninfected people touch infected people or items that infected people have touched	athlete's foot
insect vectors (I)	insects, e.g., mosquitoes, feed on the blood of an infected person and then feed on an uninfected person	malaria, dengue fever
body fluids (D)	blood from an infected person enters the blood of an uninfected person, e.g., in an unsterilised needle shared between drug users	HIV, hepatitis
sexual activity (D)	pathogens pass from infected person to sexual partner in blood, semen or vaginal fluid	HIV, non-specific urethritis (NSU), chlamydia

Defences against disease

We have three different lines of defence against disease. The first line of defence prevents pathogens entering the body; the second line destroys any pathogens that break through the first line and enter the blood. The third line of defence produces antibodies that defend us against specific pathogens.

Barriers to infection

The defences we have against entry of pathogens are mechanical and chemical barriers.

Mechanical barriers:

- The dead outer layers of the skin form a barrier to entry (Topic 14.8).

- The hairs in the nose trap larger particles that you breathe in.

Chemical barriers:

- The stomach makes hydrochloric acid that kills pathogens in food (Topic 7.6).

- Cells that line the airways (trachea and bronchi) make mucus that traps small dust particles and microorganisms. The mucus is moved away from the lungs and up to the throat by cilia (Topic 11.3). The mucus is swallowed and any pathogens are destroyed by stomach acid.

If any pathogens get through these barriers to enter the blood, then there is a second line of defence.

Blood defences

White blood cells form a line of defence against any pathogens that enter the body's tissues or the blood. **Phagocytes** engulf bacteria and viruses into vacuoles where they are digested and destroyed. The process of **phagocytosis** is described in Topic 9.7.

Lymphocytes are white blood cells that produce **antibodies**. When activated during an infection by pathogens they produce antibodies. Antibodies are proteins that have a variety of effects, such as stopping pathogens moving through the body and making it easier for phagocytes to engulf them. Vaccination is a way to make lymphocytes produce antibodies and give long-term protection against certain diseases.

Figure 10.1.3 All outdoor activities involve some risk of infection.

10.2 Defence against disease
Supplement

LEARNING OUTCOMES

- Explain that each pathogen has antigens, each with their own specific shapes
- Explain that antibodies have specific shapes that are complementary to the shapes of antigens
- Describe how antibodies destroy pathogens
- Describe the process of active immunity
- Describe the role of memory cells in long-term immunity

EXAM TIP

Do not become confused between all the words that begin with 'anti-' in this Unit. An antigen is a chemical that stimulates lymphocytes to produce antibodies. Antibodies are proteins secreted by lymphocytes that help to destroy pathogens. Antitoxins are particular antibodies that make toxins harmless.

If pathogens get through our mechanical and chemical defences, then phagocytes in the blood may engulf and destroy them. However, phagocytes need help in order to find and engulf pathogens. That help is provided by the antibodies produced by lymphocytes.

All pathogens have chemicals on their surface called **antigens**. Many of these are made of protein. When you catch a disease your lymphocytes respond by making antibodies, which can 'lock on' to the antigens on the surface of the pathogens.

There are several ways in which antibody molecules attack pathogens.

- They cause bacteria to stick together in a group. This makes it much easier for phagocytes to find them and engulf them.

- Some bacteria have flagella, which they use to move through the body. Antibodies can attach to the flagella and stop them moving. This helps phagocytes to destroy these bacteria.

- Bacteria such as those that cause tetanus and diphtheria release toxins (poisonous substances) into the blood. **Antitoxins** are a special group of antibodies that combine with the toxins. This neutralises the toxins, making them harmless.

- Some antibodies kill bacteria directly by 'punching' holes through their cells walls. This weakens the cell walls so water enters by osmosis causing the cells to burst.

Each type of pathogen has its own antigens, which have specific shapes. The antibodies that are made against each type of antigen have a shape that fits around the antigen. The antibody has a shape that is complementary to the antigen.

If you have an infection, such as the common cold, you will be ill for a while and have a variety of symptoms. You will then begin to feel better and within a week or two you will be healthy again. If the strain of the virus that causes the common cold invades again, you will not get the same symptoms. You will not be ill. Lymphocytes are responsible for this and are activated during an **immune response**.

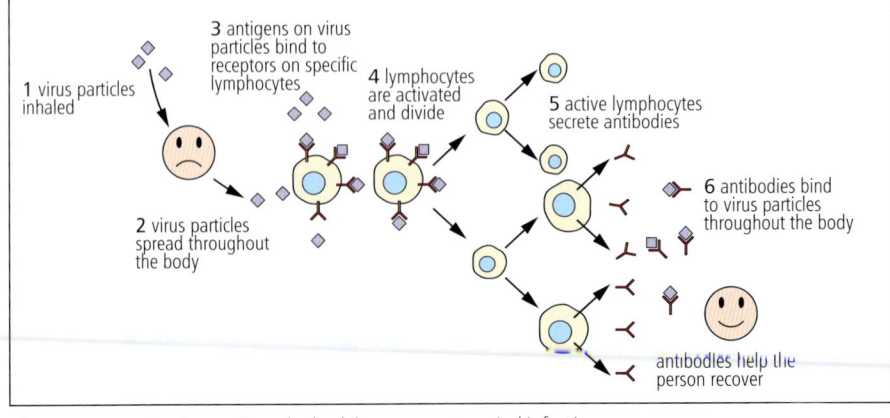

Figure 10.2.1 Active immunity – the body's response to a viral infection.

You are born with very many different types of lymphocyte. There may be as many as 100 million different lymphocytes that have the ability to become antibody secreting cells. These cells are specific to different antigens on the surfaces of bacteria, viruses and other pathogens. When a pathogen invades, some lymphocytes respond and divide to make more cells of the same type.

Many of these lymphocytes become bigger and fill with many ribosomes, which are attached to a very extensive network of endoplasmic reticulum (Topic 2.2). The antibody molecules are secreted into blood and lymph to be distributed throughout the body. The antibodies 'lock on' to the specific antigens on the pathogens that have invaded the body.

Other lymphocytes are activated to patrol the body looking for infected cells. When they find an infected cell they destroy it to stop it producing more pathogens. The reason HIV is such a serious infection is that it destroys the body's lymphocytes that are responsible for much of our defence against infectious diseases (Topic 15.11).

The response to the first infection is very slow. This is why you have fallen ill. The response to the second infection by the same pathogen is much faster; so fast that you are unlikely to have any symptoms.

This process of defence against the pathogen by antibody production is called **active immunity**. Becoming immune after catching a transmissible disease and being ill is a natural way to gain this type of immunity.

During an immune response, many lymphocytes of the specific types are produced. Many of these cells develop into antibody-producing cells, but some do not. These other cells remain in the blood and the lymphatic system circulating throughout the body. These are **memory cells**, which respond whenever there is another invasion by the same pathogen with the same antigens. Memory cells do not have a memory of the antigen in the same way that we store memories in the brain. During each immune response to specific antigens, more and more of these lymphocytes are produced to remain in the body, so that next time a pathogen invades the body there are far more lymphocytes with the ability to produce the right antibodies. The response to any subsequent infection is therefore much faster and much greater.

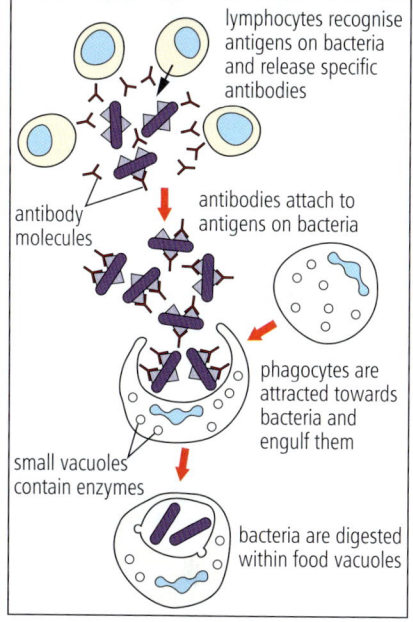

Figure 10.2.2 Lymphocytes secrete antibodies that make it easier for phagocytes to ingest bacteria and destroy them.

KEY POINTS

1 Antibodies 'lock onto' antigens leading to the destruction of pathogens.

2 Each pathogen has its own antigens, which have specific shapes that fit the shapes of specific antibodies.

3 Active immunity is a defence against a pathogen by antibody production in the body.

4 Memory cells remain in the blood and lymph after an immune response to provide a much faster and greater response to subsequent infections.

5 Active immunity is gained after an infection by a pathogen.

SUMMARY QUESTIONS

1 Explain what is meant by:
   a antigen b antibody c antitoxin
   d active immunity e memory cells

2 Outline how someone becomes naturally immune to a transmissible disease.

3 Describe how the second response to an antigen differs from the first response.

10.3 Vaccination

Supplement

LEARNING OUTCOMES

- Explain the role of vaccination in controlling the spread of disease
- Explain the control of many diseases by the mass vaccination of children

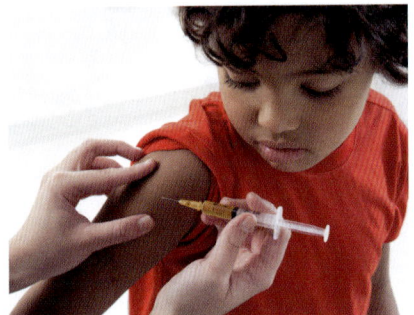

Figure 10.3.1 Some vaccines are injected.

You should have been vaccinated against most of these diseases. You may need to have boosters for some of them now or in the near future. Do you know which?

Diphtheria

Tetanus

Pertussis (Whooping cough)

Hepatitis B (HepB)

Haemophilus influenze type B (Hib)

Poliomyelitis (Polio)

Measles

Mumps

Rubella

Tuberculosis

One problem with developing natural immunity is that you have to catch the disease first. In the case of many transmissible diseases this is extremely risky.

Vaccination is an example of active immunity. The person gains immunity through being given a small quantity of the antigen or antigens required to stimulate an immune response by lymphocytes leading to the production of antibodies and memory cells. This is an artificial way to gain active immunity, which is long-lasting.

Various techniques have been devised to ensure that the vaccination results only in a mild form of the disease, or in no disease whatsoever.

- The killed organism may be used, e.g., whooping cough vaccine.

- A live non-virulent (non-dangerous) strain of the organism may be used. This is usually derived from many generations of growth in the laboratory, e.g., vaccines for tuberculosis and rubella.

- The antigens may be separated from the organism and used as a vaccine, e.g., influenza.

- The toxin (poison) produced by the organism may be chemically modified so that it is no longer toxic, but is sufficiently similar to the toxin to result in manufacture of antibodies with shapes complementary to the toxin. Examples are tetanus and diphtheria vaccines.

- Genetically modified cells, such as yeasts, are used to mass produce the relevant antigen, e.g., vaccine for hepatitis B.

Vaccines can be used to help prevent many diseases including polio, rubella, tetanus, tuberculosis, typhoid and some strains of influenza. Sometimes they need to be helped, after a number of years, by giving the person a booster injection. Some vaccines are so effective that protection lasts a lifetime without the need for any boosters. The World Health Organization (WHO) publishes a recommended schedule of vaccinations that should be given to children to protect them against vaccine-preventable diseases. The table lists some of the vaccine-preventable diseases that are part of this recommended schedule for all countries.

Each year the WHO coordinates the production of influenza vaccine because the virus that causes influenza changes as a result of mutation.

Vaccinations have saved the lives of many millions of people. They have even led to the complete eradication of smallpox in the 1970s as a result of a programme organised by the WHO.

The role of vaccination

Vaccination programmes are an important part of the health protection offered by governments to their citizens. Infants and children are vaccinated against diseases that used to be very common and were responsible for much ill health and many deaths.

Many of these diseases are now very rare in many parts of the world. The last case of polio in the Americas was in 1991, and in 1994 it was declared that transmission of polio had been successfully interrupted. But the disease still exists in other regions of the world and could be introduced into the Americas by travellers. There are still cases of polio in Pakistan, Afghanistan and some countries in Africa. You can follow the progress of the campaign to eradicate polio from the whole world by searching online for 'polio eradication'.

During eradication programmes vaccination is used in two ways.

- Mass vaccination schemes attempt to give active immunity to everyone. Some people do not respond to vaccines, but they can be protected because the chances of them coming into contact with the disease are small as most people around them have immunity and will not transmit the disease.

- Careful surveillance by health workers identifies people who have infectious diseases, and their spread can be limited or stopped by vaccinating all people in the neighbourhood including those who may have come into contact with infected people.

For some diseases there are no vaccines as yet. As of 2020, examples include HIV and malaria. Several vaccines with proven efficacy to provide resistance against COVID-19 were developed in 2020.

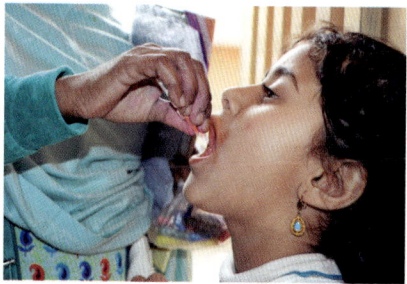

Figure 10.3.2 Vaccination against polio in Quetta, Pakistan

KEY POINTS

1 Vaccination is an artificial way of gaining active immunity.

2 Vaccination programmes are used to prevent the spread of transmissible diseases.

3 Many diseases are controlled effectively by mass vaccination of children.

SUMMARY QUESTIONS

1 Explain why vaccination is a form of active immunity.

2 Outline the events that occur in the body in response to receiving a vaccine.

3 Explain the difference between an antigen and an antibody.

4 Explain the importance of memory cells in defence against disease.

5 Explain the importance of the rubella vaccination.

6 The Americas were declared free of polio in 1994. Why do children throughout the Americas receive polio vaccinations?

7 Why is it important that as many people as possible receive the vaccinations offered by their country's health service?

10.4 Aspects of immunity

Supplement

EXAM TIP

The mother's lymphocytes do not cross the placenta or pass to the baby in breast milk. The fetus, and later the baby, 'borrow' antibodies from the mother. It has to wait until the immune system has developed before its own lymphocytes can produce antibodies.

Figure 10.4.1 There are many advantages of breast-feeding – one of the most important is providing antibodies to protect the baby against infectious diseases.

So far we have looked at active immunity, the type of immunity that occurs during the course of an infection or that follows vaccination. During an infection the pathogen enters the body and antigens stimulate an immune response. This involves the production of many lymphocytes, some of which secrete antibodies and others that become memory cells.

Passive immunity

If protection is required in a hurry, then antibodies may be given by injection. This provides only temporary immunity, but should be sufficient to give protection. Immunity gained in this way is **passive immunity**.

Antibodies pass across the placenta during pregnancy and are present in breast milk. This gives a baby protection against diseases that are in the environment and which the mother has immunity against. This is only temporary as the baby's body treats the mother's antibodies as foreign and destroys them. The immunity is gained without exposure to antigens, so is much safer than active immunity. Amongst other diseases, babies are usually immune to tetanus and measles, both of which can cause deaths of children. The protection gained from passive immunity lasts until the time when they should receive their first vaccination against these diseases.

Passive immunity is used when people are involved in nasty accidents and are at risk of tetanus bacteria entering the body through open wounds. In these cases, health workers inject tetanus antibodies as soon as possible. This neutralises the toxins produced by these bacteria that can cause muscle paralysis.

The venom from snakes and other animals, such as spiders and scorpions, acts far too quickly for active immunity to be of any use. People who have been bitten by a venomous animal need an injection of antivenom quickly. Antivenom contains antibodies to the molecules in the venom. Specialist companies collect venom from venomous animals and inject small quantities into horses or sheep. This prompts an immune response and the antibodies are collected, processed and sent to hospitals to treat bite or sting victims.

No memory cells are produced in passive immunity, so this type of immunity is short-term, but it works instantly. As soon as the antibodies enter the body they are available to be used. There is no wait between the arrival of antigens and the production of antibodies as there is in active immunity.

Malfunction of the immune system

Sometimes the immune system does not work perfectly. When this happens, it detects our own antigens as something foreign and this prompts an immune response. As a result, the immune system attacks and destroys healthy tissues by mistake.

Malfunctions of the immune system cause a variety of diseases that are collectively known as autoimmune diseases. There are many examples of these diseases. Examaples include rheumatoid arthritis, multiple sclerosis and Type 1 diabetes.

Type 1 diabetes

This type of diabetes is caused when the immune system attacks and destroys the cells of the pancreas that make insulin. This usually happens rapidly and often before the age of 20.

Insulin helps to control the concentration of glucose in the blood. It is made in response to an increase in the glucose concentration in the blood usually while a meal is being absorbed. Insulin stimulates the liver and muscles to store glucose as glycogen (Topic 14.7). When the body does not make insulin, glucose is not stored for the times when it will be needed by cells to release energy in respiration – the person has type 1 diabetes.

The symptoms of Type 1 diabetes are:

● weight loss – cells use protein and fat instead of glucose as sources of energy

● thirst – due to the increased concentration of glucose in the blood that lowers its water potential (Topic 3.2)

● tiredness due to lack of glycogen that can be converted to glucose to provide energy between meals.

One indicator of diabetes is the presence of glucose in the urine.

People with Type 1 diabetes are treated with regular injections of insulin. The drug is manufactured using techniques of genetic modification (Topic 20.4) and many different types are available so that diabetics can control their blood glucose concentrations efficiently.

Figure 10.4.2 Milking the venom from a dangerous snake. The venom will be used as an antigen to produce antibodies to give protection against the bites of this species of snake.

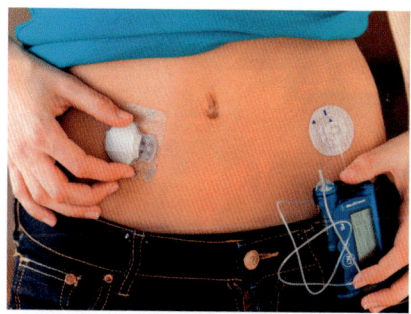

Figure 10.4.3 Many people with diabetes use programmable pumps to deliver insulin continuously into body fat through a small needle or a tube known as a cannula.

KEY POINTS

1 Passive immunity is a short-term defence against pathogens using antibodies from another individual.

2 No memory cells are produced in passive immunity.

3 Antibodies cross the placenta and are in breast milk. These protect a baby against the diseases its mother has had or been vaccinated against.

4 Type 1 diabetes is caused by the immune system destroying cells in the pancreas. It is treated by regular injections of insulin.

SUMMARY QUESTIONS

1 What is meant by the following terms?
   a passive immunity   b Type 1 diabetes

2 a Use an example to explain the advantages of passive immunity.
   b Explain why your example is only a short-term form of immunity.

3 a Explain the cause and describe the symptoms of Type 1 diabetes.
   b Suggest why insulin has to be injected or delivered by a cannula and not taken orally (by mouth).

10.5 Controlling the spread of disease

LEARNING OUTCOMES

- Explain the importance of hygienic food preparation, good personal hygiene, proper waste disposal, sewage treatment and the supply of clean drinking water in preventing the spread of disease

S
- Describe the cause and transmission of cholera

Figure 10.5.1 It is important that everyone learns to wash their hands with soap and water to prevent the spread of disease.

Figure 10.5.2 Chicken meat is often found to be contaminated by bacteria, such as *Salmonella*. It is safe to eat if it is cooked thoroughly.

Preventing infection

Personal hygiene is important in preventing the spread of some infectious diseases.

- People of all ages should wash their hands after going to the toilet to urinate or defecate and also before handling or eating food.
- Hair should be washed with shampoos to prevent dandruff and headlice.
- Everyone should wash themselves frequently, especially in hot weather.
- Dental hygiene is most important in fighting tooth decay (Topic 7.5).
- Cuts and bruises should be washed with an antiseptic and plasters applied to open wounds.

Hygienic food preparation

- Food should be covered to keep flies away.
- Kitchen surfaces should be cleaned with disinfectants to kill bacteria.
- Food should be cooked thoroughly to make sure any bacteria, such as *Salmonella*, are killed.
- Cooked food that is going to be eaten cold should be kept separate from raw food, especially meat.
- Water used for cooking and/or drinking should be boiled or sterilised by adding water purification tablets if it comes from sources that might be contaminated.

Proper waste disposal

- Household waste should be put into covered bins and collected at regular intervals, e.g., weekly.
- Garbage collected from houses and businesses should be disposed of correctly so it is not a health hazard. It should be recycled, incinerated or buried in properly regulated landfill sites. Putting garbage onto rubbish tips where it is not buried carefully means it will attract rats and flies, which spread disease. The effluent from rubbish tips may also contain harmful chemicals that cause pollution.

Sewage treatment

- Toilet waste is a serious health hazard if it is not disposed of properly through drainage pipes to a sewage treatment works.
- Human waste is broken down by microorganisms in sewage treatment works.
- The pathogens that cause typhoid and cholera are transmitted through faeces. They are transmitted to people who drink food or water contaminated with raw sewage.

- Thorough sewage treatment stops the transmission of typhoid and cholera. This means that these diseases are unknown in countries with efficient sanitation.

Drinking water treatment

- Water from the source, e.g., a reservoir, is passed through screens to remove any large floating matter

- It is then pumped to tanks where insoluble particles settle out, before filtration to remove fine particles, organic matter and bacteria

- Remaining organic matter and most of the bacteria in the water are now digested by useful bacteria, fungi and protozoa

- Chlorination kills any harmful bacteria that cause cholera, typhoid, dysentery, and food poisoning.

- In emergencies or if clean water is not available, any water to be used for drinking or cooking should either be boiled or treated with chlorine tablets to kill any harmful microorganisms.

Figure 10.5.3 A water treatment plant in the Caribbean. Water suitable for drinking is called portable water

Supplement

Cholera

Cholera is a disease caused by the bacterium *Vibrio cholerae* that is transmitted in drinking water and food contaminated by faecal matter. The bacteria produce a toxin that causes cells lining the small intestine to release chloride ions. These ions lower the water potential inside the small intestine, causing water to move from the blood into the gut. The loss of ions and water from the blood causes diarrhoea and dehydration.

Water supplies become contaminated when infected people pass out large numbers of the bacteria in their faeces. Cholera tends to occur in areas where there is a lack of proper toilet facilities, poor drainage and poor or non-existent treatment of human waste.

Figure 10.5.4 Cholera is most easily transmitted where there is a lack of clean water and sanitation is poor or non-existent

SUMMARY QUESTIONS

1. Explain why it is important to teach children to wash their hands especially before eating and after using the toilet.
2. Explain the importance of cooking chicken thoroughly, putting waste food into covered bins and connecting toilets to the main drainage.
3. **S** a State the type of pathogenic organism that causes cholera, explain how it is transmitted and describe the effects that it has on the body.
   b Suggest why cholera is a very rare disease in many countries.
   c Why is there a danger of cholera following natural disasters?

KEY POINTS

1. The spread of transmissible diseases is prevented by good personal hygiene, hygienic food preparation, proper waste disposal and sewage treatment.
2. **S** Cholera bacteria infect the small intestine causing diarrhoea that leads to the loss of ions and water from the blood.
3. Transmission of cholera is prevented by proper treatment of sewage and chlorination of drinking water.

10.6 Drugs

LEARNING OUTCOMES

- Define the term *drug*
- Describe the use of antibiotics in the treatment of bacterial infection
- State that some bacteria are resistant to antibiotics, which makes them less useful
- State that antibiotics kill bacteria but do not affect viruses
- **S** Explain the strategies that can be used to reduce the development of antibiotic-resistant bacteria

EXAM TIP

The definition at the top of this page covers all drugs, including medicinal drugs, which we describe here. You should be able to write the definition from memory.

Figure 10.6.1 These people have the bacterial disease tuberculosis. They are taking a mixture of drugs under supervision in a hospital in Tomsk, Russia.

What are drugs?

A drug is any substance that is taken into the body that alters or influences chemical reactions in the body. There are several different types of drug. Some are the active ingredients of medicines and are used to treat and cure people of disease. Others are mood-enhancing substances that alter sensory perception.

Medicinal drugs

Aspirin, paracetamol and morphine are examples of drugs that are used to suppress pain. Antibiotics are used to cure diseases that are caused by pathogenic bacteria.

Non-medicinal drugs

Alcohol, nicotine and caffeine are examples of mood-enhancing drugs that are considered socially acceptable in many countries. These are drugs that people take for their pleasurable effects, to help them to relax or to concentrate. Heroin, cocaine and amphetamines are drugs that are not socially acceptable and are illegal in many countries.

Drugs act upon the human body in various ways. Some drugs, such as nicotine and heroin, interfere with the way that the nervous system works and these are often abused by people. Many of the mood-enhancing drugs, such as heroin and nicotine, act at synapses in the nervous system and change the way in which neurones send impulses. They do this by combining with protein molecules on the cell membranes of the neurones.

Antibiotics

Antibiotics are a group of chemicals that are made by microorganisms (bacteria and fungi) or were originally derived from microorganisms but are now made artificially. Antibiotics are prescribed by doctors and vets to treat and cure human and animal diseases caused by bacteria and fungi. Antibiotics can be injected, applied to the skin or taken by mouth. Antibiotics either kill bacteria directly or stop them reproducing so that the body's defence system can kill and destroy them.

Penicillin was the first antibiotic to be discovered and mass produced. Penicillin acts on bacteria by preventing the formation of cell walls. This leads to a breakdown of the cell wall and the leakage of cell contents. Some antibiotics stop substances crossing cell membranes and others prevent enzymes catalysing important reactions, such as making proteins.

People are prescribed a course of antibiotics that has to be taken over a certain period of time. Tuberculosis is caused by a very slow-growing bacterium. The treatment of tuberculosis involves taking four drugs, including two antibiotics for a period of about six months (Figure 10.6.1). One of the antibiotics (ethambutol) prevents cell wall growth and the other (rifampicin) prevents the synthesis of mRNA so that the bacteria cannot make proteins.

Antibiotics do not affect viruses

Antibiotics have no effect on viruses so there is no point in taking antibiotics to treat viral diseases. Viruses are not cells and they do not carry out their own metabolism, but rely entirely on the cells of their host. We cannot use antibiotics to prevent processes that occur in our own cells, so the drugs used to control viruses work in different ways. An example is the anti-viral dug AZT, which is used to treat HIV/AIDS. AZT is very similar in structure to one of the bases in DNA and prevents the duplication of viral DNA so stopping the virus producing copies of itself.

Antibiotic resistance

A big problem with antibiotics is that many bacteria have become resistant to them. Antibiotic resistance occurs in a variety of ways. Some bacteria have enzymes that break down antibiotics before they can have their effect. This reduces the effectiveness of antibiotics.

Antibiotics have been in use since the 1940s. Antibiotic resistance is now widespread and increasingly there are no antibiotics available to treat some infectious diseases in some people. Some health experts have warned that we are now reaching the end of 'The age of antibiotics'.

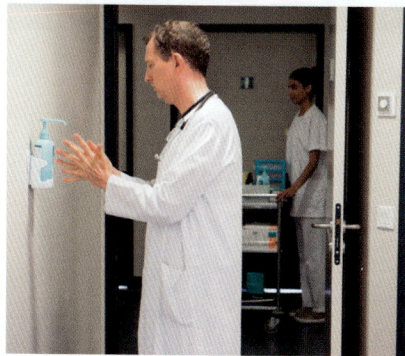

Figure 10.6.2 MRSA and other antibiotic-resistant bacteria are being controlled in hospitals by use of antibacterial gel, which this doctor is using to clean his hands.

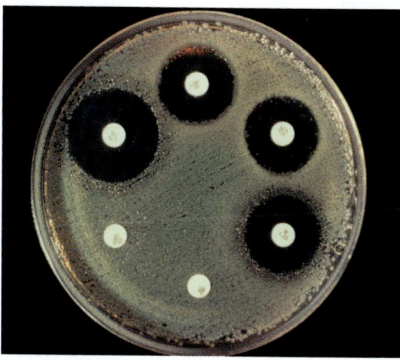

Figure 10.6.3 Before prescribing, a doctor may ask for a sensitivity test to be carried out in a hospital laboratory to find out which antibiotic to use against the pathogen that has infected a patient. The white discs on this agar plate contain different antibiotics. Here, four antibiotics would be effective because they have killed the bacteria, whereas two would not.

Supplement

Bacteria become resistant to antibiotics as a result of mutation of one or more genes. Bacteria with a mutation to a specific antibiotic survive when others without the mutation are killed when the antibiotic is used.

The overuse of antibiotics, especially in hospitals, has led to an increase in bacterial resistance. Overuse was responsible for the development of methicillin-resistant *Staphylococcus aureus* or MRSA. This has caused deaths in hospital patients with suppressed immune systems, such as those who have had organ transplants. The number of cases of MRSA in the USA and UK has decreased in recent years as a result of specific strategies such as prescribing antibiotics only when absolutely necessary. See more on antibiotic resistance on page 227.

KEY POINTS

1 A drug is a substance taken into the body that alters or influences chemical reactions in the body.

2 Antibiotics destroy pathogens by disrupting cell wall formation, inhibiting protein synthesis and metabolism in the pathogen cell.

S 3 Antibiotics cannot kill viruses since viruses have no cell wall but rather live inside host cells, taking over their metabolic processes.

SUMMARY QUESTIONS

1 Define the term *drug*.

2 Name two drugs that are used as medicines to treat bacterial infections.

3 Describe how antibiotics are used to treat bacterial infections.

S 4 Explain how the development of MRSA can be minimised.

Practice questions

1 A pathogen is an organism that:

   A feeds on dead and decaying organisms

   B lives in an organism and causes disease

   C lives inside a host organism without causing harm

   D spreads disease organisms from infected to uninfected people

   (Paper 1) [1]

2 Which defines a transmissible disease?

   A a reduction in blood flow because of a blocked blood vessel

   B a condition caused by a lack of a nutrient in the diet

   C a genetic condition inherited from one or both parents

   D a pathogen that is passed from one person to another

   (Paper 1) [1]

3 Three defences of the human body are:

   1 skin

   2 hairs in the nose

   3 mucus

   Which prevent the entry of disease-causing organisms into the lungs?

   A 1, 2 and 3   **B** 1 and 2 only

   C 1 and 3 only   **D** 2 and 3 only

   (Paper 1) [1]

4 Which is a transmissible disease?

   A cholera

   B coronary heart disease

   C rickets

   D scurvy

   (Paper 2) [1]

5 Which barrier to infection is most effective against cholera bacteria?

   A mucus   **B** skin

   C stomach acid   **D** phagocytosis

   (Paper 1) [1]

6 Which gives people long-term protection against transmissible diseases?

   A hand washing

   B garbage collection

   C sewage treatment

   D vaccination

   (Paper 1) [1]

7 Diarrhoea is a symptom of

   A anaemia   **B** cholera

   C rickets   **D** scurvy

   (Paper 1) [1]

8 An antigen is:

   A a disease-causing organism

   B a molecule produced by a lymphocyte

   C any substance that stimulates an immune response

   D a pathogen

   (Paper 2) [1]

9 During an immune response:

   A lymphocytes and specific phagocytes divide to make antibodies

   B all lymphocytes divide to produce cells that make antibodies

   C phagocytes engulf and completely digest bacteria

   D specific lymphocytes divide to produce cells that make antibodies

   (Paper 2) [1]

10 Passive immunity to snake venom is gained by:

   A being bitten by a venomous snake

   B being treated with antibodies to the venom of a spider

   C injecting a small quantity of snake venom

   D injecting snake antivenom with antibodies produced in a horse

   (Paper 2) [1]

11 The second immune response to an invasion by the measles virus many months after the first invasion is much faster because:

   A the concentration of antibodies specific to measles remains high from the first response

   B lymphocytes respond faster with age

   C lymphocytes specific to other viruses respond as well

   D memory cells are present in the blood

   (Paper 2) [1]

12 Some microorganisms, such as the bacteria that cause cholera, enter the body in our food and cause disease.

   (a) State the term given to a disease-causing organism. [1]

   (b) State two ways in which food can become contaminated with microorganisms. [2]

   (c) Explain how the stomach provides a defence against microorganisms. [2]

   (d) Some diseases are transmitted in food and water. State four other ways in which infectious diseases can be transmitted. [4]

(Paper 3)

13 (a) Explain the importance of the following in maintaining good health:
   - personal hygiene
   - sewage treatment
   - proper waste disposal [6]

(Paper 3)

14 A child received a first vaccination against measles at four months of age and then a booster at eight months. The concentration of antibodies to measles is shown in the graph.

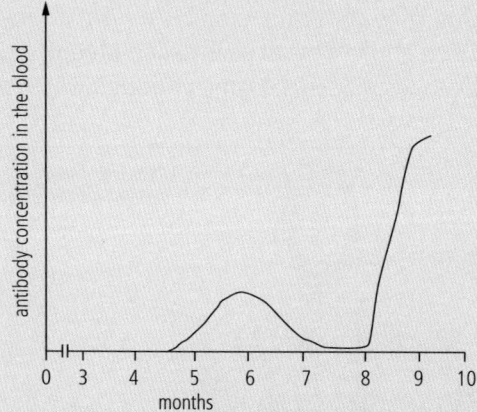

   (a) Explain why no antibodies were present in the blood for the first week. [3]

   (b) The response to the two injections of the vaccine is different. Use the information in the graph to describe how the response to the booster at eight months differs from the response to the first injection at four months. [4]

(c) Explain why the response to the second injection is different to the response to the first injection. [4]

(d) Suggest why further boosters of this vaccine may be given. [2]

(Paper 4)

15 The diagram shows four different antigens and an antibody molecule.

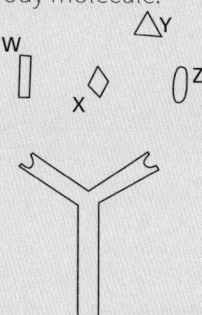

   (a) The antibody forms a complex with one of the antigens.

   (i) Which one? [1]

   (ii) Explain your answer. [1]

There is no vaccine for the viral disease chikungunya. A child has been vaccinated against several diseases including measles, tetanus and rubella. The child catches chikungunya and has the symptoms of the disease.

   (b) Use the information in the diagram to explain why it is possible to be immune to many diseases, but still be ill with another disease, such as chikungunya. [4]

   (c) Explain how vaccinations can be used to eradicate a disease from a country. [3]

   (d) Suggest why vaccination against diseases must continue to be carried out even if there are no cases of the disease for many years. [3]

(Paper 4)

16 (a) Rubella is a transmissible disease caused by a virus. Many children are vaccinated against rubella. Explain how vaccination provides protection against disease. [4]

   (b) The immune system consists of phagocytes, lymphocytes and antibodies. Explain how the immune system defends us against infection by bacterial pathogens. [5]

(Paper 4)

11.1 The gas exchange system

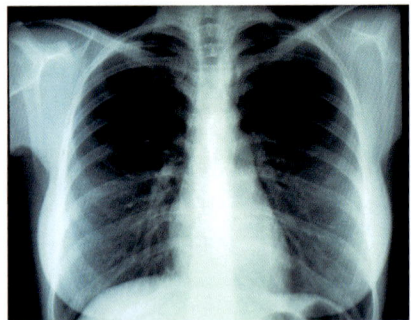

Figure 11.1.1 An X-ray of healthy lungs, which appear black.

Structure of the gas exchange system

We need to breathe air into our lungs in order to get oxygen. We breathe air out of our lungs to get rid of carbon dioxide. The lungs form part of the **gas exchange system**. The lungs are two spongy organs found inside the chest (**thorax**). They are surrounded and protected by the ribs and the sternum (breastbone). The **diaphragm** is a sheet of fibrous tissue and muscle that separates the thorax from the abdomen. Its movement up and down changes the volume of the lungs to move air when you breathe out and in. The **intercostal muscles** between the ribs move the ribs during breathing, especially during deep breathing.

Supplement

The **external intercostal muscles** (on the outside) contract to raise the ribs upwards and outwards. The **internal intercostal muscles** (on the inside) contract to lower the ribs downwards and inwards.

Air enters the mouth or nose and passes through the throat to the **larynx** (voice box). It then enters the **trachea** (windpipe), which connects the throat to the lungs. It branches to form two **bronchi**, which each enter a lung. They continue to divide to form many small **bronchioles**, which end in tiny air sacs called **alveoli**. It is here that gas exchange takes place. The tubes through which air moves are often called airways.

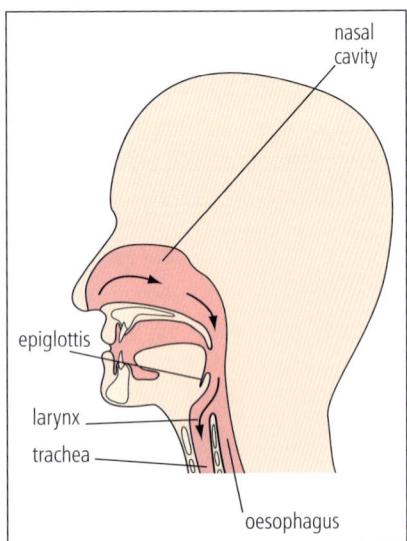

Figure 11.1.2 Entry of air into the nose and trachea.

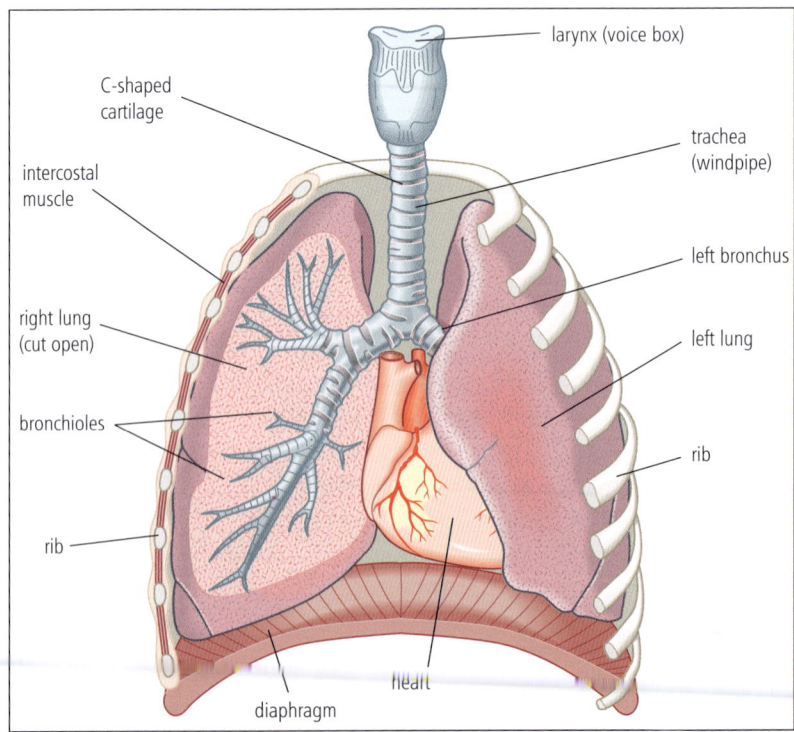

Figure 11.1.3 The gas exchange system. The lungs are surrounded by two pleural membranes.

Supplement

The larynx contains the vocal cords. When air passes over these you make sounds. You cannot breathe and swallow at the same time. This is because when you swallow, a flap called the **epiglottis** moves to cover the opening to your larynx. This stops any food from going down your trachea (see page 80).

The trachea is kept open by C-shaped rings of cartilage. The 'arms' of the C are joined by muscle at the back of the trachea. The cartilage prevents the trachea from collapsing as you breathe in when the air pressure decreases.

Deeper into the lungs

There are a very large number of alveoli in the lungs to give a huge surface area for diffusion (see page 28 to remind yourself about diffusion). Each alveolus is surrounded by a network of blood capillaries for efficient gas exchange into and out of the blood.

The alveoli have thin walls made of a single layer of cells, so there is a short distance for diffusion. They are moist so oxygen dissolves in this watery fluid before diffusing through the walls into the blood.

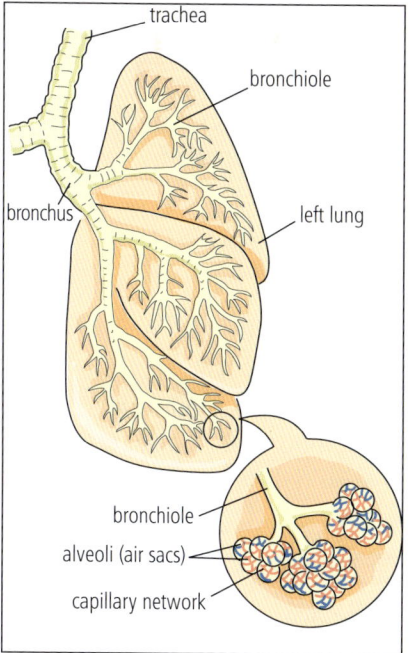

Figure 11.1.4 Structure of a lung.

1 Describe the route taken by inspired air (air we breathe in) from the nasal cavity to the alveoli in the lungs.

2 The diagram shows the human breathing system:

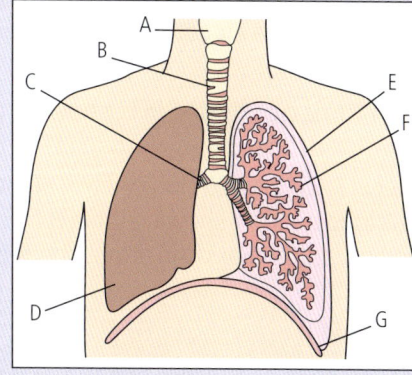

  a Name structures A to G.
  b Match parts A to G with these descriptions:
    i sheet of muscle forming the floor of the thorax
    ii one of these enters each lung
    iii contains the vocal cords
    iv flexible tube kept open by rings of cartilage
    v where exchange of gases takes place
    vi slippery membranes that surround the lungs
    vii an organ made of spongy tissue and found in the chest

Draw a simple diagram of the gas exchange system and label it. Add some notes to the labels explaining the functions of the different parts. By doing this you are <u>annotating</u> your diagram.

1 The inspired air passes down the larynx, trachea, bronchi, bronchioles to the alveoli.

2 Cartilage keeps the trachea open when air passes through it.

3 Gas exchange occurs between the air in the alveoli and the blood in the surrounding capillaries.

4 The external intercostal muscles raise the ribs; the internal intercostal muscle lower the ribs.

11.2 Gas exchange

LEARNING OUTCOMES

- Investigate and compare the composition of inspired and expired air
- List the features of gas exchange surfaces
- Describe gas exchange at the alveolus
- **S** Explain the difference in composition of inspired and expired air

EXAM TIP

See page 28 to remind yourself about diffusion and concentration gradients.

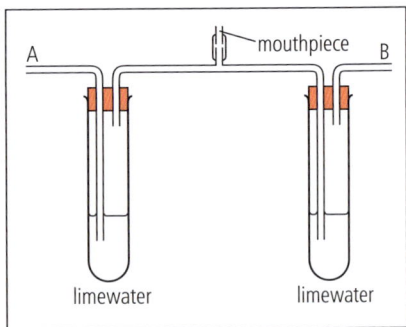

Figure 11.2.1 Apparatus for showing carbon dioxide in inspired and expired air.

EXAM TIP

Beware! We can say that alveoli and capillaries have <u>walls made of cells</u>, but not <u>cell walls</u> – remember only plant cells have cell walls, animal cells do not.

Breathing out carbon dioxide

Carbon dioxide gas is a waste product made in respiration. It can become toxic if it builds up in cells.

We breathe out in order to get rid of the carbon dioxide.

Table 11.2.1 Comparing the composition of inspired and expired air.

gas	composition / %	
	inspired air (breathing in)	expired air (breathing out)
oxygen	21	16
carbon dioxide	0.04	4
nitrogen	78	78
water vapour	variable	saturated

The table shows that we breathe out more carbon dioxide than we breathe in. We breathe in more oxygen than we breathe out, and the air that we breathe out also has a lot more water vapour.

The following practical demonstration confirms that we breathe out more carbon dioxide.

PRACTICAL

Carbon dioxide in inspired and expired air

1. Set up the apparatus as shown in Figure 11.2.1.

2. Then breathe gently in and out of the mouthpiece several times.

   - When you breathe in the air comes in through A.
   - When you breathe out the air goes through B.
   - The limewater turns cloudy first in B, showing that there is more carbon dioxide in expired air than in inspired air.
   - The limewater in A will also go cloudy if you breathe through the apparatus long enough, showing there is a lower concentration of carbon dioxide in atmospheric (inspired) air.

Gas exchange surfaces

Gas exchange surfaces, such as the gills of a fish and the alveoli (air sacs) of a human, have features in common that adapt them for efficient exchange of oxygen and carbon dioxide:

- a very large surface area for the diffusion of gases
- moist surfaces so that gases can dissolve before diffusion
- a thin surface (only one cell thick in each alveolus) so the gases do not have to diffuse very far
- a good blood supply so that lots of oxygen is removed quickly and lots of carbon dioxide is supplied quickly. This maintains the **concentration gradients** for these gases.

Ventilation of the lungs ensures that the air in the air passages is changed; this helps maintain the gas concentration gradients between air in the alveoli and that in the blood.

Gas exchange at the alveolus

When inspired (breathed in) air reaches the alveoli (air sacs) it contains a lot of oxygen. Oxygen dissolves in the water lining each alveolus. It then diffuses through the wall of the alveolus and through the capillary wall into the blood. Although this involves diffusing through two cells, the distance is very small.

Each alveolus has a network of capillaries around it. Oxygen molecules from the alveolus diffuse into the red blood cells and combine with haemoglobin. The blood cells can then transport this oxygen to the body tissues.

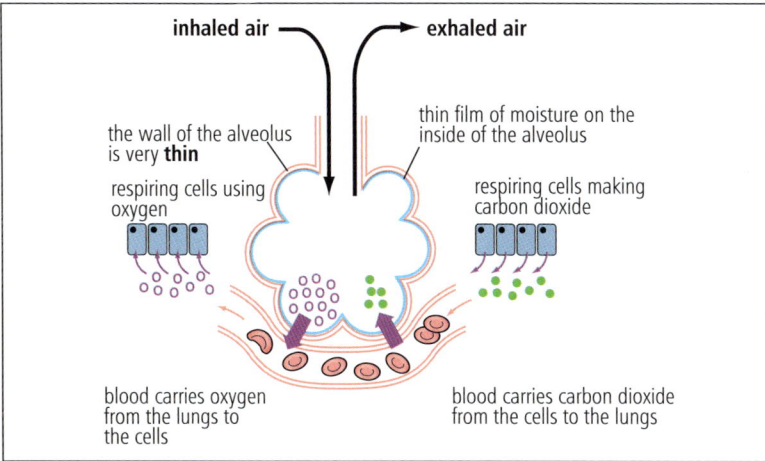

Figure 11.2.2 Gas exchange at an alveolus.

There is a lot of carbon dioxide in the capillary. It has been carried there from the respiring tissues in the blood plasma. The carbon dioxide diffuses in the opposite direction, through the capillary wall across the alveolar wall into the space inside the alveolus. From here it is breathed out.

Alveoli are surrounded by elastic tissue. This stretches when you breathe in and recoils when you breathe out to help remove air from the lungs.

Supplement

As a result of respiration, oxygen is used and carbon dioxide is produced. So expired air has less oxygen and more carbon dioxide than inspired air. The air we breathe out is also saturated with water vapour that has evaporated from the moist walls of the alveoli.

KEY POINTS

1 There is more carbon dioxide, less oxygen and more water vapour in expired air than in inspired air.
2 Gas exchange surfaces are moist, thin, have a large surface area and a good blood supply.
3 Alveoli are well adapted for gas exchange as they have these features and are surrounded by many capillaries.

SUMMARY QUESTIONS

1 List four features of the alveoli of the lungs that adapt them for efficient exchange of gases.

2 Look at this apparatus used in an experiment:

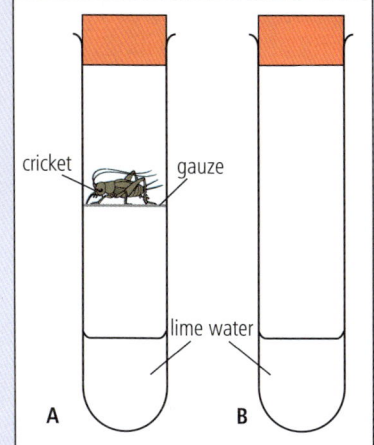

After an hour the tubes were examined.

a What result would you expect?

b Explain your answer.

c Explain why tube B is included.

3 Give an explanation for each of the following.

a When you breathe out on to a cold surface, water droplets form.

b Expired air turns limewater cloudy quicker than inspired air.

c Alveoli (air sacs) are very thin, have a large surface area and have walls with elastic fibres.

4 Explain why there is less oxygen in expired air than in inspired air.

11.3 Breathing

Supplement

PRACTICAL

A model of the chest

1 Obtain a model like the one in the diagram. The rubber sheet represents the diaphragm.

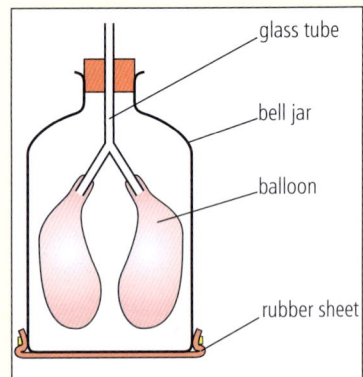

2 Pull the rubber sheet down and then push it up.

3 Do this a few times and watch what happens to the balloons.

This model shows how movement of the diaphragm changes the volume and pressure in the thorax represented by the bell jar. As the diaphragm contracts and moves downwards, the lungs inflate as a result of an increase in volume and decrease in pressure in the space around the balloons.

When you breathe it feels as if the lungs are expanding to push your ribs outwards. This is not so. When you breathe in, movements of your diaphragm and ribcage move the chest cavity and pull on the lungs so they occupy a larger volume. This decreases the air pressure inside the lungs to below the pressure of atmospheric air so that air moves in through your nose and/or mouth.

When you breathe out, the diaphragm and intercostal muscles relax and this causes the chest cavity to decrease in volume. The air pressure in the lungs increases to above atmospheric pressure so air is forced out. You can do this just by movement of your diaphragm alone (try breathing without moving your ribs).

This is fine for quiet breathing but during deep breathing and when you exercise the intercostal muscles contract to move the ribs. (Intercostal means between the ribs.) There are two layers of these muscles: the external muscles that are closer to the skin, and the internal muscles that are deeper into the chest wall (Figure 11.3.3). The diaphragm is made of tough fibrous tissue in the centre and strong muscles that connect to the backbone, the lower ribs and the sternum (breastbone).

Inspiration (breathing in)

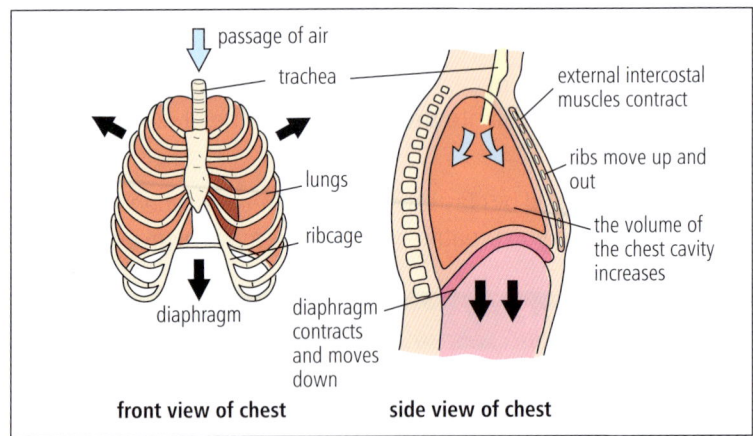

Figure 11.3.1 Increasing the volume of the thorax at expiration.

- The external intercostal muscles *contract* and the internal intercostals muscles *relax*, raising the ribs upwards and outwards.

- At the same time, the diaphragm contracts and flattens.

- Both of these actions *increase* the volume inside the thorax, causing the pressure inside the thorax to *decrease*.

- Since atmospheric pressure is greater, air moves *into* the lungs and they inflate.

Expiration (breathing out)

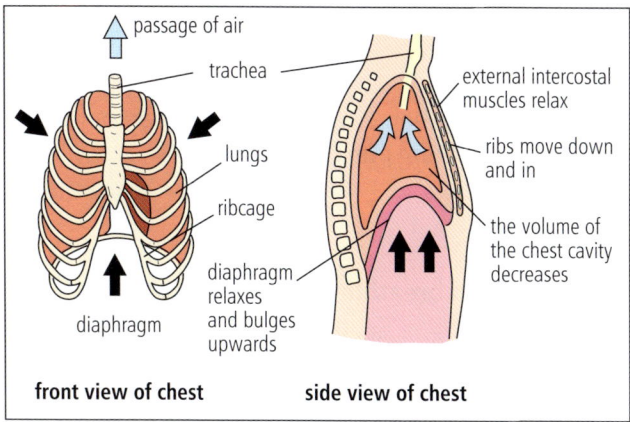

Figure 11.3.2 Decreasing the volume of the thorax at expiration.

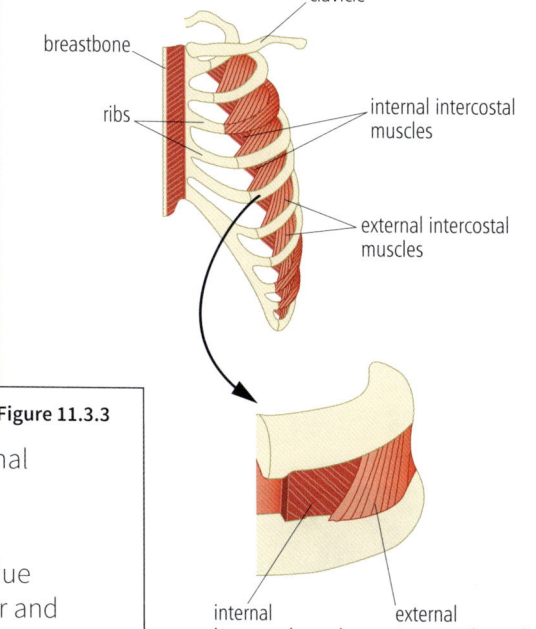

Figure 11.3.3

- The internal intercostal muscles *contract* and the external intercostal muscles *relax*.
- This lowers the ribs downwards and inwards.
- The muscle in the diaphragm *relaxes* and it bulges up due to pressure from the organs below, for example the liver and stomach, and contraction of the abdominal muscles.
- Both of these actions *decrease* the volume inside the thorax, causing the pressure inside the thorax to *increase*.
- Elastic recoil of the alveoli helps to force air *out* of the lungs.

Cleaning the air

As you breathe in through your nose, air is warmed and moistened by evaporation of water from the lining. Hairs inside the nose filter the air, removing particles and some pathogens.

The trachea, bronchi and bronchioles are lined with ciliated epithelial cells and goblet cells, which secrete mucus. Dust particles and pathogens become trapped in the slimy mucus. The **cilia** beat to carry a stream of mucus up to your nose and throat, removing the particles and pathogens, which you then swallow.

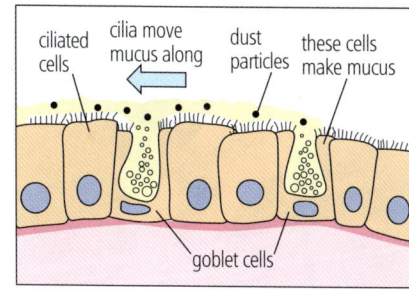

Figure 11.3.4 Goblet cells and ciliated epithelial cells keep the lungs free of dust and pathogens.

SUMMARY QUESTIONS

1 When a person breathes in, describe what happens to:
   a the intercostal muscles **b** the ribs **c** the diaphragm **d** the volume inside the thorax **e** the pressure inside the thorax
2 Look at the model of the thorax in the Practical, opposite.
   a Which part of the model represents:
      i the lungs **ii** the diaphragm **iii** the ribs?
   b Make some criticisms of the model as a way of representing the changes that occur during breathing.
3 Explain the role of mucus and cilia in keeping the air passages free from dust particles and pathogens.

KEY POINTS

1 Inspiration and expiration occur due to changes of volume and pressure in the thorax.

2 The internal intercostal muscles, external intercostal muscles and the diaphragm control these changes during ventilation.

3 Goblet cells produce mucus, which traps dust particles and pathogens in the air and cilia beat to remove mucus from the airways.

11.4 Rate and depth of breathing

LEARNING OUTCOMES

- Investigate and describe the effects of physical exercise on the rate and depth of breathing

S • Explain how the effects of exercise increase carbon dioxide concentration in the blood and how this affects the rate and depth of breathing

Figure 11.4.1 Performing step-ups.

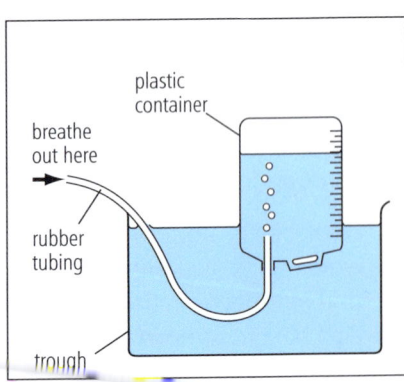

Figure 11.4.2 This apparatus can be used to measure your vital capacity (see top of next page).

Breathing and exercise

At rest you probably take between 12 and 16 breaths per minute. When you exercise this breathing rate changes.

PRACTICAL

The effects of exercise on breathing rate

You can measure your breathing rate by counting the number of times that you breathe out in a minute.

You could record your results in a table like this:

	breathing rate / breaths per min		
	at rest	light exercise	heavy exercise
1			
2			
3			
average =			

- Try doing step-ups for 1 minute (light exercise).
- As soon as you have finished, sit down and count your breathing rate.
- Wait until your breathing rate returns to its resting value.
- Now try doing step-ups as quickly as you can for 3 minutes (heavy exercise).
- Then count your breathing rate when you have finished.
- You will notice that your breathing rate and depth of breathing change after doing exercise.
- Repeat this twice and calculate averages.

The rate and depth of breathing increase with exercise. When your muscles are working hard, they need more oxygen for respiration. They also produce more carbon dioxide in respiration.

Increasing the rate and depth of breathing gets more oxygen into the blood and gets rid of more carbon dioxide from the blood. Muscles continue to respire quite fast after exercise finishes. They still need a good supply of oxygen and they still have carbon dioxide to be removed. Your pulse rate remains high after exercise because your heart is beating fast to deliver plenty of blood to your muscles so they gain this extra oxygen and have their carbon dioxide removed.

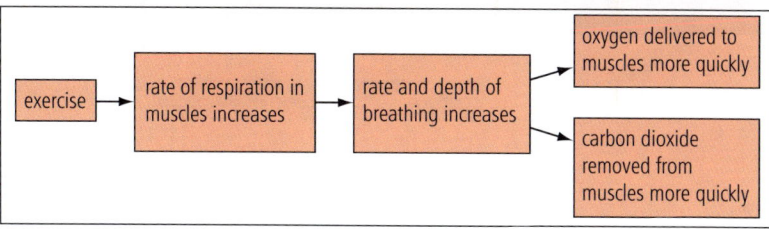

Figure 11.4.3 The effect of exercising on breathing.

Depth of breathing

An adult can take in about 5 litres of air in their deepest breath. This is their **vital capacity** – the maximum volume of air that is breathed out after breathing in as much air as possible. At rest, about half a litre of air is breathed in and out. During exercise approximately four and a half litres of extra air can be taken in during a deep breath.

Supplement

Control of breathing

You don't think about your breathing – most of the time it just happens. So what makes you breathe faster and deeper when you exercise? The brain has a special part for controlling breathing. When you exercise the tissues respire more quickly and make *more carbon dioxide*, which *lowers* the *pH* in the tissues and the blood (i.e., they become more acidic). They may also make lactic acid, which does the same (see page 140).

Your brain detects this rise in carbon dioxide and the lowering of the pH of the blood reaching it. The brain sends nerve impulses to the diaphragm and to the intercostal muscles so they contract faster and further to increase the rate and depth of breathing. By breathing deeper and more rapidly, you lower the concentration of carbon dioxide in the blood and this raises the blood pH back to normal. In turn this has the same effect on the tissues because the blood flows through them. This is an example of **homeostasis** – see page 164.

EXAM TIP

See page 28 to remind yourself about diffusion. Compare your results with the results you got when investigating pulse rate (page 103). Summarise the effects of exercise on the body including the muscles, heart and lungs. You will be able to add some more about body temperature and blood glucose in Unit 14.

EXAM TIP

S Remember the muscles! They are the organs doing the hard work. They require the extra oxygen and glucose and need to have their waste products taken away. Start any answer on this topic with the changes that occur in the muscles during exercise.

SUMMARY QUESTIONS

1 When you exercise, describe what happens to:
   a the rate of breathing   b the depth of breathing

2 Describe what happens to the concentration of carbon dioxide in the blood when you exercise.

S 3 a Explain how the brain monitors and controls the rate of breathing when you exercise.

   b Explain what happens to the following when you exercise:
      i the intercostal muscles   ii the diaphragm
      iii the heart rate

KEY POINTS

1 Exercise increases both the rate and depth of breathing.

S 2 An increase in the concentration of carbon dioxide in the blood reaching the brain triggers these changes in breathing.

Practice questions

1 Which is a pathway that air takes from the atmosphere to the gas exchange surface in the lungs?

A mouth → larynx → bronchus → bronchiole → alveolus

B mouth → nose → trachea → bronchus → alveoli → bronchiole

C nose → larynx → trachea → bronchus → bronchiole → alveolus

D nose → trachea → bronchus → bronchiole → alveolus

(Paper 1) [1]

2 A student filled a conical flask with lime water and put a drawing of a cross beneath it. The student then breathed into the flask of lime water to find out how long it took before the cross became invisible. The student repeated the test after doing three minutes of strenuous exercise.

Which describes the time taken for the cross to become invisible in the test after exercise?

A The cross became invisible in a shorter time.

B The cross took three times as long to become invisible.

C The cross took twice as long to become invisible.

D The time was the same.

(Paper 1) [1]

3 The diagram shows apparatus to investigate the production of carbon dioxide during respiration of insects. What would you expect to see in tubes **A** and **B** five minutes after setting up the apparatus?

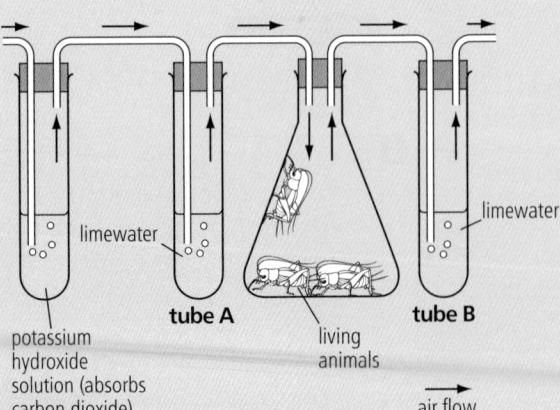

	tube A	tube B
A	cloudy	cloudy
B	cloudy	no change
C	no change	cloudy
D	no change	no change

(Paper 1) [1]

4 What are the features of a gas exchange surface?

A thick-walled, large surface area, many capillaries

B thick-walled, small surface area, few capillaries

C thin-walled, large surface area, many capillaries

D thin-walled, small surface area, few capillaries

(Paper 1) [1]

5 Which is the function of the cartilage in the trachea?

A cleans the air as it moves to the lungs

B contracts to move air into the lungs

C increases resistance to air flowing to the lungs

D prevents the trachea collapsing

(Paper 2) [1]

6 The diaphragm and the intercostal muscles act together during expiration at the end of strenuous exercise. Which of the following occurs?

	diaphragm		intercostal muscles	
			internal	external
A	contracts	moves down	relax	contract
B	contracts	moves up	relax	contract
C	relaxes	moves down	contract	relax
D	relaxes	moves up	contract	relax

(Paper 2) [1]

7 The photograph shows some cells from the gas exchange system.

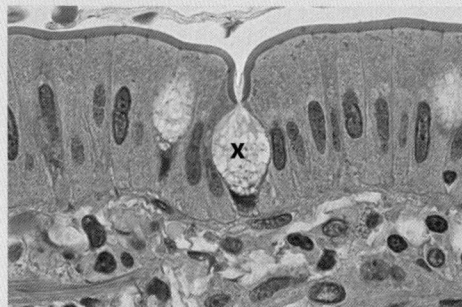

What is cell X?

A alveolar cell

B ciliated cell

C goblet cell

D muscle cell

(Paper 2) [1]

8 The percentage composition of air changes as it passes through the lungs because:

A carbon dioxide and water vapour are absorbed from the air

B carbon dioxide is absorbed and oxygen excreted

C oxygen is absorbed and carbon dioxide excreted

D oxygen is absorbed and nitrogen excreted

(Paper 2) [1]

9 Some students investigated the effect of exercise on the rate and depth of breathing. Here are the results for one of the students.

activity	volume of each breath / cm³	number of breaths taken per minute
rest	500	18
20 step-ups per minute	750	25
50 step-ups per minute	1200	34

(a) (i) Calculate the change in the volume of each breath when the student changed from 20 to 50 step-ups per minute. [1]

(ii) Describe the effect of exercise on the volume of each breath. [2]

(b) (i) Calculate the percentage change in the volume of each breath when the student changed from rest to 50 step-ups per minute. [2]

(ii) Describe the effect of exercise on the rate of breathing. [2]

(c) The students decided to investigate the effect of this exercise on the boys and girls in the class. State two features of the investigation that they would have to keep the same so that they could make valid comparisons between the boys and girls.

(Paper 3) [2]

10 The diagram shows the ribs and muscles of the thorax from the side.

(a) Name **A** to **C**. [3]

(b) (i) Describe how the structures shown bring about breathing movements during inspiration and expiration. [6]

(ii) Explain how the movements you describe cause air to move into and out of the lungs. [5]

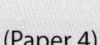

(Paper 4)

11 Look at the table of results in question 9.

(a) Describe the effect of exercise on the student's breathing. [4]

(b) Explain why the student's breathing changed during exercise. [6]

(c) A sample of air taken from the alveoli of a person had 14.5% oxygen and 5.5% carbon dioxide.

Suggest why the percentage composition of alveolar air differs from the composition of expired air. [2]

(Paper 4)

- Describe aerobic respiration and state the word equation
- State that respiration involves enzymes catalysing reactions in cells
- State the uses of energy in living organisms
- Investigate and describe the effect of temperature on respiration in yeast
- **S** State the balanced chemical equation for aerobic respiration

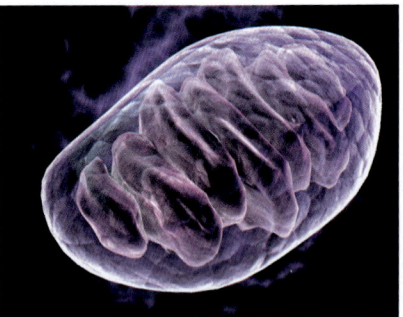

Figure 12.1.1 A computer-generated image of a mitochondrion. Most of the ATP in cells is made in these cell structures. Each is about 0.002 mm long.

Aerobic respiration

When we burn a fuel such as petrol, oxygen is used up and carbon dioxide is made. The flame gives off a lot of light and heat (thermal energy). Every cell in every living organism needs energy. A fuel used to provide energy in the cells is called glucose, but unlike burning petrol, the energy is released very gradually in a series of small, enzyme-controlled reactions.

Respiration involves a number of chemical reactions that break down nutrient molecules, such as glucose, in living cells to release energy.

In **aerobic respiration** oxygen is used in the breakdown of glucose:

glucose + oxygen $\longrightarrow$ carbon dioxide + water + energy released

Cells also respire fats and proteins to provide energy, but you only have to know about the respiration of glucose, which is a carbohydrate.

Using energy

Organisms need energy for many different things. Energy is used in humans for the following processes:

- muscle contraction
- cell division
- absorption of nutrients in the gut by active transport
- sending impulses along nerves
- protein synthesis for making enzymes, some hormones and antibodies
- making new cell membranes and cell structures like the nucleus during growth
- keeping the body temperature constant; some of the energy released is in the form of heat.

Supplement

Respiration takes place in *all* living cells *all of the time* because cells need a constant supply of energy to stay alive. The balanced chemical equation for aerobic respiration is:

$$C_6H_{12}O_6 + 6O_2 \longrightarrow 6CO_2 + 6H_2O + \text{energy released}$$

Most of the energy released in aerobic respiration is released in **mitochondria**. Cells that require a lot of energy have many mitochondria. For instance, insect flight muscle has numerous mitochondria located between the muscle fibres. The mitochondria provide the energy for the muscles to contract during flight. Liver cells have a high metabolism and have many mitochondria. The epithelial cells of the small intestine absorb glucose and other molecules by active transport, so they too have large numbers of mitochondria.

PRACTICAL

Investigating the uptake of oxygen by respiring yeast

The apparatus in Figure 12.1.2 is a respirometer and it can measure how quickly yeast uses up oxygen. The yeast takes up oxygen during aerobic respiration and releases carbon dioxide, which is absorbed by the soda lime. The decrease in volume and air pressure causes the coloured water to move along the delivery tube. The rate of movement can be measured by the scale – the faster the rate of respiration, the quicker the liquid will move. Why is it important to keep the apparatus at a constant temperature?

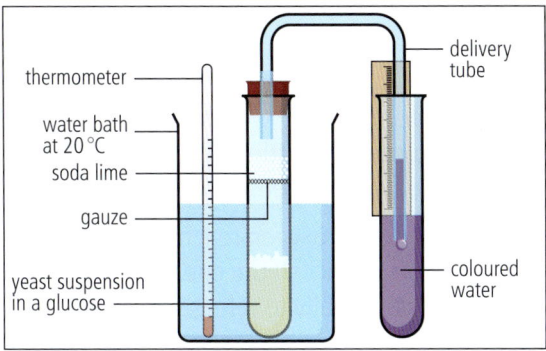

Figure 12.1.2 Testing the gas produced by a yeast suspension. The test tube contains limewater, which becomes cloudy as the bubbles of gas pass through it. This shows that carbon dioxide gas in given off during aerobic respiration.

Figure 12.1.3 A respirometer.

SUMMARY QUESTIONS

1 Copy and complete the sentences using these words:

> **aerobic   glucose   oxygen   mitochondria**
> **energy   carbon dioxide   cells**

Respiration takes place in all living _____. A fuel for respiration is _____. It is broken down to release _____ Also produced is water and the waste gas _____. When it takes place in the presence of _____ we call it _____ respiration. These reactions take place in every living cell, inside tiny structures called _____.

2 **a** Write down the word equation for aerobic respiration.

  b Energy is released during aerobic respiration.

  Make a list of five processes that use this energy.

3 A student used the apparatus in Figure 12.1.2 to investigate the effect of temperature on respiration of yeast. The table shows the results.

temperature / °C	5	10	20	30	40	50	60	70
mean number of bubbles per minute	0	3	7	12	7	4	1	0

  a Use the figures in the table to draw a graph.

  b Describe the effect of temperature on respiration of yeast.

  c Describe how the student could use the apparatus to confirm that yeast is the source of carbon dioxide in the bubbles.

  d Describe three steps that the student should take in the investigation to make sure that the results are as accurate as possible.

  e Another student criticised the investigation because it relies on counting bubbles. Explain this criticism and suggest how to improve the investigation.

KEY POINTS

1 In aerobic respiration, glucose is broken down to release energy in the presence of oxygen, forming carbon dioxide and water.

2 Respiration involves the action of enzymes to catalyse reactions in cells.

3 The word equation for aerobic respiration is:

   glucose + oxygen $\longrightarrow$ carbon dioxide + water

S 4 The balanced chemical equation for aerobic respiration is:

   $C_6H_{12}O_6 + 6O_2 \longrightarrow 6CO_2 + 6H_2O$

5 The effect of temperature on the respiration of yeast can be investigated by measuring the carbon dioxide produced by suspensions of yeast in glucose solutions.

12.2 Anaerobic respiration

LEARNING OUTCOMES

- Describe anaerobic respiration
- State the word equations for anaerobic respiration in muscle during vigorous exercise and in yeast
- State that anaerobic respiration releases much less energy per glucose molecule than aerobic respiration
- **S** State the balanced chemical equation for anaerobic respiration in yeast
- Describe the build-up of lactic acid in muscle to cause an oxygen debt and how it is removed during recovery from exercise

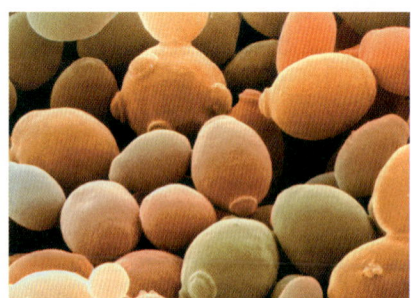

Figure 12.2.1 Yeast cells under an electron microscope.

Figure 12.2.2 Hurdlers build up lactic acid in their muscles.

Muscles without oxygen

Your muscles need oxygen and glucose to respire aerobically. These substances are brought to your muscles by your blood system. During vigorous exercise, your heart and lungs cannot get enough oxygen to your muscles quickly enough. When this happens, your muscles start to carry out anaerobic respiration.

Anaerobic respiration

It is possible to release energy from glucose without using oxygen. **Anaerobic respiration** is the release of a little energy from each molecule of glucose in the *absence* of oxygen.

During strenuous exercise, not enough oxygen reaches the body muscles for aerobic respiration to supply all the energy the muscles need. Muscle tissue respires anaerobically to release energy. Most of the enzyme-catalysed reactions of aerobic respiration do not happen without oxygen. As a result, the glucose is not broken down to carbon dioxide and water, but to **lactic acid** instead:

$$glucose \longrightarrow lactic\ acid + \text{energy released}$$

Other tissues, such as cardiac muscle in the heart, do not normally respire anaerobically as this would not release enough energy to keep the heart beating properly. Some bacteria also respire anaerobically to make lactic acid in the same way, for example those that we use to make yoghurt.

Other microorganisms, such as yeast, respire anaerobically when oxygen is absent or in short supply in their surroundings, to make alcohol and carbon dioxide:

$$glucose \longrightarrow alcohol + carbon\ dioxide + \text{energy released}$$

Supplement

$$C_6H_{12}O_6 \longrightarrow 2C_2H_5OH + 2CO_2 + \text{energy released}$$

Plant roots respire anaerobically when land is flooded and soils become saturated with water so little or no oxygen is available.

Far less energy is released from each molecule of glucose in anaerobic respiration compared to aerobic respiration. This is because glucose is not completely broken down and a lot of energy remains stored in lactic acid and in alcohol in the form of chemical bond energy. You can see this if you set alcohol alight.

Table 12.2.1 Comparing energy released in different types of respiration.

	energy released / kJ g⁻¹ glucose
aerobic respiration	16.1
fermentation by yeast	1.2
anaerobic respiration in muscle	0.8

Supplement

Lactic acid

Lactic acid can slowly poison muscles and cause cramps, so it must be removed from the body. The build up of lactic acid in the muscles and blood, during vigorous exercise, causes the **oxygen debt**.

Our oxygen debt builds up when we exercise hard, and it has to be 'paid back' straight away. We carry on breathing faster and deeper after vigorous exercise in order to supply more oxygen for aerobic respiration to break down lactic acid. The oxygen debt is removed during recovery by aerobic respiration of lactic acid in the liver.

After exercise, the heart continues to beat at a fast rate in order to transport lactic acid in the blood from the muscles to the liver. Sprinters build up lactic acid in their muscles. They often hold their breath during a 100 metre race. Afterwards, they need about $7 \, dm^3$ of oxygen to get rid of the lactic acid. That is why they breathe deeply after the race in order to 'repay their oxygen debt'.

The enzyme-catalysed reactions that release most of the energy in aerobic respiration occur in mitochondria. These cell structures do not function without a supply of oxygen. During anaerobic respiration, glucose is partly broken down in the cytoplasm and alcohol or lactic acid is produced from the end product of this partial breakdown.

Figure 12.2.3 This runner is repaying her oxygen debt by breathing deeply after her training.

SUMMARY QUESTIONS

1   a   Define the term *anaerobic respiration*.
    b   Give two examples of anaerobic respiration.
    c   Compare the energy released from a molecule of glucose during aerobic and anaerobic respiration.

2   Describe and explain the differences between the energy released in aerobic and anaerobic respiration. (See the table opposite for some figures to use in your answer.)

S 3   The table below shows the units of lactic acid produced in the leg muscles of an athlete during a race:

time / minutes	0	10	20	30	40	50	60	70	80
units of lactic acid	0	1	7	12	9	6	3	1	1

   a   Draw a graph using the data.
   b   When did the lactic acid reach a maximum?
   c   Explain why the muscles produced lactic acid.
   d   Suggest why the muscles produced less lactic acid towards the end of the race.

4   Explain why we carry on breathing heavily after we finish a session of hard exercise.

5   Explain how an oxygen debt develops and how it is repaid.

KEY POINTS

1   Anaerobic respiration is the chemical reaction in cells that break down glucose to release energy without using oxygen.

2   In anaerobic respiration, glucose is broken down to release energy in the absence of oxygen, forming lactic acid in muscle tissue, or alcohol, plus carbon dioxide in yeast and plants.

3   Far more energy is released from each molecule of glucose in aerobic than in anaerobic respiration.

S 4   The build up of lactic acid in muscles during hard exercise produces an *oxygen debt* that has to be paid back by breathing in extra oxygen.

Practice questions

1 Which process does *not* require energy from respiration?

   A diffusion of oxygen across the alveolar wall

   B contraction of muscle during exercise

   C passage of nerve impulses

   D synthesis of protein molecules

   (Paper 1) [1]

2 When yeast respires without oxygen, the cells produce:

   A alcohol

   B carbon dioxide and alcohol

   C lactic acid

   D lactic acid and carbon dioxide

   (Paper 1) [1]

3 During aerobic respiration, enzymes catalyse chemical reactions to breakdown nutrient molecules to:

   A produce glucose

   B produce lactic acid

   C produce oxygen

   D release energy

   (Paper 1) [1]

4 During vigorous exercise, muscle respires without oxygen. The result of this respiration is the release of

   A carbon dioxide

   B carbon dioxide and water

   C little energy from each molecule of glucose

   D much energy from each molecule of glucose

   (Paper 1) [1]

5 Which does not have mitochondria?

   A ciliated epithelial cell

   B liver cell

   C muscle cell

   D red blood cell

   (Paper 2) [1]

6 The equations for respiration show the overall change that occurs to glucose as a result of enzyme-catalysed reactions. Which term describes the role of glucose for the first enzyme in respiration?

   A activator

   B active site

   C product

   D substrate

   (Paper 2) [1]

7 After vigorous exercise, a student's heart rate is high. Which is not a reason for this?

   A carbon dioxide from anaerobic respiration needs to be removed from the muscles

   B lactic acid from anaerobic respiration needs to be removed from the muscles

   C oxygen is required in the liver for metabolism of lactic acid

   D waste products of respiration need to be removed

   (Paper 2) [1]

8 The oxygen debt is the additional volume of oxygen absorbed by the lungs after exercise. This oxygen is required for

   A anaerobic respiration in muscle tissue

   B metabolism of lactic acid in the liver

   C removal of carbon dioxide

   D respiration of muscle tissue to release energy for contraction

   (Paper 2) [1]

9 **(a)** Write out and complete the word equation for aerobic respiration:

   ____ + oxygen $\longrightarrow$ ____ + ____ + ____

   [4]

A student set up the apparatus shown opposite to find the rate of respiration of crickets.

The student took several readings from the apparatus. After each reading the clip was opened for several minutes before being sealed again.

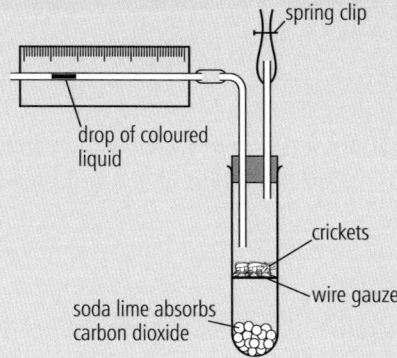

spring clip

drop of coloured liquid

crickets

wire gauze

soda lime absorbs carbon dioxide

(b) State the direction in which the coloured drop will move as the crickets respire in the apparatus. Explain your answer. [4]

(c) Explain why the clip is opened after taking the readings. [2]

(d) State one factor that should be kept constant during the investigation. [1]

(Paper 3)

10 Copy and complete the table with ticks and crosses. The first row has been completed.

feature	photosynthesis	respiration
occurs only in plant cells	✓	✗
produces oxygen		
consists of a series of enzyme-catalysed reactions		
produces carbon dioxide		
releases energy		
takes place all the time (night and day)		

(Paper 3) [5]

11 The apparatus shown in question 9 was used to determine the effect of temperature on the rate of respiration of Pinto beans.

The apparatus was put into a water bath at 27 °C with the clip open. After 10 minutes the clip was closed and the position of the droplet recorded over time. The results are as follows.

time / minutes	0	5	10	15	20	25	30	35
position of droplet / mm	0	0	0	31	65	95	130	162

(a) Explain why
  (i) a water bath was used
  (ii) the apparatus was left for 10 minutes before closing the clip. [2]

(b) The diameter of the capillary tube is 0.8 mm. Use the results in the table to calculate the rate of oxygen uptake in mm³ per hour. [3]

(c) Explain how the results would differ if the investigation was repeated at 17 °C and at 50 °C. [4]

(Paper 6)

12 (a) Write out the balanced equation for aerobic respiration in animal cells. [3]

(b) Name the structures inside animal cells that carry out aerobic respiration. [1]

The graph shows the oxygen consumption of an athlete who exercised for 10 minutes after a short period of rest.

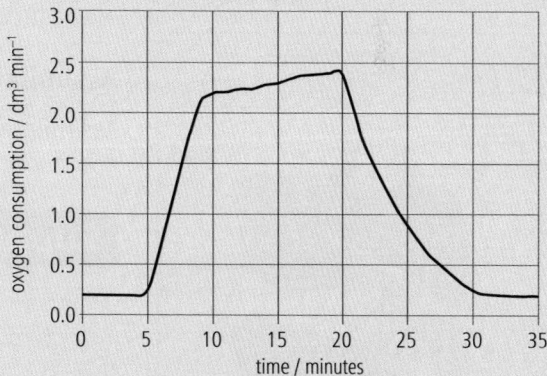

(c) Describe the change in oxygen consumption over the 35 minutes that recordings were taken. [4]

(d) Explain the changes in oxygen consumption between
  (i) 10 and 20 minutes [6]
  (ii) 20 to 30 minutes. [5]

(Paper 4)

13.1 Excretion

Excretion is the removal from the body of the waste products of metabolism, toxic materials and substances in excess of requirements. The waste substances are removed from the body by the excretory organs, which are the lungs, liver and kidneys (Figure 13.1.1).

Excretory products

Humans have two main excretory products.

- **Carbon dioxide** is made in body tissues during respiration. It is transported to the lungs in the blood plasma. Here, it diffuses out of the blood into the air in the alveoli and is breathed out (see page 130).

- **Urea** is made in the liver from excess amino acids. It is carried to the kidneys in the plasma where it is filtered out and leaves the body dissolved in the form of urine.

Excretion is a vital process in the metabolism of the body. Without it, substances like carbon dioxide and urea would build up in the tissues. These substances would reach a toxic level and destroy the tissues.

Do not confuse 'excretion' with 'egestion'. Substances that are **excreted** are produced by body cells in metabolism. Substances that are **egested** have been eaten and have passed through the alimentary canal without being digested and absorbed into the bloodstream. Fibre is an example.

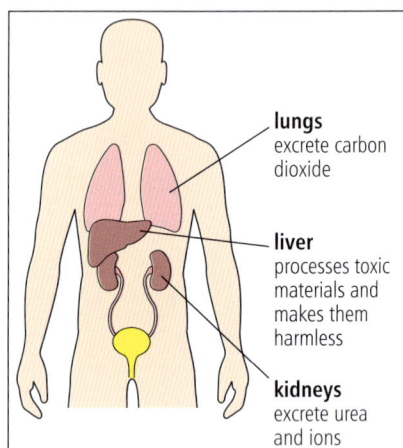

Figure 13.1.1 The excretory organs.

lungs
excrete carbon dioxide

liver
processes toxic materials and makes them harmless

kidneys
excrete urea and ions

Assimilation

Assimilation means that the food molecules that have been absorbed now become part of the cells or are used by the cells.

Supplement

The liver carries out a number of important functions as part of assimilation:

- stores glucose by removing it from the blood and storing it as glycogen. This helps to regulate the concentration of glucose in the blood (see pages 164–165)
- uses amino acids to make proteins, such as plasma proteins, e.g., fibrinogen, involved with blood clotting
- breaks down excess amino acids
- converts fatty acids and glycerol into fat, which is stored around the body, e.g., under the skin
- produces cholesterol from fats (see page 71).

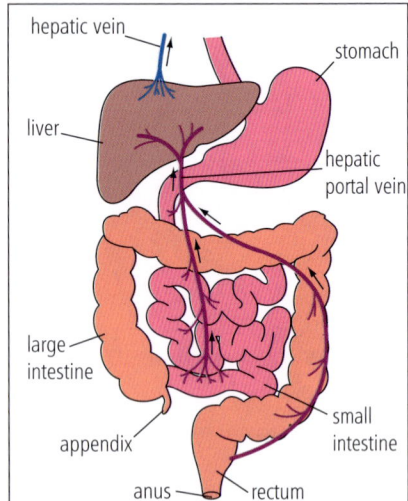

hepatic vein

stomach

liver

hepatic portal vein

large intestine

small intestine

appendix

anus   rectum

Figure 13.1.2 The position of the hepatic portal vein.

Roles of the liver

The liver carries out over 200 different roles in the body. As you can see from Figure 13.1.2, all the blood from the digestive system flows to the liver before going into the rest of the circulation. The amino acids produced in the alimentary canal by the digestion of protein are absorbed into the blood in the small intestine. Your body cannot store the amino acids that it cannot use and so each molecule is broken down into two new molecules by the process of **deamination**.

One molecule is converted to carbohydrate or fat and used as a source of energy. The other molecule is ammonia (NH_3), which combines with carbon dioxide to form the excretory product **urea**. This is carried to the kidneys in the blood where it is filtered out and is excreted in urine (see page 146). The part of the amino acid molecule that contains nitrogen is known as the amino group – and the process is called deamination because this group is removed.

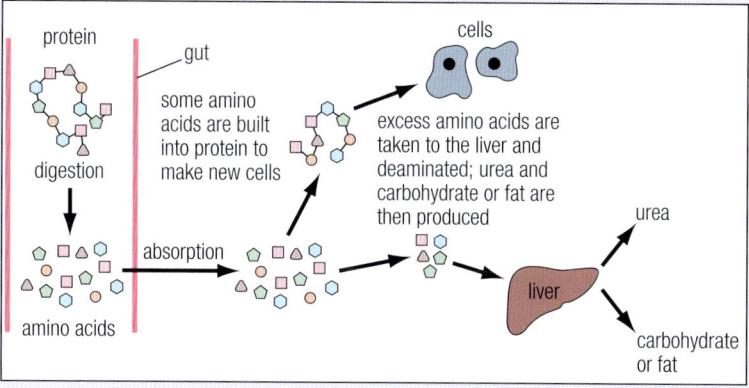

Figure 13.1.3 This shows what happens to amino acids in the body.

Hormones that have circulated in the blood for a while are broken down by the liver. The liver also breaks down harmful substances such as alcohol and other drugs like paracetamol.

Besides assimilation and breaking down unwanted substances, the liver also makes bile. bile. Bile neutralises the acid contents of the small intestine and also emulsifies fats (see page 82).

KEY POINTS

1 Excretion is the removal from the body of waste chemicals made in tissues during metabolism.

2 Excretory products include carbon dioxide and urea.

3 The lungs, liver and kidneys are excretory organs. Carbon dioxide is excreted by the lungs.

4 The liver makes urea from excess amino acids. The kidneys excrete urea in urine.

S 5 The liver uses amino acids to make proteins such as plasma proteins for example fibrinogen.

6 Deamination is the breakdown of excess amino acids in the liver to form ammonia, which is made into urea.

SUMMARY QUESTIONS

1 a Explain how each of these excretory products are produced and then removed from the body:
   i carbon dioxide   ii urea

2 Distinguish between excretion and egestion.

S 3 Describe what happens during deamination.

13.2 Kidney structure

LEARNING OUTCOMES

- Describe the structure of the urinary system
- Identify the parts of the urinary system on diagrams and images
- Describe the role of the kidneys in excretion
- **S** Describe the internal structure of the kidney

EXAM TIP

<u>Urine</u> is a body fluid that consists of waste substances dissolved in water. It is yellow because one of the waste substances is a yellow pigment. <u>Urea</u> is a colourless compound that dissolves in water, in blood plasma and in urine.

The kidneys

The kidneys make up part of the **urinary system**. They are responsible for the excretion of urea and excess ions from the body. They also control the water and ion content of the blood.

Your kidneys are at the top of your abdominal cavity just underneath the diaphragm. They are protected by the backbone, lower ribcage and the fat that surrounds them.

Waste chemicals like urea diffuse from cells into the blood and are transported to the kidneys. Blood enters the kidneys through the **renal arteries**. Inside each kidney there is a complex network of filtering units called kidney tubules. As the blood flows through these filtering units, small molecules such as glucose, ions, water and urea are forced out of the blood plasma to form a fluid known as filtrate. As the filtrate passes along the tubule, the useful substances (glucose and ions) are reabsorbed into the blood.

The volume of water reabsorbed depends on the body's water content. Sometimes, you can get dehydrated (short of water) perhaps because you have been sweating a lot during exercise, or because it is very hot. Then the kidneys will take back as much water as they can so you produce concentrated urine and conserve water. If you have taken in a lot of water, or it is a cold day and you have not been sweating, the kidneys reabsorb less water. In this case, you will produce pale-coloured, dilute urine.

At the end of each tubule the urine is released and flows into the **ureter** and to the **bladder** where it is stored. From here it leaves the body, at intervals, through a shorter tube called the **urethra**. The **renal vein** carries blood with a low concentration of waste chemicals away from the kidney.

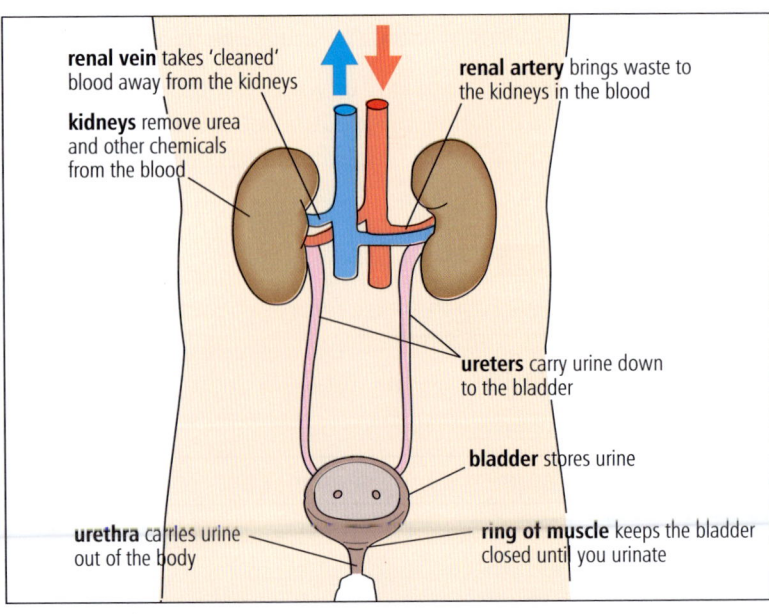

Figure 13.2.1 The urinary system.

Supplement

Inside the kidneys

If you cut open a kidney lengthways, you can see three areas: the **cortex** – a brown outer area, the **medulla** – a reddish inner area, and the **pelvis** – a white area.

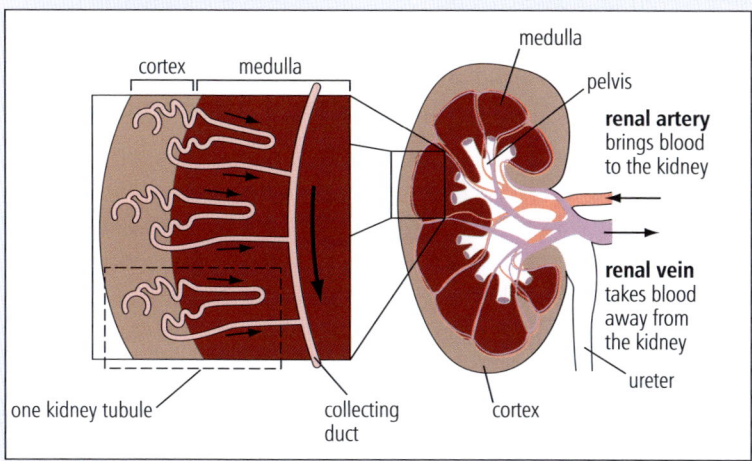

Figure 13.2.2 A vertical section of a kidney (right hand side). Detailed view of part of the kidney (left hand side).

Inside each kidney are about a million tiny tubes called **kidney tubules**, which you can see only with a microscope. The function of the kidney tubules is to filter the blood and remove waste chemicals and determine how much water is excreted. The filtering is carried out in the cortex. The waste chemicals and excess water are removed from the body in the urine, which flows from the kidneys down the ureter and is stored in the bladder.

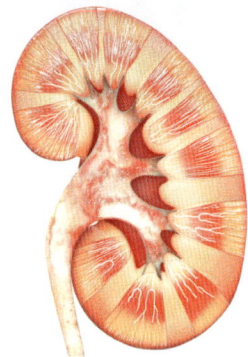

Figure 13.2.3 A kidney with part of the ureter attached.

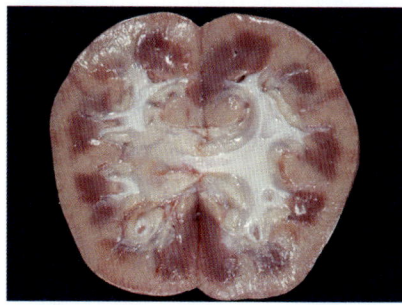

Figure 13.2.4 A kidney cut open to show the cortex, medulla, pelvis and ureter.

SUMMARY QUESTIONS

1 Match each structure in the urinary system in the left column with one of the functions in the right column.

renal artery — stores urine
ureter — filters urea and other waste chemicals out of the blood
kidney — carries blood with a high concentration of urea
bladder — carries urine down to the bladder
renal vein — carries blood with a low concentration of urea

2 Sketch a vertical section of the kidney and label:

  A cortex   **B** medulla   **C** renal artery

  D renal vein   **E** ureter

3 **a** Distinguish between *excretion* and *egestion*.

  b What is the main nitrogenous waste product excreted by the kidneys and where is it produced?

  c Distinguish between the ureter and the urethra.

  d Name two substances found in the blood plasma that are not found in the urine of a healthy person.

KEY POINTS

1 The urinary system consists of the kidneys, ureters, bladder and urethra.

2 The internal structure of a kidney consists of an outer cortex and an inner medulla.

3 The kidneys filter materials out of the blood, take back useful substances into the blood, e.g., glucose, control the quantity of water lost in the urine and excrete urea in the urine.

4 The ureters carry urine to the bladder from the kidneys.

5 The volume and concentration of urine depends on the intake of water and how much water has been lost in sweat during exercise and in hot weather.

13.3 Kidney function

Supplement

LEARNING OUTCOMES

- Describe the structure of a kidney tubule
- Describe the role of the kidney tubule in filtration, removal of urea and excess water and the reabsorption of glucose and some ions

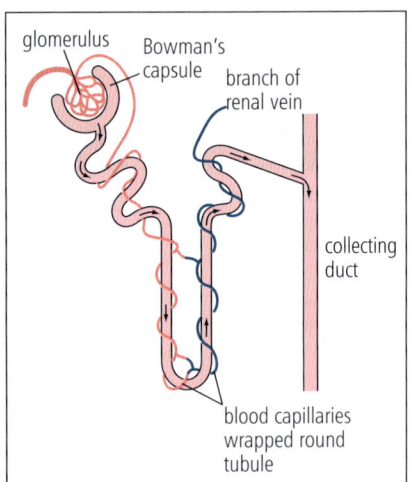

Figure 13.3.1 Kidney tubule.

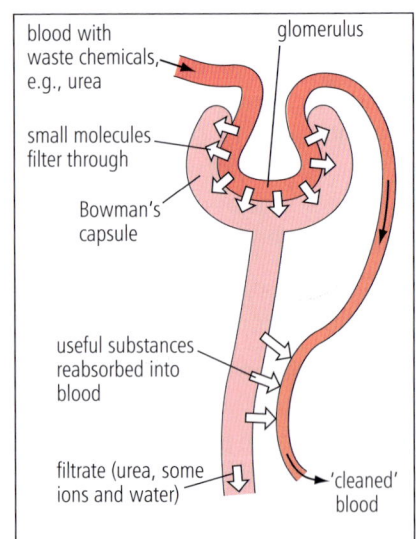

Figure 13.3.3 Filtration and reabsorption.

Structure of a kidney tubule

Blood containing waste chemicals flows into the kidney in the renal artery. Inside the kidney, the renal artery branches many times to give arterioles. Each of these arterioles supplies blood to a closely packed group of capillaries called a **glomerulus**.

After filtration, the blood flows out of the glomerulus into another arteriole. It then flows into capillaries around the rest of the tubule, which eventually join to form the renal vein. Each kidney tubule starts with a Bowman's capsule. This is formed by the end of the tubule being pushed in like the finger of a rubber glove. The glomerulus is found inside the Bowman's capsule.

Figure 13.3.1 shows that the tubule forms a loop. As you can see in Figure 13.2.2 (page 147), the loop goes deep into the medulla and back to the cortex. The end of the kidney tubule drains into a **collecting duct**. This passes through the medulla and empties urine in the pelvis and thus into the ureter.

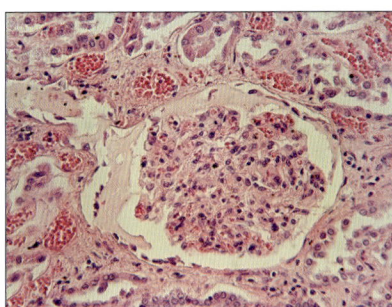

Figure 13.3.2 Photomicrograph of a single glomerulus in the kidney cortex surrounded by kidney tubules.

Filtration and reabsorption

Filtration

Look at Figure 13.3.3 showing how a kidney tubule works.

The kidneys are close to the heart so the blood pressure in the renal artery is high. You can see that the blood vessel entering the glomerulus is wider than the one leaving it. So there must be more blood entering the glomerulus than there is leaving it. This causes pressure to increase inside the glomerulus (it's rather like a bottleneck situation). It is this pressure that causes the blood to be filtered.

The lining of the capillaries is like a net with tiny holes in it. Blood cells and large molecules, like blood proteins, are too big to pass through the capillary lining and so stay in the blood. Small molecules, like urea, glucose, ions and water, pass out of the glomerulus and into the Bowman's capsule. It is in this space that the filtrate collects.

Reabsorption

All of the glucose, some ions and much of the water are needed by the body. They are **reabsorbed** back into the blood from the kidney tubule. This reabsorption involves active transport (see pages 34–35).

If you look at the structure of the cells in the wall of the tubule, you will see adaptations associated with active transport:

- **microvilli** provide a large surface area for absorption.
- numerous mitochondria provide energy for active transport.

After the process of reabsorption, what is left is urea and excess ions dissolved in water. As this fluid flows on through the tubule, some water may be reabsorbed if the body is low in water. The fluid that enters the collecting ducts is urine. It flows down the collecting duct and then to the ureter and the bladder. Urine collects and is stored in the bladder. From here it is expelled through the urethra at intervals. The blood leaving the kidney in the renal vein has a much lower concentration of waste chemicals.

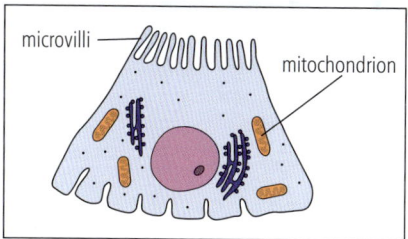

Figure 13.3.4 An epithelial cell from the lining of the kidney tubule.

EXAM TIP

Reabsorption of glucose and salts occurs by diffusion and active transport. Water is reabsorbed by osmosis. Expect to explain reabsorption in the kidney by referring to these processes (see pages 30 and 34).

SUMMARY QUESTIONS

1 Sketch a diagram of the kidney tubule and label:
  a glomerulus **b** Bowman's capsule **c** collecting duct
  d branch of the renal vein **e** branch of the renal artery

2 Describe what happens in kidney tubules to form urine.

3 The table shows the concentrations of five substances in three body fluids.

substance	concentration / g dm^{-3}		
	blood in renal artery	filtrate in kidney tubule	urine
urea	0.2	0.2	20.0
glucose	0.9	0.9	0.0
amino acids	0.05	0.05	0.0
ions	8.0	8.0	16.5
protein	82	0	0

Use the information in the table to:
a state the substances that pass from the blood into the kidney tubule
b explain how the substances you have named pass into the kidney tubule
c state the substances that are reabsorbed into the blood from the kidney tubule and explain why they are reabsorbed
d explain the results for protein.

KEY POINTS

1 Blood is filtered as it passes through the glomerulus removing urea, ions, glucose and water.

2 The useful materials such as glucose, some ions and most of the water are reabsorbed into the blood capillaries along the tubule.

3 Urine, consisting of urea, some ions and water, passes out of the kidneys and along the ureters, to be stored in the bladder until urination.

Practice questions

1 Which organ is responsible for the excretion of carbon dioxide?

   A   bladder

   B   kidney

   C   liver

   D   lung

   (Paper 1)   [1]

2 Which substances are excreted in the urine of a healthy person?

   A   glucose and proteins

   B   glucose and water

   C   proteins and ions

   D   ions and water

   (Paper 1)   [1]

3 The urinary system makes urine, stores it and removes it from the body. Which organ is *not* part of the urinary system?

   A   bladder

   B   kidney

   C   liver

   D   ureter

   (Paper 1)   [1]

4 Which is used to make urea?

   A   excess amino acids

   B   excess fatty acids

   C   excess glucose

   D   excess hormones

   (Paper 2)   [1]

5 Which is the blood vessel with the lowest concentration of urea?

   A   aorta

   B   renal artery

   C   renal vein

   D   vena cava

   (Paper 2)   [1]

6 Which process reabsorbs glucose from the filtrate in the renal tubules?

   A   active transport

   B   cohesion

   C   diffusion

   D   osmosis

   (Paper 2)   [1]

7 Which is an excretory function of the liver?

   A   formation of glycogen

   B   deamination

   C   respiration of lactic acid

   D   protein synthesis

   (Paper 2)   [1]

8 As part of an investigation, samples of urine were collected from a person throughout a 24 hour period. The urine sample collected early in the morning was more concentrated than the samples collected later in the day. Which is the most likely explanation for this?

   A   less urea was produced during the day

   B   less water was reabsorbed from the urine in the renal tubules at night

   C   the blood glucose concentration decreased during the day

   D   the concentration of the blood plasma increased at night

   (Paper 2)   [1]

9 The diagram shows the urinary system and its blood supply.

   (a) (i)   Name parts **A** to **D**.   [4]

       (ii) State the functions of structures **A** to **D**.   [4]

(b) (i) The following is a list of components of the blood. Identify those that are found in the urine of healthy people.

blood proteins, urea, glucose, platelets, red blood cells, ions, water [1]

(ii) State three factors that affect the volume and concentration of urine. [3]

(Paper 2)

10 Ammonia is produced by all animals as a waste product of their metabolism.

(a) Ammonia produced by humans is converted to urea. Explain why this is necessary. [3]

The table shows the composition of blood plasma in the renal artery, filtrate in Bowman's capsule and urine.

substance	concentration / g dm⁻³		
	blood plasma in renal artery	filtrate in Bowman's capsule	urine
urea	0.2	0.2	20.0
glucose	0.9	0.9	0.0
amino acids	0.05	0.05	0.0
mineral ions	8.0	8.0	16.5
protein	82	0.0	0.0

(b) Explain why

(i) four of the substances shown in the table are present in the filtrate, but protein is not [2]

(ii) glucose and amino acids are not present in the urine [2]

(iii) the concentration of urea and mineral ions is higher in the urine than in the filtrate [3]

(iv) the blood in the renal artery has a higher concentration of urea than blood in the renal vein. [2]

(c) Explain how filtrate is formed in the kidney. [4]

(Paper 4)

11 (a) The diagram shows a section through a kidney with its blood supply.

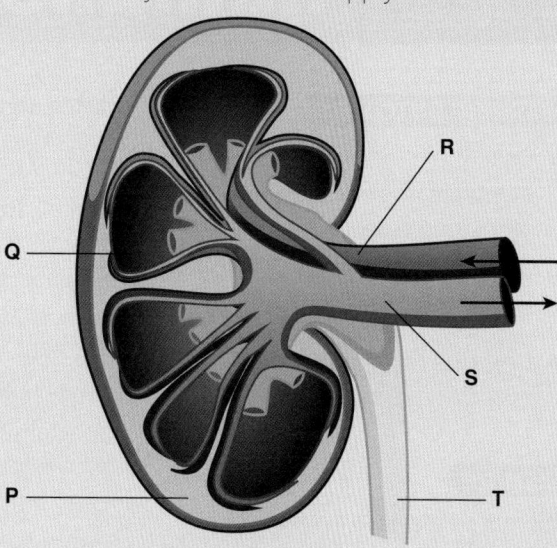

(i) Name parts **P** to **T**. [5]

(ii) State the functions of structures **R** and **S**. [2]

(b) The diagram shows a kidney nephron and some of the blood vessels associated with it.

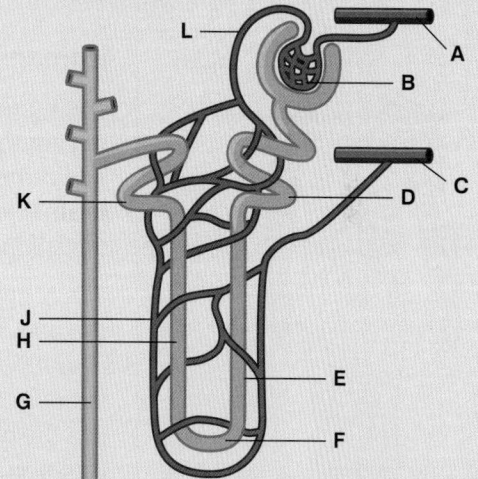

Use the letters on the diagram of the nephron to identify the parts that carry out the functions listed.

1. Transport of blood at high pressure.
2. Filtration of the blood.
3. Reabsorption of glucose.
4. Reabsorption of water to produce a concentrated urine. [4]

(c) Explain why the body produces urea. [3]

(Paper 4)

14.1 Nervous control in humans

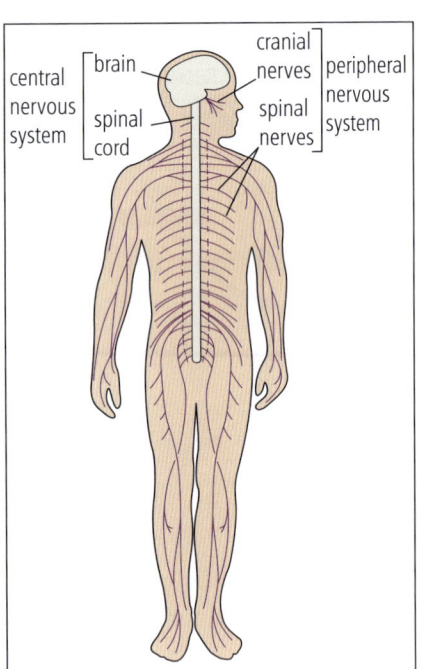

Figure 14.1.1 The main parts of the human nervous system.

All living organisms are sensitive to changes in their environment; the changes in daylight, wind speed, temperature, movement, availability of food that they detect are called **stimuli** (singular: stimulus). **Receptors** are the cells that detect the stimuli; for example, cells in your taste buds detect chemicals in your food. **Effectors** are the organs, such as glands and muscles, that bring about responses. Glands secrete useful substances and muscles contract to bring about movement. The actions they take as a result are called **responses**.

Animals have nervous systems, which at their simplest are networks of specialised nerve cells, as in jellyfish and sea anemones. At their most complex, they are organs with powerful processing skills, such as the human brain. Two organ systems are involved:

- the sensory system for detecting changes in our environment
- the nervous system for coordinating our responses and initiating actions.

Here we discuss the nervous system. The nervous system is involved in the coordination and regulation of body functions. You will find information about the sensory system in Topic 14.4.

Human nervous system

The human nervous system consists of two parts: the **central nervous system** (CNS) and the **peripheral nervous system** (PNS). The central nervous system is made up of only two organs: the **brain** and the **spinal cord**. The brain is protected by the skull and the spinal cord is protected by the vertebral column.

The central nervous system is connected to different parts of the body by nerves that make up the peripheral nervous system (Figure 14.1.1). **Cranial nerves** and **spinal nerves** are in pairs on either side of the brain and spinal cord respectively. Cranial nerves link the brain with all the organs in the head and also some in the thorax and abdomen. Spinal nerves link the brain to the arms, thorax, abdomen and legs. Spinal nerves leave the spinal cord in pairs through spaces between the vertebrae.

Each nerve is made up of lots of nerve cells or **neurones** surrounded by a protective fibrous tube. Nerves are visible to the naked eye, but you need microscopes to see neurones. Neurones transmit information in the form of nerve impulses, which are like pulses of electricity passing along a wire.

Voluntary actions are under conscious control and involve the brain in decision making. A footballer carries out a voluntary action when he or she kicks a ball. Turning over the pages of a book is an example of a voluntary action. These actions are under the conscious control of the brain.

Reflex actions

We make two types of action. Some happen automatically without us having to think about them. These are involuntary actions that occur unconsciously. Swallowing, blinking, breathing and the beating of the heart are four such actions that we are not conscious of most of the time, but we can choose to control occasionally. Movement of food in the ileum (see page 82) and of urine in the ureter (see page 146) are other involuntary actions you have studied that we can't control by thinking about them. Voluntary actions are those we choose to make and the decisions to make them occur in our brains.

When a person sits down on a sharp object such as a pin without realising it, there is a very quick, automatic response that the person does not need to think about.

This is an example of a **simple reflex**, which is an involuntary action.

The stimulus is the pin and the receptors are pain sensors in the skin. The effectors are the muscles in the legs that cause the person to get up quickly.

The sequence of events is:

stimulus ⟶ receptor ⟶ coordinator ⟶ effector ⟶ response

The **coordinator** is the part of the body that connects information about the stimulus to the effector. This always happens inside the central nervous system to allow connections with other parts of the body and for the brain to be aware of stimuli. Inside the spinal cord, neurones from pain receptors connect with neurones to the muscles. Some reflexes involve the brain as we will see in the section on the eye (see page 158).

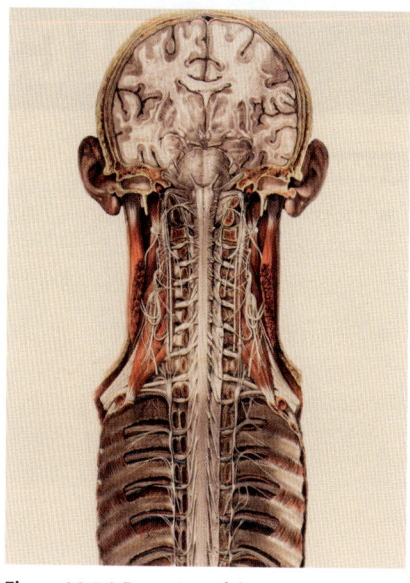

Figure 14.1.2 Front view of the top part of the nervous system.

EXAM TIP

You can see more about the neurones that control simple reflexes on page 155. They are arranged into reflex arcs involving peripheral nerves and the CNS. All responses are coordinated through the CNS.

SUMMARY QUESTIONS

1 Copy and complete the sentences using these words:

   cranial   peripheral   spinal   neurones   central

   The brain and spinal cord make up the _____ nervous system. All the nerves of the body make up the _____ nervous system. _____ nerves are connected to the brain and _____ nerves are connected to the spinal cord. The nervous system is composed of many specialised cells known as _____ .

2 a Put the following words into the correct sequence:

   coordinator   stimulus   effector   receptor   response

   b Define each of the above words.

3 a Give an example of a simple reflex that involves the spinal cord.

   b Outline what happens in a simple reflex.

   c Give advantages of having simple reflexes.

KEY POINTS

1 The human nervous system is made up of the central nervous system (brain and spinal cord) and the peripheral nervous system (cranial and spinal nerves).

2 Neurones or nerve cells transmit nerve impulses.

3 A simple reflex is an automatic response to a stimulus coordinated by the nervous system.

14.2 Neurones and reflex arcs

Neurones

Neurones are highly specialised cells. Their structure allows them to transmit information as nerve impulses over long distances. The cell body contains the nucleus and almost all of the cytoplasm. However, there are thin extensions of cytoplasm that may be quite short or extend a long way through the body.

Sensory neurones transmit impulses from sense organs to the brain and spinal cord.

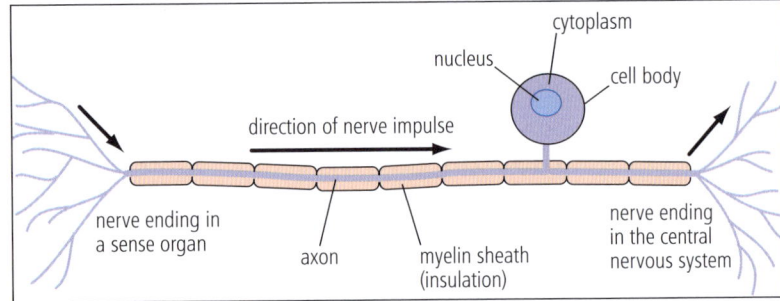

Figure 14.2.1 A sensory neurone.

Motor neurones transmit impulses away from the brain and spinal cord to effector organs – muscles and glands.

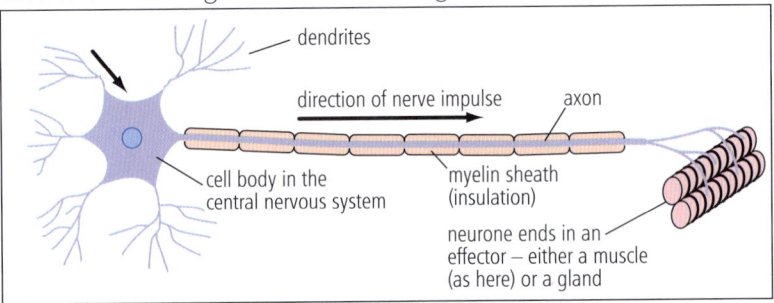

Figure 14.2.2 A motor neurone.

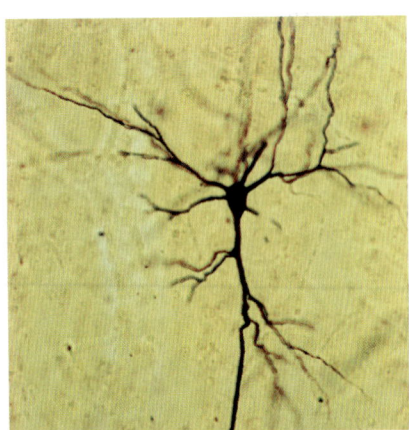

Figure 14.2.3 This neurone has a long axon

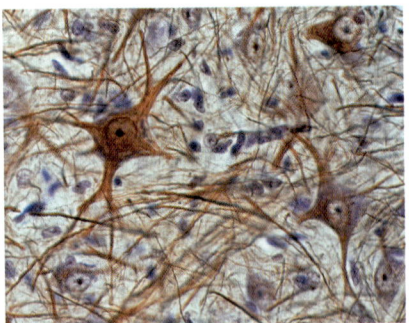

Figure 14.2.4 Relay neurones in the brain. Notice the cell bodies with their nuclei.

Figures 14.2.1 and 14.2.2 show a sensory neurone and a motor neurone surrounded by insulation. This insulation is known as **myelin**. It is formed by separate cells that grow around the neurones to form a layer rich in fat. This insulation makes impulses travel very quickly. Some myelinated neurones, which are used in our fast reflexes, transmit impulses at 100 metres per second.

A third type of neurone is the **relay (connector) neurone**. These are short and pass on impulses from sensory neurones to motor neurones inside the brain and spinal cord.

When any two neurones meet, they do not actually touch each other. There is a tiny gap between them called a **synapse**. When an impulse reaches a synapse a **chemical transmitter substance** is released from the first neurone. It diffuses across the synapse and triggers an impulse in the second neurone. Since the chemical transmitter is only produced on one side of the synapse, it ensures that impulses travel in one direction through the nervous system. Many drugs produce their effects by acting at synapses (see page 159).

Reflex arcs

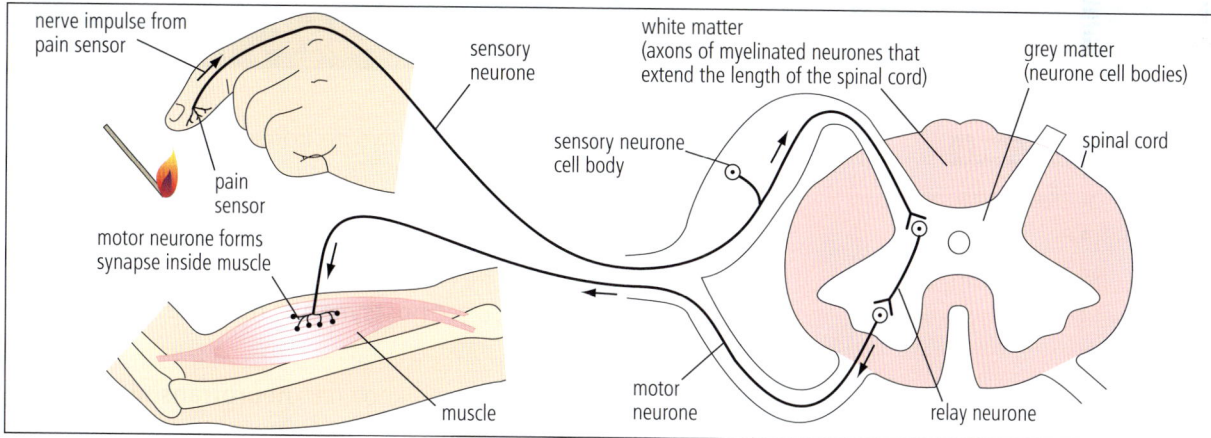

Figure 14.2.5 A reflex arc.

If you touch a hot object, you remove your hand very quickly and automatically without thinking. This is an example of a simple reflex. Your brain receives impulses about the pain but not until you have already moved your hand. If you relied on the brain to make a decision you would be burnt in the time it took for the impulses to travel to the brain and back.

Look at Figure 14.2.5 and trace the pathway of the impulse along the neurones. This pathway is called a **reflex arc**. The **stimulus** in the example is the hot flame. The **receptor** is the heat sensor in the skin. When it is stimulated the heat sensor generates an impulse. The impulse travels to the spinal cord along a sensory neurone.

Inside the grey matter of the spinal cord the impulse passes across a synapse to a relay (connector) neurone. The relay neurone passes the impulse across a second synapse to a motor neurone.

The motor neurone transmits the impulse to a muscle in the arm. The muscle is the **effector** and it contracts to remove the hand from the hot object. This action is the **response**.

Many reflexes are protective; they happen very quickly, so you do not harm yourself.

1 Copy and complete the sentences using these words:

> synapse   grey   arc   relay   motor
> sensory   effectors   transmitter

In a reflex _____ nerve impulses are transmitted to the spinal cord by _____ neurones. Inside the _____ matter of the spinal cord the impulses are passed on to _____ neurones. The impulses leave the spinal cord along _____ neurones to go to _____. There is a gap between two neurones called a _____ where chemical _____ substances are released to pass the impulse to the next neurone.

2 Look at Figure 14.2.6:

   a Name the parts labelled A to D.

   b Give the functions of parts **B**, **C** and **D**.

3 a Define each of the following:

   i synapse   **ii** chemical transmitter substance
   iii myelin   **iv** relay neurone

   b Explain the function of **i** – **iv** in the nervous system.

4 Make a drawing of a relay (connector) neurone.

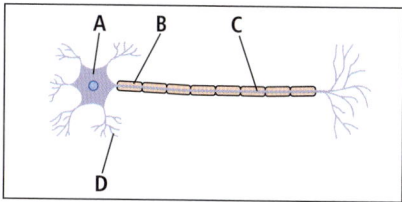

Figure 14.2.6 Diagram of a motor neurone.

1 A reflex arc is made up of a receptor, three types of neurone (a sensory neurone, a relay neurone and a motor neurone) and an effector.

2 Reflex actions are rapid, automatic and often protective. They interpret and coordinate stimuli with responses by effectors.

14.3 Synapses

Supplement

A **synapse** is a junction between two neurones. Where two neurones meet, they do not touch. There is a small gap between them, about 20 nm wide. The gap is called the **synaptic gap**. The neurone that carries the impulse to the synapse is called the **presynaptic neurone**. The neurone that carries the impulse away from the synapse is called the **postsynaptic neurone**.

How does a nerve impulse transfer across the synapse from one neurone to the next? Chemicals known as **neurotransmitters** are released by the presynaptic neurone and diffuse across the synaptic gap to trigger an impulse in the postsynaptic neurone.

Structure of the synapse

The axons of neurones end in swellings called **synaptic bulbs**. The surface of the synaptic bulb is called the **presynaptic membrane**. It is separated by the synaptic gap from the **postsynaptic membrane** of the cell body or axon of the next neurone.

In Figure 14.3.1 you see that the synaptic bulb has many vesicles containing neurotransmitter molecules. It also contains many mitochondria. This suggests that energy is required in synaptic transmission. The postsynaptic membrane has a number of large protein molecules on its surface, which act as receptor sites for the neurotransmitter substance.

Estimates for the number of neurones in the adult human brain reach 86 billion. As each neurone has many synapses with other neurones, the estimate for the number of these is therefore far greater: 0.15 quadrillion (0.15×10^{15}).

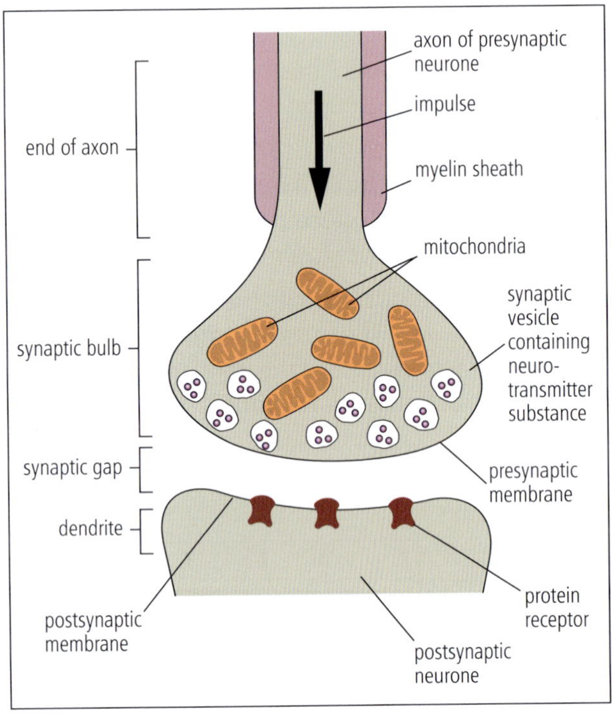

Figure 14.3.1 The structure of a synapse.

Labels in figure:
- axon of presynaptic neurone
- impulse
- myelin sheath
- end of axon
- mitochondria
- synaptic vesicle containing neurotransmitter substance
- synaptic bulb
- presynaptic membrane
- synaptic gap
- dendrite
- protein receptor
- postsynaptic membrane
- postsynaptic neurone

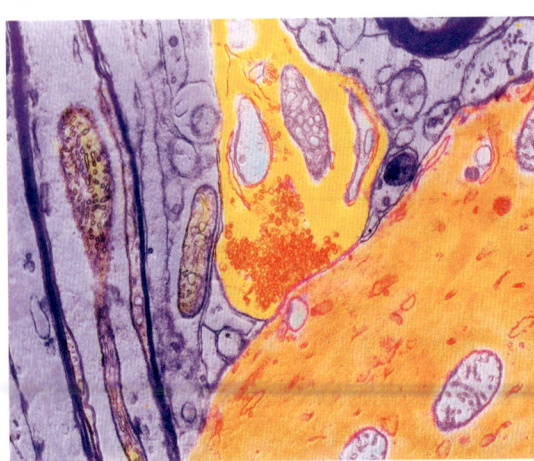

Figure 14.3.2 In this image the cytoplasm of the neurones is coloured orange. The image shows the synapse between a presynaptic neurone (above) and a postsynaptic neurone (below). The synaptic gap is the tiny space between them. You can see lots of vesicles clustered together at the bottom left of the presynaptic neurone. (×14 000).

Synaptic transmission

1 When an impulse arrives at the synaptic bulb it causes vesicles containing the neurotransmitter to move towards the presynaptic membrane.

2 The vesicles fuse with the presynaptic membrane, releasing the neurotransmitter into the synaptic gap.

3 The neurotransmitter diffuses across the synaptic gap and attaches to specific receptor sites on the postsynaptic membrane. These receptor sites have a complementary shape to the neurotransmitter, but the binding is only temporary.

4 The binding of the neurotransmitter triggers an impulse in the postsynaptic neurone. Once this has happened the neurotransmitter is broken down by an enzyme in the synaptic gap.

5 The mitochondria provide energy to reform the neurotransmitter.

In a reflex arc synapses ensure that nerve impulses can only travel in one direction, i.e., from presynaptic neurone to postsynaptic neurone. This happens because vesicles of neurotransmitter are only present in the presynaptic bulb and the receptors are only found on the postsynaptic membrane.

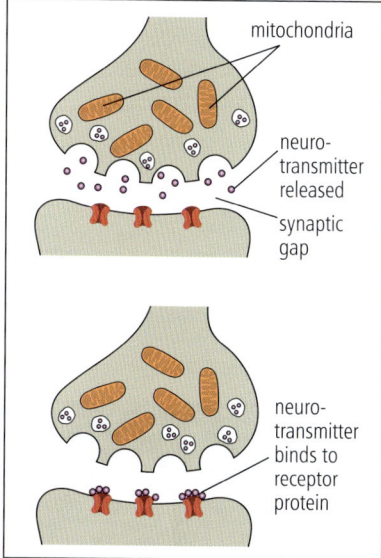

Figure 14.3.3 Action of a neurotransmitter at a synapse.

SUMMARY QUESTIONS

1 What is meant by each of these terms?
   a synapse b neurotransmitter c receptor molecules

2 Explain the function of each of the following in the transmission of an impulse across synapses:
   a synaptic vesicles   b neurotransmitter
   c protein receptor molecules d mitochondria

3 Make a flow chart diagram to summarise the events that occur between the arrival of an impulse at the end of the axon of a presynaptic neurone and the transmission of an impulse in the postsynaptic neurone.

4 a Find out what is meant by the following terms.
      i excitatory drug
      ii inhibitory drug
   b Give an example of each different type of drug.

KEY POINTS

1 A synapse is a junction between two neurones.

2 The synaptic bulb contains many mitochondria and many vesicles containing molecules of neurotransmitter.

3 An impulse triggers the release of a neurotransmitter, which diffuses across the synaptic gap to bind with the receptor molecules on the postsynaptic membrane and trigger an impulse.

4 In a reflex arc, synapses ensure that impulses travel in only one direction.

14.4 Sense organs

LEARNING OUTCOMES

- Describe sense organs as groups of receptor cells responding to particular stimuli
- Describe the structure and function of the eye
- Explain the pupil reflex in terms of light intensity and pupil diameter

Figure 14.4.1 Musicians use a variety of their senses when performing.

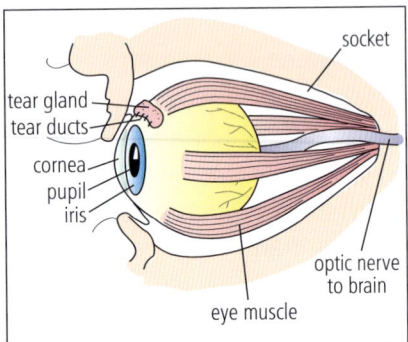

Figure 14.4.2 The eye in its socket.

EXAM TIP

The sensory system also has receptors inside the body for detecting changes in things like blood temperature and the tension in our muscles. This information helps with homeostasis.

We are aware of our surroundings and use our sense organs to detect stimuli. We have receptor cells that are specialised to detect certain stimuli, such as light and chemicals. These cells generate nerve impulses when stimulated and these impulses travel along sensory neurones to the central nervous system.

Some receptors are found concentrated in groups in certain organs. The eye and the ear have receptors arranged in special tissues. They also have other tissues to modify the information from the environment before it reaches the receptors. In the eye there are structures for focusing light rays; in the ear there are structures for amplifying sound.

Sense organs are groups of receptor cells that respond to specific stimuli.

- The eyes respond to light rays and give us our sense of sight.
- The nose responds to chemicals in the air and gives us our sense of smell.
- The tongue responds to chemicals in our food and drink and gives us our sense of taste.
- The ears respond to sound vibrations and give us our sense of hearing. The ear also detects movement and position of the body, so provides information about our balance.
- The skin responds to pressure and gives us our sense of touch. It also detects pain and temperature (see page 168).

Eye structure

Your eye sits inside a socket in the skull and is moved about by three pairs of eye muscles . In Figure 14.4.2, you can see the median pair of muscles and one of each of the upper and lower pairs. The eye is a complex organ composed of different tissues, such as fibrous tissue, muscle tissue, sensory tissue and blood. Each tissue has a function to help with the detection of light. Figure 14.4.3 shows the structures in the eye and their functions.

At the front of the eye is the transparent **cornea**, through which light enters the eye. Light then passes through the **pupil**, which is a hole in the centre of the pigmented (coloured) **iris**, through the lens to the **retina** at the back of the eye. The retina is the tissue with light-sensitive receptor cells.

At the very front of the eye is a delicate, transparent layer, which provides protection. It is kept moist by the tear glands that make the tears that wash your eye clean every time you blink. Tears also contain **lysozyme**, an enzyme that kills bacteria.

In humans, the eyes are set side by side at the front of the head as opposed to at the side, like a rabbit. What you see in front of you is called your visual field. As the eyes are a few centimetres apart, each eye views the same visual field but from a slightly different

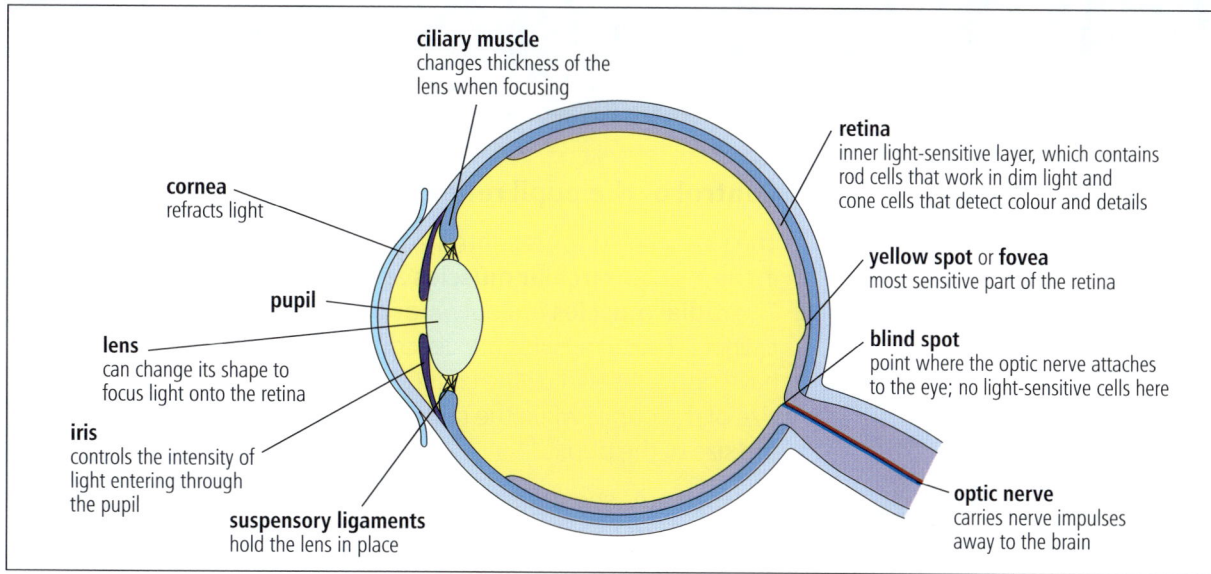

Figure 14.4.3 A horizontal section of the eye.

angle. Just try shutting your eyes alternately to experience this. The two fields of view overlap to give stereoscopic vision, which gives us the ability to judge distances accurately. Stereoscopic vision is an essential adaptation for tree-dwelling animals such as monkeys, and predators such as lions and tigers, all of which must judge distances accurately to survive.

The pupil reflex

Look in a mirror and shine a bright light near your eyes and then turn it off. Wait a few minutes then turn the light back on. You will see that your pupils decrease in size. This lets less light into your eyes and protects the retina from damage. When you turned the light off, your pupils increased in size to let more light into your eyes so you could still see clearly. This is another example of a simple reflex. Receptor cells in the retina send impulses along sensory neurones in the optic nerve. These connect with motor neurones that pass impulses to the muscles in the iris to stimulate them to contract.

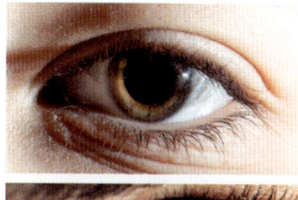

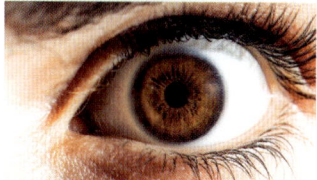

Figure 14.4.4 When a bright light is shone into the eye the pupil constricts to protect the retina from damage.

KEY POINTS

1 Sense organs are groups of receptor cells that respond to specific stimuli: light, sound, touch, temperature and chemicals.

2 The cornea, iris, pupil, lens, retina, optic nerve and blind spot are parts of the eye that have functions described in Figure 14.4.3.

3 In bright light, the diameter of the pupil decreases. When the light intensity decreases the diameter of the pupil widens to let in more light. This is controlled by a simple reflex.

SUMMARY QUESTIONS

1 Give the name in each case for the part of the eye that:
   a is sensitive to light
   b controls the intensity of light entering the eye
   c holds the lens in place within the eye
   d refracts the light after it has entered the eye
   e transmits nerve impulses to the brain
   f controls the shape of the lens
   g secretes the enzyme lysozyme
   h allows light to pass through the iris

2 Describe and explain the changes in the diameter of the pupils when the surrounding light intensity changes.

14.5 More about the eye

Supplement

LEARNING OUTCOMES

- Explain the pupil reflex in terms of light intensity and the antagonistic action of muscles in the iris
- Explain how the eye focuses on near and far objects
- State that rods and cones are light sensitive cells in the retina
- Describe the distribution of rods and cones and the way they function
- Identify on diagrams and images the position of the fovea

EXAM TIP

Here is a way to remember action of the circular muscles. The three Cs: circular, contract, constrict

EXAM TIP

In Unit 2 we dealt with cells, tissues, organs and organ systems. The eye is a very good example of an organ where you can identify tissues (e.g., muscle tissue) and cells (e.g., rod and cone cells). The eye is part of an organ system: the sensory system.

Control of the pupil reflex

The size of the pupil is controlled by a pair of antagonistic muscles in the iris. The **circular muscles** are arranged around the pupil. The **radial muscles** run outwards from the pupil like the spokes in a wheel. These muscles are described as antagonistic because they have opposite effects: the circular muscles contract to reduce the size of the pupils while the radial muscles contract to increase the size of the pupils.

The diagrams show what happens in dim light when the light intensity is low and in bright light when the light intensity is much higher and has the potential to damage the retina.

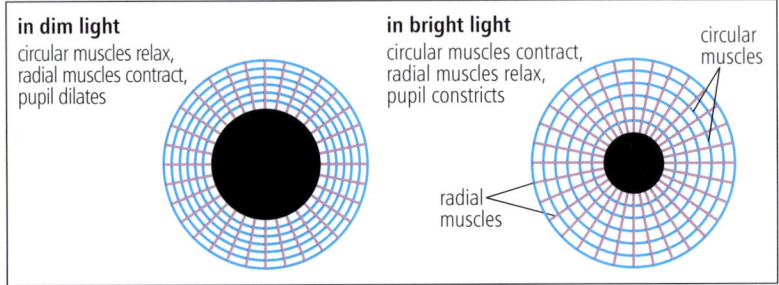

Figure 14.5.1 The iris muscles control the diameter of the pupil.

Focusing

Figure 14.5.2 shows how light is focused in the eye. Light enters the eye through the transparent cornea, passes through the lens to be focused on the retina. Notice that the image on the retina is inverted (upside down) because of the way the light rays reflected from the top and bottom of the object enter the eye and cross over behind the lens. Information about this inverted image goes to the brain where it is interpreted the right way up. We learn how to do this in the first few months of life.

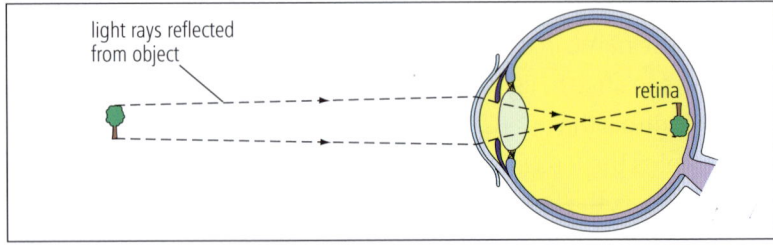

Figure 14.5.2 This shows light rays refracted by the cornea and the lens.

Rods and cones

Rods and cones are the light-sensitive receptor cells in the retina. Rod cells are sensitive to light of low intensity and send impulses only when it is dark, for example at night. They are responsible for our night vision. Cone cells are only sensitive to light of high intensity.

There are three types of cone cell, each sensitive to different wavelengths of light (different colours). The brain interprets the impulses from cone cells to give us our colour vision. The **fovea** in the centre of the retina contains cones and no rods. Each cone has its own neurone to the brain so this area in the middle of our visual field gives us a very detailed image. The rest of the retina contains rods and few cones. This area gives us our peripheral vision, which is not as detailed. The position of the fovea is in the centre of the retina on the same horizontal level as the blind spot. The fovea in your right eye is to the right of your blind spot.

Accommodation is the term used to describe the changes that occur in the eye when focusing on far and near objects. As light enters the eye it must be refracted (bent) so that we can see the image clearly. About 60% of the refraction of the light rays is done by the cornea and the rest is done by the lens. The lens is surrounded by elastic tissue that can be stretched and can recoil. The shape of the lens is controlled by the **ciliary muscles**.

If you are looking at a distant object:

- the ciliary muscles relax
- the pressure inside the eye pulls the suspensory ligaments tight (or taut) so the lens is pulled into an **elliptical** (thin) shape
- Light rays are refracted as they pass through the lens and focused on the retina; the distant object is in focus.

If you are looking at a near object:

- the ciliary muscles contract to counteract the pressure inside the eye
- the suspensory ligaments become slack so the elastic tissue around the lens recoils and the lens becomes more **spherical** (fatter)
- Light rays are refracted more than they were when looking at the distant object; the near object is in focus.

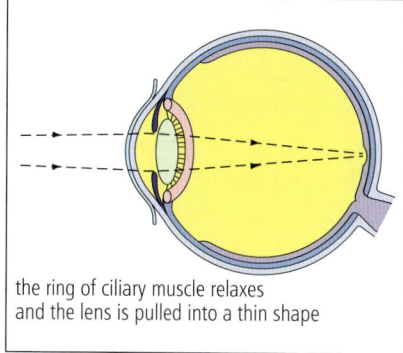

the ring of ciliary muscle relaxes and the lens is pulled into a thin shape

Figure 14.5.3 Focusing on a distant object.

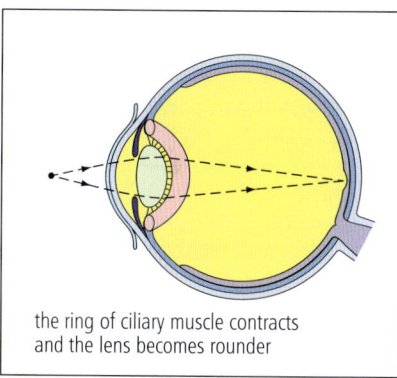

the ring of ciliary muscle contracts and the lens becomes rounder

Figure 14.5.4 Focusing on a near object.

KEY POINTS

1 The pupil reflex involves the coordination of radial muscles and circular muscles in the iris to control the intensity of light entering the eye through the pupil.

2 The ciliary muscles, suspensory ligaments and the lens are all involved in accommodation (focusing) in the eye.

3 In the retina, rod cells are distributed around the periphery and respond to light of low intensity. Cone cells respond to high light intensity and detect colour. The fovea is made entirely of cones.

SUMMARY QUESTIONS

1 Describe and explain the changes that occur to the ciliary muscles, the suspensory ligaments and the shapes of the lens, when focusing on a distant object.

2 a State what happens to each of the following in bright light:
   i circular muscles of the iris   ii radial muscles of the iris
   iii the size of the pupil
   b State what happens to each of the above in dim light.
   c Explain how the pupil reflex is controlled.

3 Describe the distribution of rods and cones in the retina and explain their functions.

14.6 Hormones

LEARNING OUTCOMES

- Identify the adrenal glands, pancreas, testes and ovaries and name the hormones that they release
- Define the term *homeostasis*
- State that insulin decreases blood glucose concentration
- Describe the effects of adrenaline on the heart, lungs and pupils
- S Describe the roles of adrenaline in controlling metabolic activity
- Compare the endocrine and nervous systems

EXAM TIP

The term 'secretion' means that cells make a substance that is useful for the body and release it into the blood or some other part of the body, such as the gut. Do not confuse secretion with 'excretion', which is the removal of metabolic waste (see page 144).

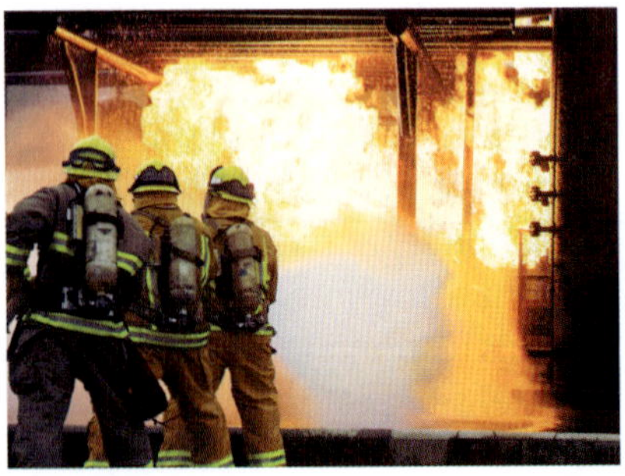

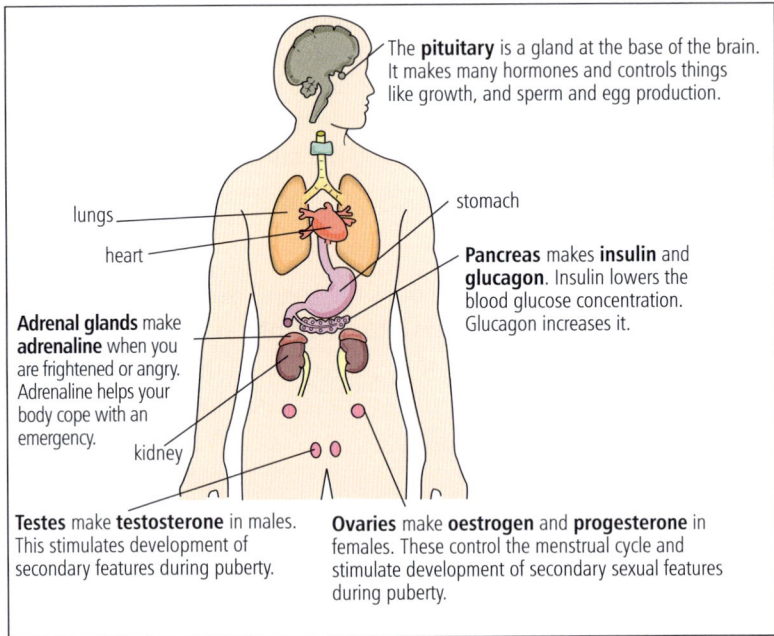

Figure 14.6.2 Adrenaline is released during times of excitement, fear or stress.

The endocrine system

A **hormone** is a chemical substance produced by an **endocrine gland** and transported in the blood. Endocrine means that the gland produces hormones that are secreted into the bloodstream rather than into a duct. Hormones travel throughout the body but only certain organs or tissues recognise each hormone and respond to it. These are referred to as **target organs**. Each hormone alters the activity of its target organs.

Hormones coordinate the activities of the body but in a different way to the nervous system. The main endocrine glands, the hormones that they secrete and their actions are shown in Figure 14.6.1.

The **pituitary** is a gland at the base of the brain. It makes many hormones and controls things like growth, and sperm and egg production.
lungs
heart
stomach
Pancreas makes **insulin** and **glucagon**. Insulin lowers the blood glucose concentration. Glucagon increases it.
Adrenal glands make **adrenaline** when you are frightened or angry. Adrenaline helps your body cope with an emergency.
kidney
Testes make **testosterone** in males. This stimulates development of secondary features during puberty.
Ovaries make **oestrogen** and **progesterone** in females. These control the menstrual cycle and stimulate development of secondary sexual features during puberty.

Figure 14.6.1 The organs of the endocrine system.

Controlling conditions in the body

During metabolism, many chemical reactions occur inside the body. In cells, some molecules, such as proteins, are built up; some molecules, such as glycogen, are broken down. Enzymes control these reactions and work best when conditions, such as temperature, pH and concentration of water, are kept suitable and constant. see Unit 5 to see how temperature and pH influence enzyme activity.

Any slight change in the conditions in the body can slow down or stop enzymes from working. **Homeostasis** is the maintenance of constant internal conditions in the body, so that enzymes control metabolism efficiently. The role of insulin in the body is to decrease the concentration of glucose in the blood after a meal so that plenty is stored in the liver and muscles for when it is needed, e.g., during exercise.

Adrenaline

Adrenaline is secreted by the adrenal glands. If you realise that you are in a situation that puts you at risk of injury or even death, you will immediately become much more aware and ready to take action. The responses necessary to survive in these 'fight or flight' situations are coordinated by adrenaline, which gets you ready for action by:

- increasing your breathing rate
- increasing your heart (pulse) rate
- widening your pupils so that more light enters your eyes.

Supplement

Adrenaline also stimulates an increase in the rate of chemical reactions within the body. To provide enough energy for increased levels of activity, adrenaline stimulates cells in the liver to convert glycogen to glucose, which diffuses into the blood. This increases the concentration of glucose in the blood, so that there is more available to the muscles as a source of energy for muscle contractions needed for sudden action.

Adrenaline coordinates the increased uptake of oxygen and changes in the supply of blood so that oxygen and glucose reach the muscles that need them for respiration.

- The air passages (trachea, bronchi and bronchioles) widen to allow more air into the alveoli in the lungs. This increases the volume of oxygen that can be absorbed.
- The breathing rate increases to increase the uptake of oxygen and excrete carbon dioxide at a faster rate.
- The heart pumps harder to force out more blood with each beat and the heart rate (beats per minute) also increases.

Comparing the endocrine and nervous systems

The endocrine system and the nervous system act to control and coordinate the activities of the body.

Table 14.6.1 Comparing the endocrine and nervous systems

feature	nervous system	endocrine system
structures	nerves	secretory cells in glands
forms of information	electrical impulses	hormones (chemicals)
pathways	along neurones	in the blood
speed of information transfer	fast	slow
longevity of action	short-lived, e.g., muscle contracts for a short time	usually slow and longer lasting
target area	only the area at the end of a neurone	whole tissue or organ
response	muscle contraction or secretion by glands	e.g., conversion of glucose to glycogen, protein synthesis, rate of respiration

SUMMARY QUESTIONS

1 What is meant by the following terms?
 a endocrine system
 b hormone
 c target organ

2 State the gland that secretes each of the following:
 a testosterone b insulin
 c oestrogen

3 Suggest four situations in which the secretion of adrenaline increases.

4 State four ways in which the endocrine system differs from the nervous system.

S 5 Explain how adrenaline prepares the body for action.

The brain has overall control of our body processes. The control of body temperature is an example of **homeostasis**.

In humans, the normal temperature of the blood is 37 °C. When blood reaching the brain is warmer than this temperature, the brain sends impulses along nerves to the skin to promote heat loss. For example, our sweat glands produce sweat and this helps us to cool down, returning body temperature to normal.

We feel cold when our blood temperature falls below 37 °C. Then the brain will stimulate the production of more heat in the muscles by shivering and by increasing the rate of respiration in the liver. It also stimulates the body to conserve heat by decreasing the production of sweat by sweat glands.

The normal body temperature (37 °C) is described as the **set point**. The hypothalamus in the brain continually monitors the temperature of the blood and compares it with the set point. Whenever the temperature increases above the set point or decreases below the set point, it directs changes to occur that return the temperature to 37 °C. If you made a continuous record of your body temperature you would see it fluctuating within a very narrow range, for example between 36.1 °C and 37.2 °C.

Negative feedback

The control of body temperature is an example of **negative feedback**. This is the same sort of control system as in an oven to keep the temperature constant at the pre-set temperature. Negative feedback acts to ensure that the actual temperature is as close to the pre-set temperature as possible. If our body temperature increases above or falls below the normal 37 °C, then the body responds by taking actions that will bring the temperature back to the set point. Our temperature may not be exactly 37 °C all the time, but will be close to that and within narrow limits. There are also control systems for blood glucose concentration, the water content of the blood, blood pH and the oxygen and carbon dioxide concentrations of the blood.

Controlling blood glucose

Cells need glucose for energy and therefore need a constant supply from the blood.

When you eat a high carbohydrate meal, your blood glucose concentration can increase by a factor of 20. However, it does not stay high because cells in the **pancreas** detect the high glucose concentration in the blood. These cells secrete the **hormone insulin** into the blood. Insulin stimulates liver cells to convert glucose into the storage compound **glycogen** (see page 38). This stimulates the liver cells to absorb lots of glucose from the blood so the concentration decreases and returns to normal.

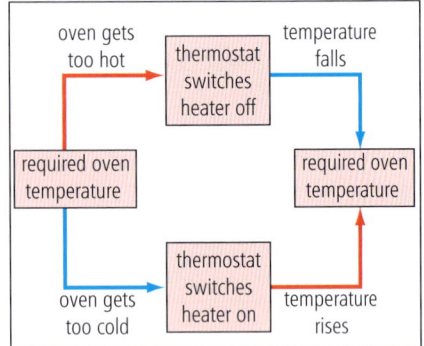

Figure 14.7.1 Controlling the temperature of an oven is an example of negative feedback.

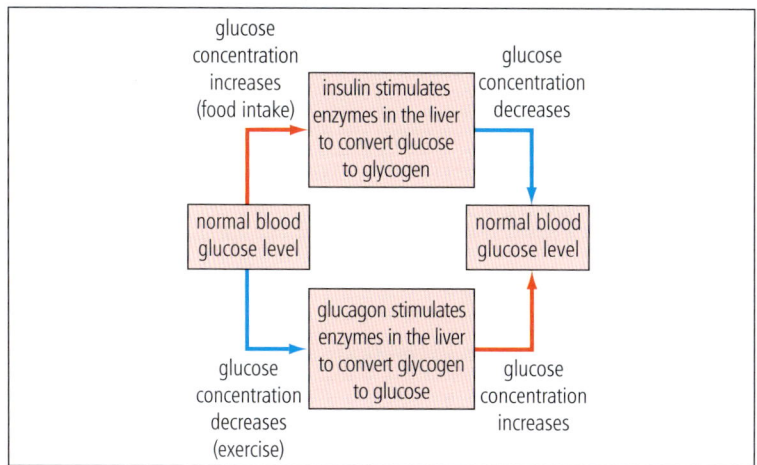

Figure 14.7.2 Controlling blood glucose concentration by negative feedback.

When you run a race or carry out some other type of exercise, your muscles take up lots of glucose from the blood to provide the energy they need. Your blood glucose concentration decreases, but it does not keep decreasing. Other cells in the pancreas detect this decrease and secrete the hormone **glucagon** into the blood. Glucagon stimulates liver cells to break down glycogen to glucose, which diffuses into the blood so the blood glucose concentration increases to normal.

Blood glucose concentration is controlled by the pancreas, which *monitors* the blood glucose concentration and releases hormones to instruct liver cells to remove glucose molecules from the blood or add more.

This is an example of negative feedback. The pancreas:

- makes insulin to stimulate a decrease in blood glucose concentration
- makes glucagon to stimulate an increase in blood glucose concentration.

This ensures that the blood concentration of glucose stays within limits and does not increase or decrease too much. (See also Type 1 diabetes in Topic 10.4.)

SUMMARY QUESTIONS

1 a Define the term *homeostasis*.
  b Explain why is it important that conditions in cells are kept constant.
2 a Explain the term negative feedback using the control of blood glucose concentration as an example.
  b Explain the roles of the pancreas and the liver in maintaining a constant blood glucose concentration.

EXAM TIP

Insulin and glucagon are <u>hormones</u>. They are not enzymes. So always say that insulin and glucagon <u>stimulate</u> the changes to glucose. Do not say that 'insulin converts glucose to glycogen'. The conversion is done by enzymes inside cells, such as liver cells and muscle cells.

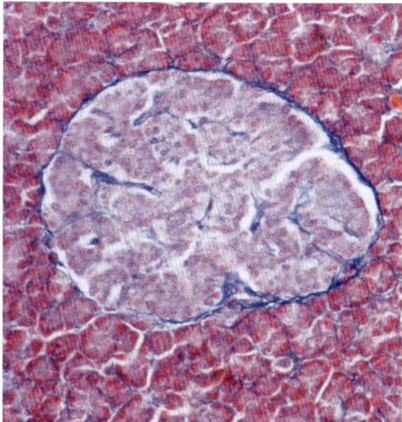

Figure 14.7.3 The tissue in the centre of this photo of the pancreas secretes insulin and glucagon.

KEY POINTS

1 Homeostasis is the maintenance of a constant internal environment.

2 Homeostasis involves negative feedback. It operates to maintain constant conditions of glucose, water, carbon dioxide and oxygen in the blood as well as its pH.

3 The pancreas controls the concentration of glucose in the blood by producing the hormones insulin and glucagon, which stimulate the liver.

14.8 Controlling body temperature

Supplement

LEARNING OUTCOMES

- Identify different structures on a diagram of the skin
- Describe how body temperature is kept constant

EXAM TIP

We <u>produce heat</u> in respiration and by shivering; we <u>lose heat</u> to our surroundings, e.g., by evaporation when we sweat; we <u>conserve heat</u>, e.g., by having fat beneath the skin.

Birds and mammals maintain a constant body temperature. They are very successful animals as they can live in cold environments as well as hot ones and be active at night as well as during the day. The skin provides ways to adjust the body temperature in different conditions.

The skin

The skin is the largest organ in the body and has several important functions It:

- protects the body from damage
- stops pathogens from entering
- prevents too much water loss
- detects changes in temperature
- detects pressure (touch) and pain
- loses heat by conduction, convection, radiation and evaporation.

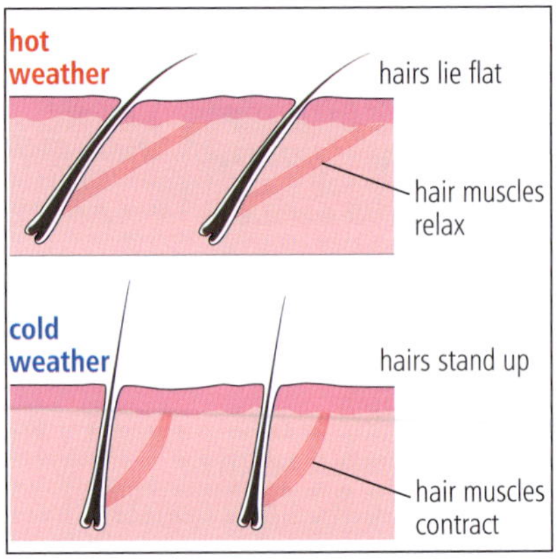

Figure 14.8.2 Hairs are involved in maintaining body temperature.

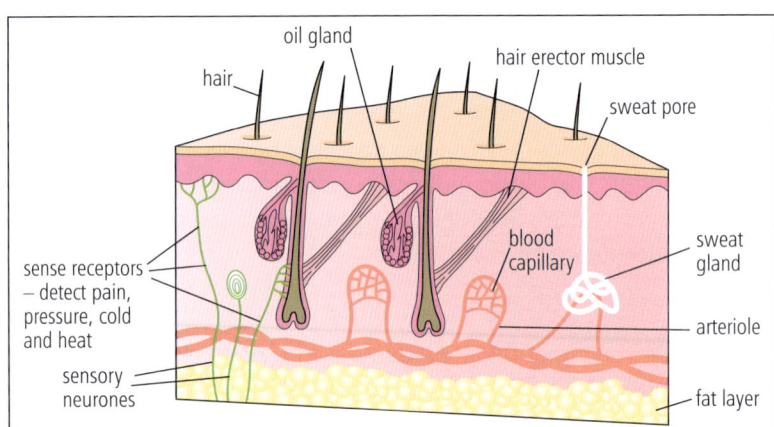

Figure 14.8.1 The structure of human skin.

Many of the structures in the diagram help in controlling body temperature. We lose and absorb heat through our skin – mostly by radiation. We use our skin for heat loss when we are hot and to conserve heat when we are cold.

Mammals have fur or hair, which traps a layer of air close to the skin. Air is a poor conductor of heat. So air pockets in the fur or hair reduce the loss of heat to the atmosphere by keeping convection currents of air away from the warm surface of the skin. When it is cold, hair erector muscles *contract* so that hairs stand erect to trap a thick layer of air so reducing heat loss.

When it is hot, hair erector muscles *relax* so that hairs lie flat. Less air is trapped close to the skin so more heat can be lost as convection currents flow closer to the surface of the skin.

Figure 14.8.3 Polar bears have a thick layer of insulating fat to conserve their body heat.

Mammals also have a layer of fatty tissue beneath their skin. Fat is a good insulator because it does not conduct heat well. Mammals that live in cold environments have very thick layers of fat.

Controlling body temperature

No matter what the weather is like, your body temperature stays very close to 37 °C unless you have a fever. There are receptors in the brain that detect changes in the temperature of the blood. When the brain detects any change in the blood temperature, it coordinates the action of effectors to restore the blood temperature to normal.

In the heat, the brain detects an increase in the blood temperature. The brain sends impulses to the skin to increase the rate of sweating so that more energy is lost by evaporation from the skin surface.

In the cold, the brain detects a decrease in the blood temperature. The sweat glands stop producing sweat and the hair erector muscles contract to raise the hairs. The body starts to shiver as some of the body muscles contract spontaneously and release heat from respiration. Blood flows through the muscles and is warmed by this heat.

A very effective way to regulate body temperature is to change how much blood flows through the capillaries near to the surface of the skin. Arterioles supply these capillaries with blood. The muscle around the arterioles can contract and relax to change the blood flow. To cool down these muscles in the walls of the arterioles relax. The vessels widen and there is an increase in blood flow through the capillaries. This is known as **vasodilation**. More energy is now transferred to the surroundings by convection and radiation.

In the cold, the muscles in the walls of the arterioles contract so the vessels become narrower. This reduces the blood flow through the capillaries and the blood is diverted to stay beneath the fat layer in the skin. This is known as **vasoconstriction**.

The control of body temperature is by negative feedback as shown in Figure 14.8.4. The brain coordinates the responses by effector organs to lose, conserve or gain heat in order to maintain a body temperature that stays within set limits.

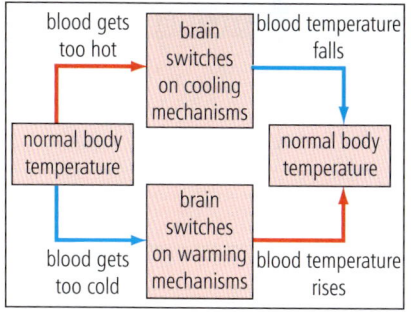

Figure 14.8.4 Control of body temperature

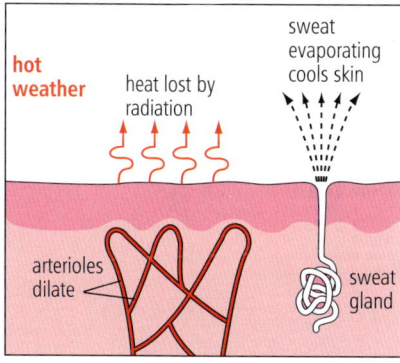

Figure 14.8.5 Vasodilation in skin

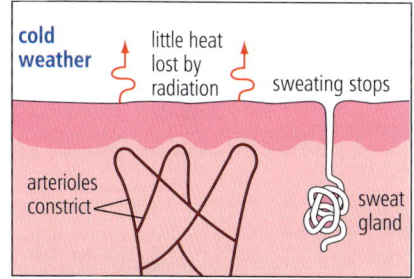

Figure 14.8.6 Vasoconstriction in skin

KEY POINTS

1 The brain controls body temperature by using the many structures in the skin.

2 Insulation conserves heat, sweating loses heat and shivering generates heat.

3 Vasoconstriction and vasodilation control how much energy is lost from the blood in the skin.

SUMMARY QUESTIONS

1 Explain how the brain controls body temperature.

2 Describe what happens in the skin:
   a to cool down when we are hot
   b to conserve heat when we are cold.

3 How do we produce heat in the body if we are cold?

4 Explain how fur and fat tissue beneath the skin help in temperature control in cold environments.

5 Explain how vasodilation and vasoconstriction are involved in maintaining body temperature.

14.9 Tropic responses

LEARNING OUTCOMES

- Describe gravitropism and phototropism as responses of plants to gravity and light
- **S** • Explain the chemical control of plant growth by auxins

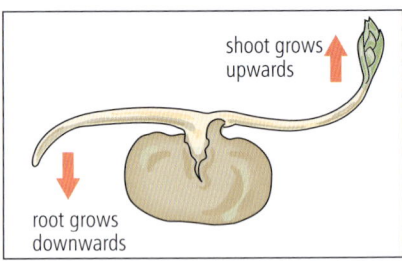

shoot grows upwards

root grows downwards

Figure 14.9.1 Which way is up?

EXAM TIP

A seed germinates. A root and then a shoot appear. Which way is up? Gravitropic responses ensure the survival of the seedling as it establishes itself in competition with other plants nearby.

EXAM TIP

Phototropism and gravitropism are examples of the chemical control of plant growth. The chemical auxin is an example of a plant hormone.

Tropisms

Plants respond to stimuli such as light and gravity. Different parts of the plant either grow towards or away from the different stimuli. These growth responses are much slower than the responses coordinated by the nervous systems in animals. They are called **tropisms**.

Gravitropism

Whichever way a seed is planted, the root always grows downwards in the direction of gravity. The shoot always grows upwards against the direction of gravity.

A growth response to gravity is called a **gravitropism**.

Roots are **positively gravitropic** because they grow in the same direction as the stimulus, gravity. They grow downwards into the soil to absorb water and mineral ions.

Shoots, on the other hand, are **negatively gravitropic**. They will always grow in the opposite direction to the direction of gravity. This ensures that shoots grow upwards getting the leaves into sunlight.

Supplement

Auxin

A plant hormone called **auxin** controls this growth. Auxin in shoots stimulates growth by causing the cells to elongate, but in roots auxin inhibits growth by slowing down cell elongation.

If a seedling is put on its side, the auxin builds up on the lower side of the shoot and root. In the shoot, the auxin on the lower side stimulates the cells there to elongate. This means the cells grow more on the lower side. This causes the shoot to bend upwards.

In the root, the auxin also builds up on the lower side. But auxin slows down growth in a root by inhibiting cell elongation. As a result, cells on the upper side of the root elongate more than those on the lower side so the root bends downwards.

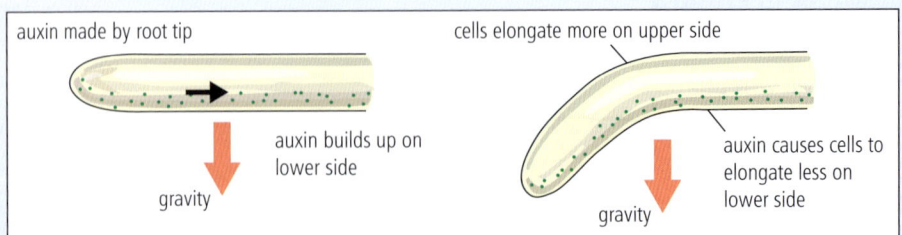

auxin made by root tip

cells elongate more on upper side

auxin builds up on lower side

gravity

auxin causes cells to elongate less on lower side

gravity

Figure 14.9.2 Auxin controls the positive gravitropism of roots.

Phototropism

The cress seedlings in the photo have been put in a window for some time and their shoots have grown towards the light.

This sort of growth response is called a **phototropism**. Because shoots grow *towards* the light they are **positively phototropic**. The advantage of this tropic response is that the leaves expose the maximum surface area to the light. They can then absorb light more efficiently for photosynthesis.

Figure 14.9.3 Roots grow downwards for anchorage and to absorb water and mineral ions.

Supplement

This response is controlled by auxin made in the shoot tip. The auxin diffuses away from the shoot tip down the stem. When a shoot receives light only from one side the auxin is unequally distributed so most is found on the shaded side of the stem. The auxin stimulates cells on the shaded side to elongate more than the cells on the side in direct light. The shoot grows more on the shaded side and bends towards the light. Roots are not sensitive to light.

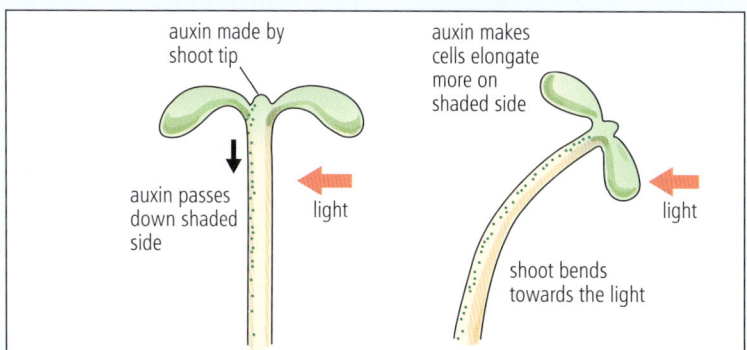

auxin made by shoot tip

auxin makes cells elongate more on shaded side

auxin passes down shaded side

light

light

shoot bends towards the light

Figure 14.9.4 Auxin controls the positive phototropism of shoots.

EXAM TIP

Cells become longer, or elongate, when they absorb water by osmosis. The vacuole swells, increasing the turgor pressure inside the cell, which causes the cell wall to stretch. Auxins work by making it easier for the walls to stretch.

Figure 14.9.5 These cress seedlings have grown towards the light.

SUMMARY QUESTIONS

1 Describe what is meant by each of the following:
   a a tropic response **b** phototropism **c** gravitropism

2 State what growth response will happen in each of the following.
   a Some seedlings are placed in a window, in sunlight for 6 hours.
   b Some germinating seeds are pinned to a board with their roots lying horizontally for a few hours.

S 3 Explain the observed growth responses in question **2** in terms of the action of auxin.

KEY POINTS

1 A gravitropism is a growth response to the stimulus of gravity. Roots are positively gravitropic and shoots are negatively gravitropic.

2 A phototropism is a growth response to the stimulus of light. Shoots are positively phototropic and grow towards the light.

S 3 Auxins are plant hormones that control growth.

14.10 Investigating tropic responses

- Investigate gravitropism and phototropism in shoots and roots

Figure 14.10.1 Tomato plant left on its side for 24 hours.

The potted plant in the photograph was placed in a horizontal position. You can see what has happened after a period of 24 hours. The plant stem has grown upwards away from the stimulus of gravity. The plant stem is said to be **negatively gravitropic**.

If the roots of this plant were examined, you would see that they will have changed their direction of growth to grow downwards. So the plant's roots are said to be **positively gravitropic**.

We can carry out some simple experiments to investigate tropic responses in plants.

Investigating gravitropism in pea radicles

1 Soak some peas in water to get them to germinate. Collect about six pea seedlings with straight radicles.

  Grow them in a vertical roll of moist filter paper so that the radicles grow straight downwards.

2 Pin the six pea seedlings in a horizontal position to the cork turntable of a **clinostat**.

3 Then pin another six pea seedlings in a horizontal position to the cork lid of a glass jar. Use some modelling clay to keep the jar in a horizontal position.

4 A clinostat is a piece of apparatus that works by electricity or clockwork. It has a turntable that rotates four times every hour so it will neutralise the effect of gravity on the seedling radicles. That is, the sides of each radicle will be exposed to gravity for the same period of time.

5 The clinostat and the glass jar are then left in darkness for two days.

6 You can see from the diagram that the pea radicles in the clinostat continued to grow horizontally. This is because the rotation of the clinostat allowed gravity to act equally on all sides of the radicles.

  The radicles in the glass jar have grown downwards towards gravity so they are said to be **positively gravitropic**.

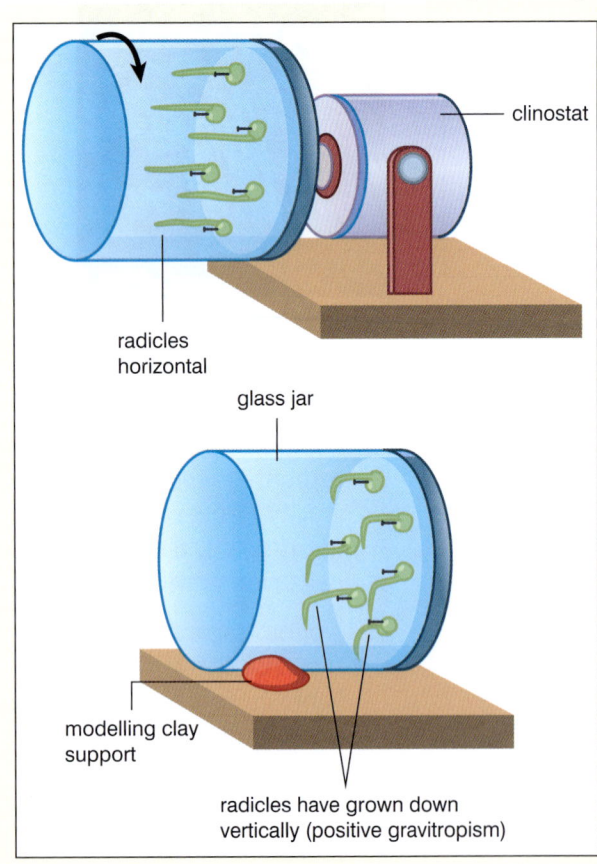

Figure 14.10.2 Apparatus for investigating gravitropism in pea radicles

Investigating phototropism in cress seedlings

1 Grow two sets of seedlings on wet cotton wool in Petri dishes in the dark.

2 Cover the seedlings in diagram A with a cardboard box with a slit in the side that allows the light to enter from only one direction so the seedllings are exposed to **unilateral light**.

3 Place the cress seedlings on a clinostat arranged in a vertical position. The clinostat rotates four times every hour, exposing each side of the cress seedlings to the light.

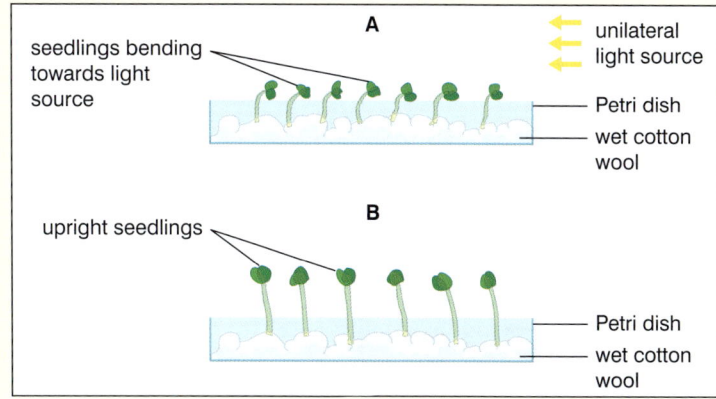

Figure 14.10.3 Apparatus for investigating phototropism in cress seedlings

4 After two days you can see that cress seedlings exposed to unilateral light grew towards the light source.

They were positively phototropic.

The seedlings on the clinostat were not exposed to unilateral light so continued to grow vertically

Advantages of tropic responses

Positive phototropism in stems means that the leaves are brought into the best position to absorb light for photosynthesis. Also the flowers are brought into a position where they are more likely to be pollinated by flying insects or their pollen can be carried by the wind.

Positive gravitropism in roots ensures that the roots grow towards gravity so penetrating the soil and providing anchorage for the plant. It also allows the roots to absorb water and mineral ions in the soil.

SUMMARY QUESTIONS

1 a What is a clinostat?
  b Explain how it can be used to study **i** gravitropic response in roots and **ii** phototropic response in stems.

2 Name the scientific term that describes:
  a a stem growing towards light
  b a root growing towards gravity
  c a stem growing away from gravity.

3 Explain how phototropic responses and gravitropic responses benefit the growth, development and reproduction of a flowering plant.

Practice questions

1. In which pathway do nerve impulses travel in a spinal reflex?

   A effector $\longrightarrow$ sensory neurone $\longrightarrow$ relay neurone $\longrightarrow$ receptor $\longrightarrow$ motor neurone

   B motor neurone $\longrightarrow$ relay neurone $\longrightarrow$ sensory neurone $\longrightarrow$ effector

   C receptor $\longrightarrow$ sensory neurone $\longrightarrow$ effector $\longrightarrow$ relay neurone $\longrightarrow$ motor neurone

   D receptor $\longrightarrow$ sensory neurone $\longrightarrow$ relay neurone $\longrightarrow$ motor neurone $\longrightarrow$ effector

   (Paper 1) [1]

2. Where are relay neurones situated?

   A effector   C spinal nerve

   B spinal cord   D receptor

   (Paper 1) [1]

3. A nerve impulse travels from a fingertip to the spinal cord in 0.02 s. This is a distance of 1.4 metres. The speed of the impulse is:

   A $0.7 \, \text{m s}^{-1}$ C $70 \, \text{m s}^{-1}$

   B $7 \, \text{m s}^{-1}$ D $170 \, \text{m s}^{-1}$

   (Paper 1) [1]

4. Which shows the correct match between the hormone and its effect?

   A adrenaline – decrease in heart rate

   B insulin – increase in blood glucose concentration

   C oestrogen – decrease in the volume of urine

   D testosterone – stimulates development of sperm

   (Paper 1) [1]

5. Which occur when in a hot environment?

	arterioles in the skin	sweat glands	hair erector muscles
A	constrict	active	relax
B	constrict	not active	contract
C	dilate	active	relax
D	dilate	active	contract

   (Paper 2) [1]

6. The diagram shows the response of a plant shoot to receiving light from one side only.

   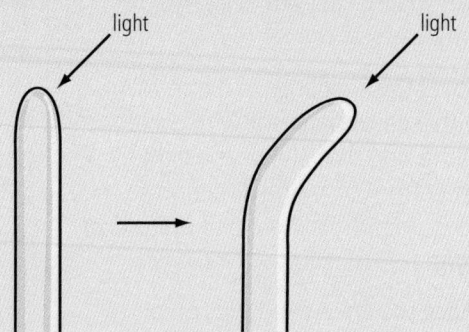

   Which statement explains how this response is controlled?

   A Auxins stimulate a gravitropic response.

   B Auxins stimulate a phototropic response.

   C Auxins stimulate cells on the side closest to the light to grow faster than the cells on the opposite side.

   D Auxins stimulate cells on the side away from the light to grow faster than the cells on the opposite side.

   (Paper 2) [1]

7. Which events occur in the eye when focusing on an object near to the face?

   A suspensory ligaments slacken and the lens becomes thicker

   B suspensory ligaments slacken and the lens becomes thinner

   C suspensory ligaments tighten and the lens becomes thicker

   D suspensory ligaments tighten and the lens becomes thinner

   (Paper 2) [1]

8. A man injures his arm in an accident. Afterwards, he can feel objects touching his hand, but he cannot move his hand away from them. The cause of this could be that:

   A all nerves between his hand and the spinal cord are damaged

   B all nerves between his spinal cord and the effectors in his arm are cut

C all nerves between the receptors in his hand and his spinal cord are cut

D all the receptors in his hand are damaged

(Paper 2) [1]

9 The diagram shows a horizontal section of the eye.

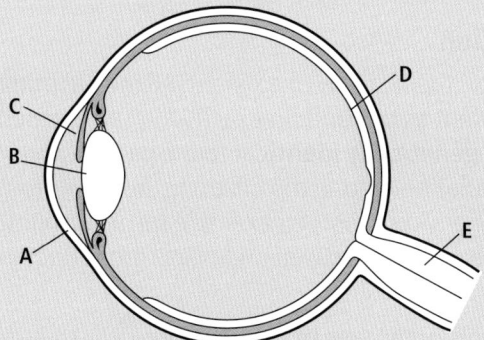

(a) Name the parts labelled **A** to **E**. [5]

(b) Explain why the eye is a sense organ. [2]

(c) A person walks from a dark room into a brightly lit place.

   (i) Describe the change that you would see happening to the person's pupils. [1]

   (ii) Suggest the advantage of the change you described in **(i)**. [1]

(Paper 3)

10 (a) Copy and complete the flow chart that shows part of the control of the concentration of glucose in the blood.

> increase in blood glucose concentration
>
> ↓
>
> detected by …
>
> ↓
>
> insulin secreted
>
> ↓
>
> liver cells respond by …

[2]

(b) Write out a similar flow chart to show the body's response to a decrease in blood glucose concentration. [3]

(c) Explain the principle of negative feedback, using the control of glucose in the blood as an example. [5]

(Paper 4)

11 A scientist investigated the distribution of rods and cones across the retina of a mammal as shown in the diagram. The graph shows the distribution of cones in the retina.

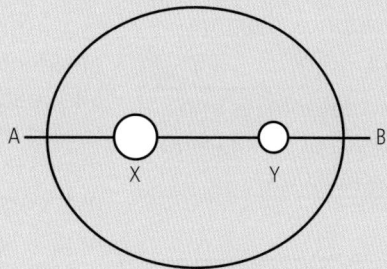

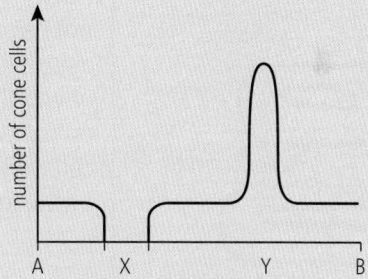

(a) Name the parts of the retina **X** and Y. Give reasons for your answers. [4]

(b) Copy the graph and show on it the distribution of rods in the retina. [3]

(c) (i) Describe the roles of rods in the eye. [3]

   (ii) There are three types of cone in the eye. Explain the functions of these different types of cone. [3]

(d) Explain the advantages to an animal of having two coordination systems: the nervous system and the endocrine system. [3]

(Paper 4)

12 (a) Describe the events that occur at a synapse when a nerve impulse arrives at the end of a sensory neurone. [4]

(b) There are huge numbers of synapses in the spinal cord and the brain. Explain why this is so. [3]

(Paper 4)

15.1 Asexual and sexual reproduction

LEARNING OUTCOMES

- Define *asexual* and *sexual reproduction*
- Identify asexual reproduction from descriptions, diagrams and photographs
- **S** Discuss the advantages and disadvantages of asexual and sexual reproduction

Reproduction means producing new living organisms. Animals and plants reproduce to make new individuals of the same species. There are two types of reproduction: **asexual** and **sexual reproduction**.

Asexual reproduction

In asexual reproduction there is only *one* parent. All the offspring are identical to the parent as they inherit exactly the same genetic information. They are **genetically identical** to the parent, which means there is little variation among the offspring. Any variation is due to the effect of the environment, for example the availability of nutrients and water determine how well organisms grow. Bacteria, fungi and plants reproduce asexually.

Bacteria are microscopic organisms made up of one cell. They do not have a nucleus, but a loop of DNA. When bacteria reproduce asexually, the DNA loop is copied so that there is some for each new cell. You can find some information about how DNA is copied on page 204. The bacterial cell divides into two by making a new cell wall. This type of asexual reproduction is called **binary fission**, because the parent cell splits into two.

Pin mould is a fungus that grows on bread (see pages 6 and 7). It reproduces asexually by making **spores**. These are small and light, like specks of dust, and they float through the air. When a spore lands on a damp surface, it splits open and a thread grows out. This thread or hypha, grows over the surface of bread forming a dense network of threads called a mycelium. Eventually, short hyphae grow upwards and produce spore cases or sporangia at their tips. Inside each sporangium hundreds of new spores are formed asexually by division of the nuclei. Each nucleus gains a small quantity of cytoplasm and a protective spore case. When ready, the sporangia break open and the spores are dispersed.

Potatoes reproduce asexually by means of stem tubers, which are swollen underground stems that grow from the parent plant. Sucrose is transported in the phloem from the leaves into these underground stems that swell as they convert the sucrose into starch. The parent plant dies at the end of the growing season leaving the tubers in the ground to survive over the winter.

However, farmers dig them up and sell them. Some are kept as 'seed' potatoes to plant the next year. All the tubers produced from one parent plant are genetically identical. When they start growing, new shoots and roots emerge from the growing points of the potato, which are known as 'eyes'.

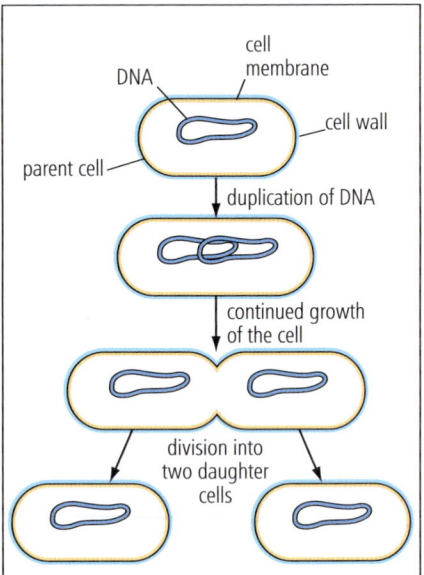

Figure 15.1.1 Binary fission in bacteria. Each new bacterial cell is genetically identical to each other and to the parent cell.

Figure 15.1.2 A seed potato showing growth points (one is circled).

Sexual reproduction

In sexual reproduction there are two parents. The parents have **sex organs**. The sex organs make sex cells or **gametes**. In animals, the male gametes are **sperm cells**. In flowering plants, the male gametes are nuclei inside **pollen grains**. In animals, the female gametes are **egg cells**. In flowering plants, the female gametes are inside structures called **ovules**. During sexual reproduction, the nuclei of the gametes fuse together at **fertilisation**. The fertilised egg or **zygote** divides to form an **embryo**, which may grow into a new individual plant.

At fertilisation, half the genetic material comes from the male gamete and half comes from the female gamete. The nuclei in the gametes each contain one set of chromosomes. A zygote has a nucleus that contains two sets of chromosomes – one from each gamete. The offspring are not genetically identical to the parents. Each zygote receives half its genes from its male parent and half from its female parent. This means that sexual reproduction brings about **variation** in the offspring.

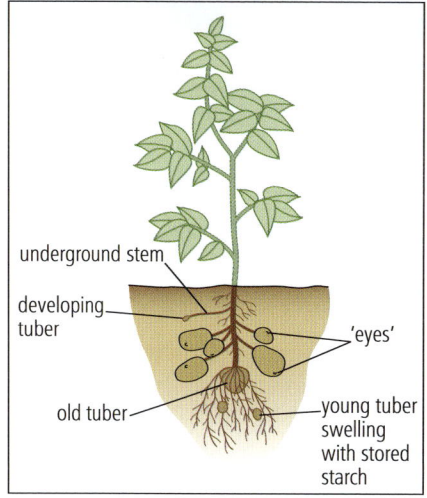

Figure 15.1.3 The potato plant.

Supplement

Advantages and disadvantages of asexual reproduction in plants

An advantage of asexual reproduction is that it is fast. Organisms can reproduce rapidly to spread and colonise areas where the parent plants grow. Plants grow more rapidly from tubers than from seeds.

If left in the ground, potato tubers grow into new plants in the same place as the parent, but compete with each other for resources. Seeds produced in sexual reproduction are dispersed to new areas where there may not be as much competition. However, many seeds land in unsuitable places for growth so many are wasted.

There is very little, if any, variation as a result of asexual methods of reproduction. This can be a disadvantage if these individuals are affected by a certain disease. Since they are all genetically identical, none may have resistance to the disease and they may all be killed. Disease could spread very quickly through a crop monoculture, in which all the plants are genetically identical. If environmental conditions change, it is unlikely that there will be individuals with features adapted to the new conditions. Sexual reproduction gives rise to variation. In the wild, many individuals may not be as well adapted to the existing conditions as their parents, but may be well adapted if the environment changes. If there is plenty of variation, species are better able to survive and evolve.

Figure 15.1.4 Potato plants reproduce asexually by producing tubers, as well as sexually with flowers.

SUMMARY QUESTIONS

1 Explain each of the following:
   a gamete   b fertilisation
   c zygote   d embryo

2 Describe each of these examples of asexual reproduction:
   a binary fission in bacteria
   b spore production in fungi
   c tuber formation in potatoes

S 3 Make a table to compare the advantages and disadvantages of asexual reproduction and sexual reproduction in plants.

KEY POINTS

1 Asexual reproduction results in the production of genetically identical offspring from one parent.

2 Sexual reproduction involves the fusion of male and female nuclei to form a zygote, producing offspring that are genetically different from each other and their parents.

3 Binary fission in bacteria, spore production in fungi and tuber formation in potatoes are all examples of asexual reproduction.

S 4 Asexual reproduction may produce many individuals rapidly, but it does not result in variation of offspring as in sexual reproduction.

15.2 Flower structure

LEARNING OUTCOMES

- Describe the structure of an insect-pollinated flower
- State the functions of the sepals, petals, stamens and carpels
- Describe the structure of a wind-pollinated flower
- Compare the adaptations of different types of pollen grain

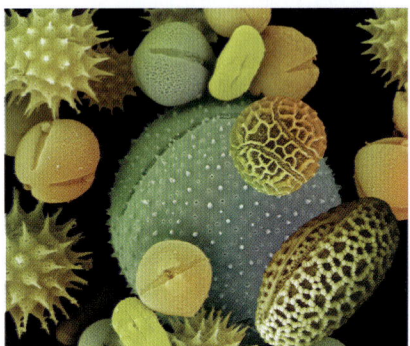

Figure 15.2.2 The spiky pollen grains are from insect-pollinated flowers and the smooth grains are from wind-pollinated flowers.

Figure 15.2.3 The Titan arum from Sumatra in Indonesia produces a smell of rotting flesh to attract flies – the flies jump in, slide down the sticky walls and are covered in pollen; after a few days the walls of the plant change, allowing the flies to climb out to visit another flower of the same species.

Flowering plants carry out sexual reproduction by producing flowers, which have male and female parts. The male parts make pollen grains to carry male gametes to the female parts. The transfer of pollen grains is **pollination**. Female gametes are deep inside the female part and pollen grains grow a tube to reach them. Each pollen tube delivers a male gamete so it can fuse with the female gamete to form a **zygote**. Fusion of male and female gametes is **fertilisation**.

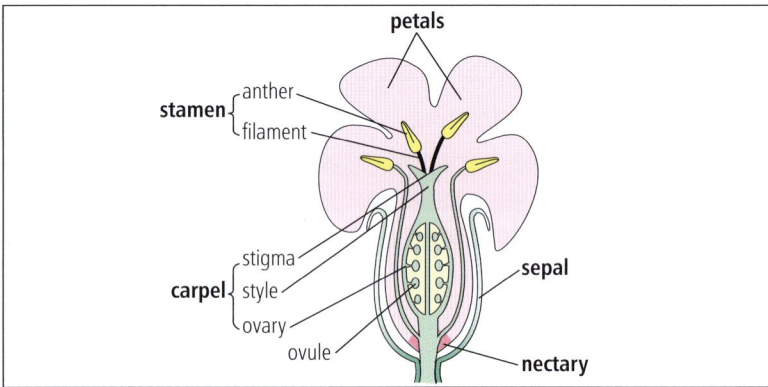

Figure 15.2.1 A half flower drawing of an insect-pollinated flower.

Flowers vary in structure depending upon their method of pollination.

Insect-pollinated flowers

Flower parts are often arranged in rings attached to the end of a swollen flower stalk. The sepals form the outer ring and the carpels form the inner ring.

- Sepals are leaf-like structures that protect the flower when it is a bud.
- Petals are brightly coloured and scented. Many have a nectary at the base, which makes sugary nectar. Visiting insects land on the petals to feed on nectar.
- Stamens are the male sex organs. Each one is made up of two parts: the anther, where the pollen is made, and the filament, a stalk, which holds the anther. In the flower shown in Figure 15.2.1, there are four stamens (eight in the full flower).
- Carpels are the female sex organs. Each carpel is made up of a stigma, a style and an ovary.
- The ovules are inside the ovary. Each ovule has a female gamete – the egg cell.

The pollen grains of insect-pollinated flowers are often spiky like those in Figure 15.2.2. They are also often sticky. The grains are adapted to attach to hairs on the surfaces of insect bodies. The grains become dislodged onto a stigma when an insect enters another flower of the same species. The flower in Figure 15.2.1 has a stigma in the centre of the flower, which an insect will push past on its way towards the nectary.

Wind-pollinated flowers

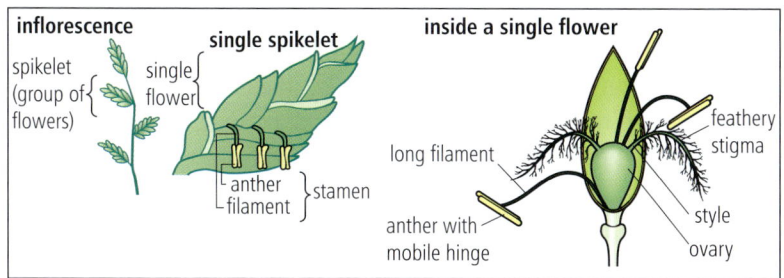

Figure 15.2.4 This species of rye grass is an example of a wind-pollinated flower.

Grasses and cereals are all pollinated by the wind. Rye grass, *Lolium perenne*, has many features typical of a wind-pollinated flower. The flowers are small, green and inconspicuous compared with insect-pollinated flowers. They have no scent and no nectar.

The anthers produce pollen grains that are small, smooth and light so that they are easily carried on the wind. These flowers make huge numbers of pollen grains as the chances of pollen grains landing on a stigma are very small. The anthers hang outside the flower so the wind can blow away pollen. The feathery stigmas are also positioned outside the flower. They act like a net, providing a large surface area for catching pollen grains that get blown into them. Wind pollination can waste a lot of pollen, which is why the anthers produce so much. But the fact that plants such as grasses and cereals grow close to one another means that pollen is more likely to be transferred than if they were separated by large distances.

Figure 15.2.5 Long anthers hanging out of rye grass flowers.

Figure 15.2.6 Two flowers of *Bromus*, showing large, loosely attached anthers and feathery stigmas.

KEY POINTS

1 Insect-pollinated flowers have sepals to protect the flower in the bud and bright petals to attract insects. The stamens consist of anthers and filaments. Each carpel is made up of a stigma, style and ovary.

2 Wind-pollinated flowers are inconspicuous, with feathery stigmas and anthers that hang outside the flower.

3 Wind-pollinated flowers produce large quantities of light, smooth pollen that is easily carried on the wind. Insect-pollinated flowers produce smaller quantities of sticky and spiky pollen that can attach to insects.

SUMMARY QUESTIONS

1 Copy and complete the sentences using these words (you may use some more than once):

| carpels | pollen | stigma |
| anther | ovary | stamens |

The male parts of a flower are called the _____ . Each is made up of an _____ and a filament. The _____ grains are made inside the _____ . The female parts of a flower are called the _____ . Each is made up of a _____ , style and _____ .

2 State the functions of the parts of an insect-pollinated flower: sepals, petals, stamens, filaments, anthers, carpels, stigma, style, ovary and ovules.

15.3 Pollination

LEARNING OUTCOMES

- Define the term *pollination*
- Name the agents of pollination
- Compare the different structural adaptations of insect-pollinated and wind-pollinated flowers
- **S** Distinguish between self-pollination and cross-pollination and discuss their significance

Pollination is the transfer of pollen grains from the anther to a stigma. Pollination may be carried out by insects or by the wind, or by other animals such as birds and bats. Flowers are adapted to one of these methods.

When an anther is ripe, it splits open along its length and releases its pollen so that pollination can occur. Pollination is needed in order to bring the male gamete (inside a pollen grain) near to the female gamete so that **fertilisation** can occur.

Insect pollination

Look at Figure 15.3.1. The flower has a number of adaptations for insect pollination.

The flower is pollinated by bees, which land on the petals. Bees have long tongues and are able to reach the nectaries at the base of the petals.

Figure 15.3.2 Honey bee with pollen on its legs.

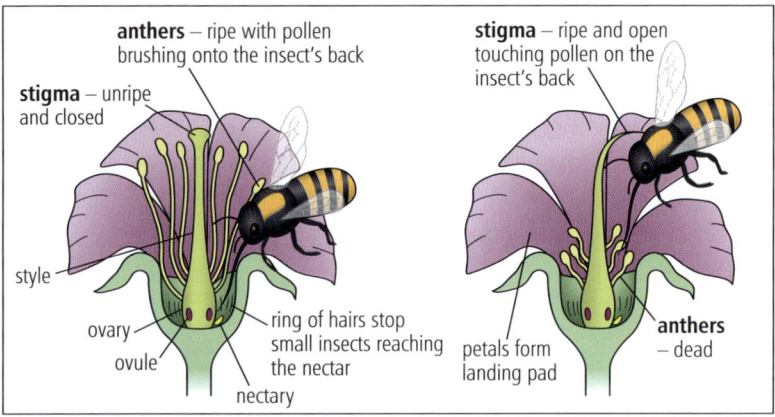

anthers – ripe with pollen brushing onto the insect's back

stigma – unripe and closed

style

ovary

ovule

ring of hairs stop small insects reaching the nectar

nectary

stigma – ripe and open touching pollen on the insect's back

petals form landing pad

anthers – dead

Figure 15.3.1 A bee transfers pollen from one flower to another of the same species.

The bees feed on the sugary nectar made in the nectaries. The anthers are positioned in such a way that sticky pollen from them will brush against the bee's back as it pushes its head down to the base of the petals. When the bee enters another flower, it brushes some of the pollen against the ripe stigma and pollination is achieved.

Wind pollination

Look at Figure 15.3.3, a diagram of wind pollination. The anthers hang outside the flower so that they release their pollen when the wind blows them.

The stigmas are feathery and are also found outside the flower, where they act as a net to catch pollen grains in the air. Wind-pollinated flowers have light, smooth pollen grains that can easily be carried by the wind to another flower. The anthers produce many pollen grains so that some, by chance, will land on the stigma of a flower of the same species. Most of the pollen produced will be lost.

EXAM TIP

When you look at insect-pollinated flowers with a hand lens, identify how their structure helps them to be pollinated by insects. You should be able do this for a flower you have not seen before.

The structural features of insect- and wind-pollinated flowers are compared in the table.

Table 15.3.1 Comparing insect-pollinated and wind-pollinated flowers.

feature	insect-pollinated flowers	wind-pollinated flowers
petals	present – colourful and scented to attract insects	absent or very small and difficult to see
nectaries	present – make nectar, which is a sugary liquid food for pollinating insects	absent
stamens	present – usually with short filaments; anthers attached firmly to filaments; inside the flower for insects to rub against	long filaments so anthers hang outside the flower; anthers loosely attached to the filaments so pollen is easily blown away
pollen	small quantities of sticky, spiky pollen grains that stick easily to insects' bodies	large quantities of smooth, light pollen that can easily be carried by the wind
carpels	sticky, small stigmas usually inside the flower for insects to rub against	large, feathery stigmas to catch pollen grains in the air

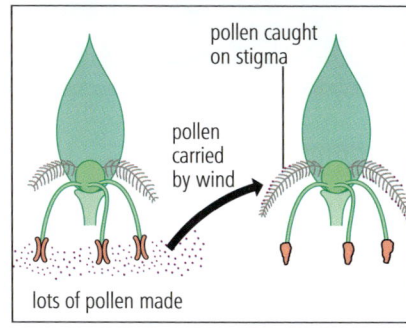

Figure 15.3.3 Wind pollination.

Figure 15.3.4 Hazel catkins release pollen to be carried by the wind.

Supplement

Pollination and variation

Self-pollination occurs when pollen is transferred from the anther to a stigma of the same flower or to a different flower but on the same plant.

Cross-pollination occurs when pollen is transferred to a stigma of another plant of the same species. Cross-pollination is of a greater benefit to a species of plant, since it ensures exchange of genetic material between different plants and results in greater **variation** upon which **natural selection** can operate (see Unit 17).

Self-pollination results in much less variation, since genetic material is not exchanged with different plants but with the same one. However, self-pollination is an advantage if there are no pollinating insects, and for plants growing in isolation from others of the same species.

KEY POINTS

1 Pollination is the transfer of pollen grains from the anther of the male part of the flower to the stigma of the female part of the flower.

2 Insect-pollinated flowers have bright petals, scent and nectar. Wind-pollinated flowers are inconspicuous with feathery stigmas and anthers that hang outside the flower.

S 3 Self-pollination involves the transfer of pollen from the anthers to the stigma of the same flower or a different flower on the same plant. Cross-pollination involves the transfer of pollen from the anthers of one flower to the stigma of another flower on a different plant of the same species.

SUMMARY QUESTIONS

1 a Give two ways in which pollen of insect-pollinated flowers is different from the pollen of wind-pollinated flowers.

b Write down three other differences between insect-pollinated and wind-pollinated flowers.

S 2 a What is the difference between self-pollination and cross-pollination?

b State the advantages of cross-pollination over self-pollination.

15.4 Fertilisation and seed formation

LEARNING OUTCOMES

- State that fertilisation in flowering plants occurs when a pollen nucleus fuses with a nucleus in an ovule

S • Describe the events that lead to fertilisation

- State the requirements for germination of seeds

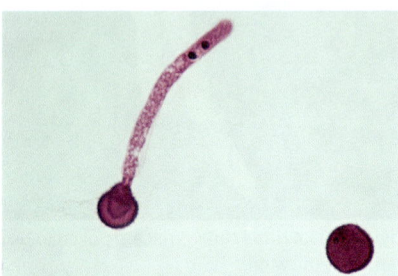

Figure 15.4.1 A pollen grain and a germinated pollen grain (left) with its pollen tube and male gamete nuclei.

EXAM TIP

Remember that the male gamete is inside the pollen grain and is delivered to the female gamete by the pollen tube. The pollen grain is <u>not</u> the male gamete.

Figure 15.4.3 Scanning electron micrograph of pollen grains and pollen tubes of lily, <u>Lilium</u> sp.

Fertilisation

Pollination is complete when the pollen grains land on the stigma of the female part of a flower. If a pollen grain lands on a ripe stigma it starts to grow a pollen tube to take the male nucleus to the female nucleus. Fertilisation occurs when the male nucleus fuses with the female nucleus. This occurs inside the ovule.

Supplement

Each pollen grain grows a **pollen tube** to take the male nucleus to the ovule. As it grows, the pollen tube gains nutrition from the tissues of the style and carries the male gamete nucleus with it. The first pollen tube to reach the ovary enters the ovule through a small hole – the **micropyle**. The male gamete nucleus then fuses with the egg cell nucleus. This is fertilisation in which a zygote is formed by fusion of the two nuclei.

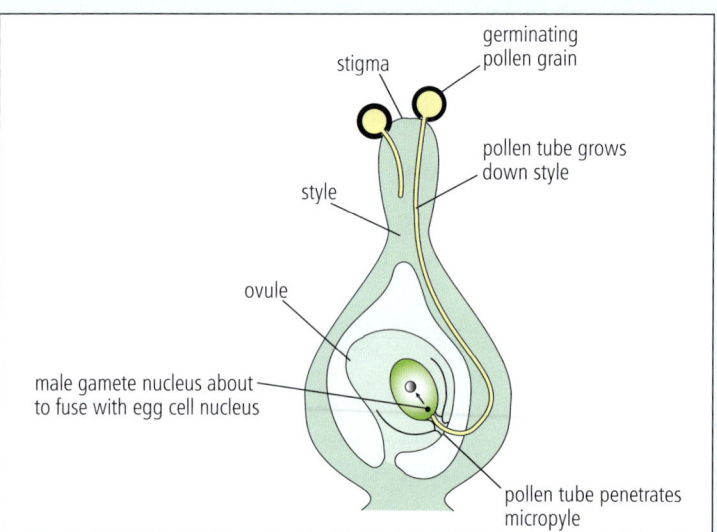

Figure 15.4.2 Fertilisation occurs within the ovule.

PRACTICAL

Growing pollen tubes

Find some flowers with anthers covered in pollen. Lily flowers are particularly suitable.

Cut some small squares of damp Visking (dialysis) tubing and transfer them to filter paper soaked in a nutrient solution rich in sucrose. Use a paintbrush to transfer some pollen grains from the anthers to the squares of dialysis tubing. Leave in a covered Petri dish for 24 to 48 hours. Remove the squares carefully with a pair of forceps, put on a microscope slide and observe.

Seed formation

After fertilisation, the zygote divides and grows into the embryo. The ovule forms the seed with the embryo inside it.

The ovary forms the fruit with the seeds inside it. Many of the parts of the flower are not needed when fertilisation has occured, so the sepals, petals and stamens wither and fall off. They have completed their functions.

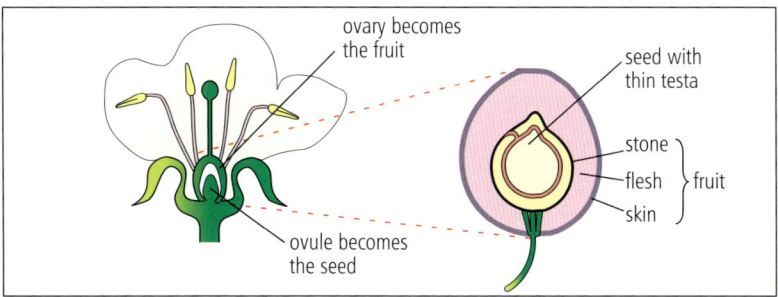

Figure 15.4.4 How a plum flower grows into a plum fruit.

Each fertilised ovule grows to form a seed. Each seed is made up of:

- the embryo
- an energy store
- a seed coat or testa.

The ovary forms the **fruit**.

Conditions for germination

Three conditions are needed for seeds to germinate.

- Water is needed for seeds to swell. This swelling breaks the seed coats of some seeds. Cells absorb water, develop vacuoles and expand. Cell expansion makes the radicle (embryonic root) grow out from the seed. Water is also needed to dissolve the soluble products made by enzymes from the food stores. Water also activates these enzymes.
- Oxygen is needed for aerobic respiration to provide the embryo with energy.
- A suitably warm temperature is needed so that enzymes can work efficiently.

PRACTICAL

Conditions for germination

You can investigate the conditions necessary for germination with mung beans. Set up the tubes as shown and leave them for a week, checking them every day to see if germination has started.

Only the seeds in **A** have all the conditions needed for germination.

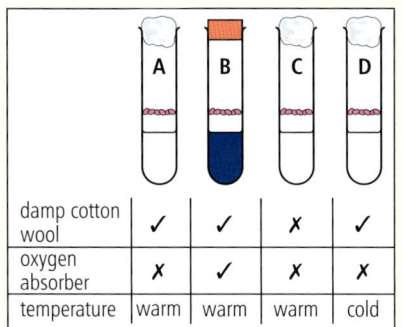

	A	B	C	D
damp cotton wool	✓	✓	✗	✓
oxygen absorber	✗	✓	✗	✗
temperature	warm	warm	warm	cold

Figure 15.4.5 Apparatus for investigating germination conditions.

KEY POINTS

1 Fertilisation occurs when the male gamete nucleus from the pollen fuses with the female gamete nucleus inside an ovule.

2 A fertilised egg grows into an embryo inside an ovule, which forms the seed. The ovary forms the fruit containing many seeds.

3 Germinating seeds require water, oxygen and a suitably warm temperature for growth and development.

S 4 Pollen grains that land on the stigma produce a pollen tube that grows down the style to reach an ovule in the ovary.

SUMMARY QUESTIONS

1 **a** State what happens when fertilisation occurs in flowering plants.

b Describe the role of the pollen tube.

c What happens to each of the following after fertilisation?

  i petals

  ii ovule

  iii ovary

2 **a** State the three environmental conditions that affect the germination of seeds.

b Explain how each condition given in part **a** affects germination.

c Explain why the seeds in tubes **B**, **C** and **D** in the Practical box (left) did not germinate.

15.5 The male reproductive system

- Identify on diagrams the parts of the male reproductive system and state their functions

- Compare male and female gametes

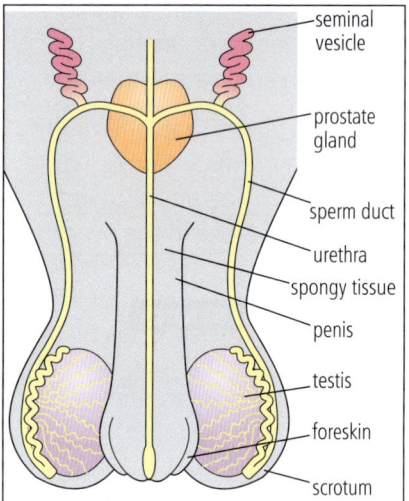

Figure 15.5.1 Male reproductive system: front view.

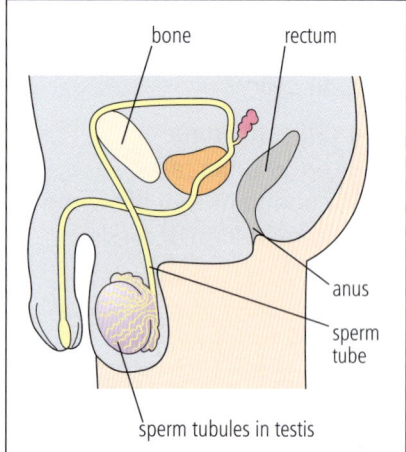

Figure 15.5.2 Male reproductive system: side view.

The **testes** are the male sex organs. They produce the male gametes or **spermatozoa** – a word that is usually abbreviated to **sperm** or sperm cells. The testes also make the male hormone **testosterone**. This stimulates changes in a boy's body as he develops into an adult during puberty. This happens between about 10 and 16 years of age.

The testes are located inside a sac called the **scrotum**, which hangs outside the body. This keeps the testes at a cooler temperature as sperm cells need a temperature cooler than 37 °C to develop properly and be stored. After puberty, the testes constantly produce sperm cells, which are stored in small tubules just outside the testes where they mature. A much wider tube called the **sperm duct** connects these tubules to the urethra. The **prostate gland** and other glands secrete fluids in which the sperm cells can swim. The prostate secretes mucus and other glands secrete sugars, which sperm cells use as a source of energy for their respiration. Sperm cells and the fluid together form **semen**, also known as seminal fluid.

The two sperm ducts join with the urethra, which runs down the centre of the **penis**. Urine and semen never pass down the urethra at the same time. A ring of muscle around the urethra contracts to prevent the loss of any urine from the bladder during sexual intercourse.

Male and female gametes

Gametes are sex cells. The male gamete is called the **sperm** and the female gamete is called the **ovum** or **egg**. The sperm cell is a specialised cell that is adapted for swimming by having a flagellum (tail) that lashes from side to side. Sperm cells carry genetic information from the male parent to the egg of the female parent. The nucleus contains the father's chromosomes. An egg is much bigger than a sperm because it provides the food store that supports the embryo after fertilisation.

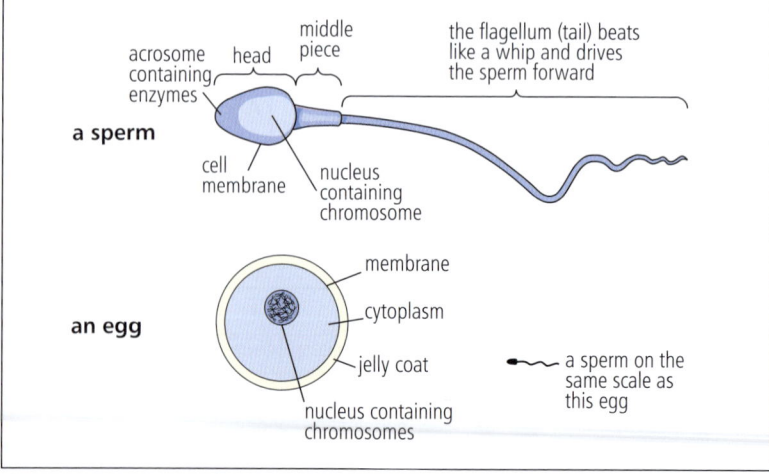

Figure 15.5.3 Structure of a human sperm (top) and an egg (bottom).

The nucleus of the egg cell contains the mother's chromosomes. Gametes contain half the number of chromosomes as body cells. They are described as haploid cells. Body cells have the normal number of chromosomes and are described as diploid.

An egg is fertilised by a sperm cell. The head of the sperm contains enzymes to help it reach the surface of the egg cell. The cell membrane of the sperm cell fuses with the egg cell membrane. The sperm nucleus moves through the cytoplasm of the egg and the two nuclei fuse together to form a **zygote** (fertilised egg). After fertilisation, the jelly coat changes so that no more sperm can enter.

Comparing male and female gametes

Table 15.5.1 Comparing male gametes and female gametes.

feature	sperm cell	egg cell
size	small	much larger than sperm cell
movement (mobility)	swims using its flagellum (or tail) that lashes from side to side	does not move itself – is moved along the oviduct by cilia and peristalsis
energy store	has very little – uses sugar in seminal fluid for respiration	protein and fat in cytoplasm – enough to last until implantation in uterus
number of chromosomes	23 (haploid number)	23 (haploid number)
number produced	millions constantly produced (after puberty) often throughout life	one a month after puberty until menopause, except when pregnant or taking a contraceptive pill

Each sperm cell has **mitochondria**, which release energy to power swimming by the tail. The **acrosome** contains enzymes that digest a pathway through the jelly coat surrounding the egg.

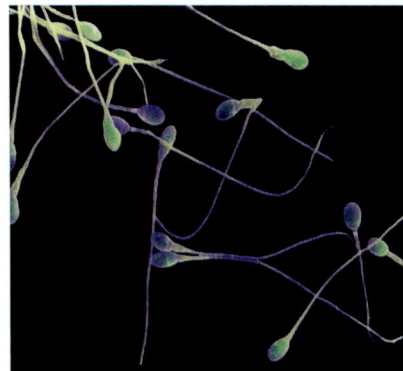

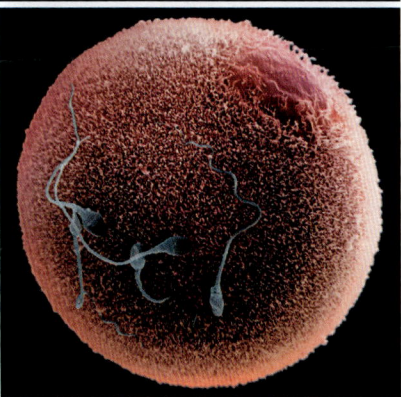

Figure 15.5.4 Human sperm (top) and sperm fertilising an egg.

SUMMARY QUESTIONS

1 Copy and complete the sentences using these words:

semen   urethra   testes   ejaculation   prostate
sperm cells   penis   sperm duct   scrotum

The male gametes are called _____ and are made in the male sex organs or _____ which are contained within a sac called the _____. A tube called the _____ carries the sperm away from each testis. The _____ gland adds fluid to the sperm cells to form _____. During an _____ the semen passes along the _____ and out of the erect _____.

2 a Give three differences between the structure of sperm cells and egg cells.

  b Explain how sperm cells and egg cells are adapted for their functions.

KEY POINTS

1 The male reproductive organs consist of the testes, scrotum, sperm ducts, prostate gland, urethra and penis.

2 The sperm cells are much smaller than the female egg cell. They are produced in huge numbers and are able to swim using a flagellum (tail).

15.6 The female reproductive system

Ovaries are female sex organs that produce the female gametes called **ova** or **eggs**. The ovaries also make the female hormones **oestrogen** and **progesterone**, which starts to happen with girls between 10 and 15 years of age during puberty.

Oestrogen stimulates the development of the sex organs and secondary sexual characteristics in a girl's body as she starts to develop into an adult. Progesterone prepares the uterus so that it is ready to receive an embryo in the case of a pregnancy.

The ovaries are attached to the inside of the abdomen just below the kidneys. After puberty, an egg is released from an ovary about every 28 days. The ovaries tend to release an egg (or eggs) on alternate months. The egg passes out of the ovary and into the funnel-shaped openings of the oviduct in a process called **ovulation**. The egg moves slowly down the oviduct towards the **uterus** (womb).

If sperm cells are present in the oviduct the egg will be fertilised. If the egg is not fertilised it will die after about a day. If the egg is fertilised it will divide to form an embryo, which may attach itself to the lining of the uterus where it develops into a fetus.

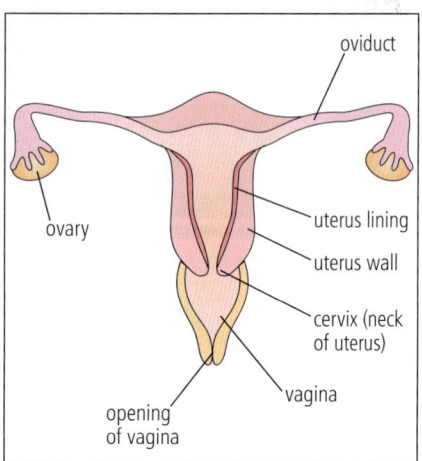

Figure 15.6.1 Female reproductive system: front view.

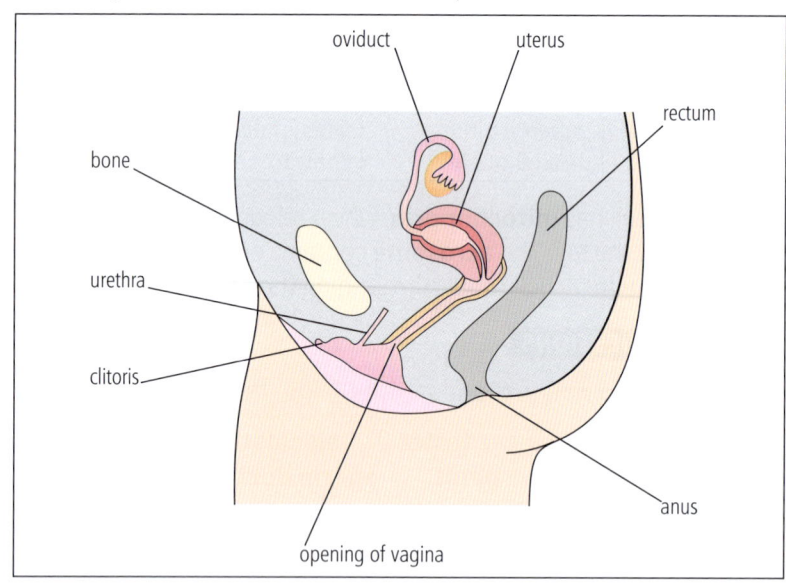

Figure 15.6.2 Female reproductive system: side view.

The lower end of the uterus has a ring of muscle called the **cervix**. It leads to a muscular tube called the **vagina** that opens to the outside of the body. Above the opening of the vagina is the opening of the **urethra** through which urine from the bladder passes out. Above the urethra is the sensitive **clitoris**. The outer opening of the vagina is called the **vulva**. The vagina is sometimes known as the birth canal because a baby passes through it at birth.

Sexual intercourse

During sexual intercourse males and females stimulate each other. Blood is pumped into special spongy tissue in the penis so that it becomes erect. The erect penis is placed into the vagina during sexual intercourse. Fluid made by the walls of the vagina lubricates movements of the penis within the vagina. This movement stimulates the penis and contractions begin in the sperm ducts to move sperm cells from the tubules around the testes towards the penis.

As they flow, secretions from the glands including the prostate gland are added to form semen or seminal fluid. Contractions of the urethra move the seminal fluid through the penis into the vagina. This is an **ejaculation** and at this time the man experiences feelings of pleasure called an **orgasm**. Repeated movements of the erect penis against the clitoris or against the vagina walls may also produce orgasms for the woman. Each ejaculation contains between 2 to 5 cm^3 of semen with up to 500 million sperm cells.

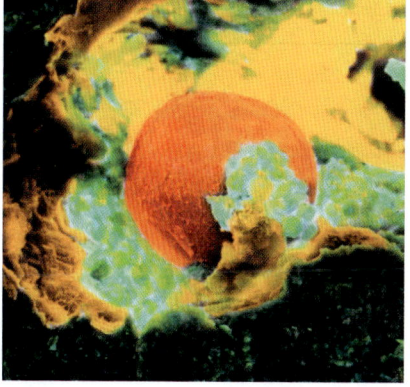

Figure 15.6.3 An egg cell leaves the ovary at ovulation. If sexual intercourse happens at around the time of ovulation there is a chance that fertilisation will occur.

SUMMARY QUESTIONS

1 Copy out the list of organs on the left and match each with its correct function or description from the list on the right.

ovary	carries urine out of the body
vagina	a ring of muscle at the opening of the uterus
oviduct	receives the penis during sexual intercourse
cervix	carries the egg from the ovary
uterus	produces eggs and female hormones
urethra	the place where a fetus develops

2 Look at the diagram of the female reproductive system below:

  a Name the parts labelled **A** to **G**.

  b Where on the diagram do each of the following take place?

    i the production of oestrogen

    ii fertilisation

    iii development of an embryo

    iv the passage of the baby during birth

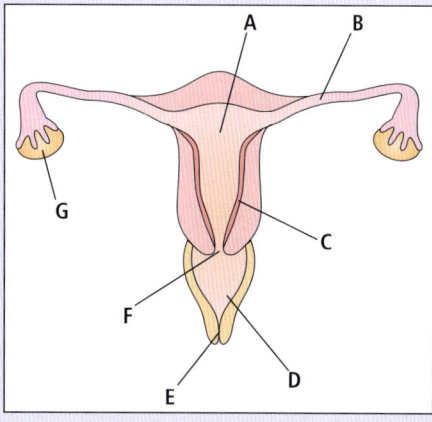

3 Some women's oviducts can get blocked. Explain why a woman with blocked oviducts would be unable to have a baby, whereas a woman with only one blocked oviduct may be able to have a baby.

KEY POINTS

1 The female reproductive system consists of the ovaries, oviducts, uterus, cervix and vagina.

2 The ovaries produce eggs, which are released into oviducts at ovulation.

3 The oviduct is the organ where fertilisation occurs. The uterus is the organ in which the embryo and later the fetus develop.

4 The cervix is a ring of muscle that separates the uterus from the vagina, which is also called the birth canal.

15.7 Fertilisation and implantation

LEARNING OUTCOMES

- Describe fertilisation in humans
- Describe the early development of the zygote to form a ball of cells that is implanted into the lining of the uterus

Human fertilisation

There are millions of sperm in a man's semen. If one of them is to meet an egg it must swim from the vagina to the oviduct – the muscular tube that links the ovary to the uterus.

After intercourse, sperm cells swim through the mucus in the cervix into the uterus and then all the way to the oviduct. Many sperm cells do not survive this difficult journey, which is why so many sperm cells are produced – to increase the chance of some of them reaching the oviduct.

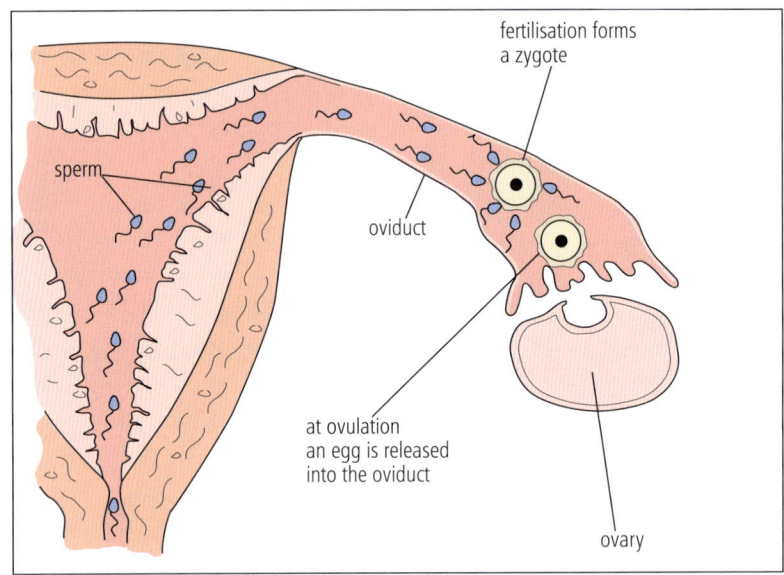

Figure 15.7.1 Ovulation and fertilisation.

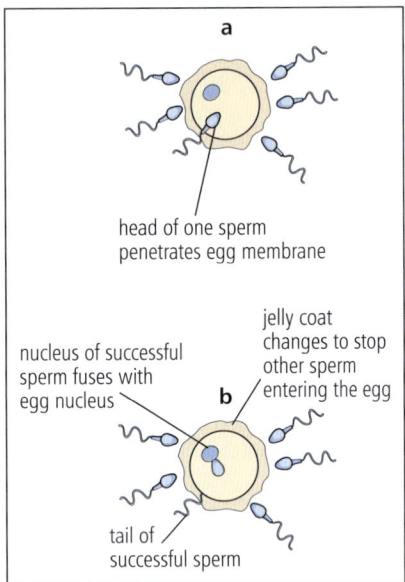

Figure 15.7.2 Events at fertilisation.

If there is an egg in the oviduct, a sperm cell may succeed in penetrating it. Enzymes released by the acrosome on the head of the sperm digest a pathway through the jelly coat surrounding the egg.

After the sperm membrane has fused with the egg membrane, the nucleus enters the egg cytoplasm and the flagellum is left outside. The sperm nucleus fuses with the egg nucleus to form the zygote nucleus, which is diploid as it contains two sets of chromosomes – one from the mother and one from the father. The zygote nucleus now contains 46 chromosomes and these start being copied in readiness for the first cell division, which occurs shortly after fertilisation. The jelly coat changes to stop other sperm cells from entering so only *one* sperm is able to fertilise the egg.

If there is no egg in the oviduct, no fertilisation can take place. However, the sperm can stay alive for 2 or 3 days. So if intercourse happened just before ovulation, the sperm can fertilise an egg if it is released during this time.

EXAM TIP

Every time a cell divides the nucleus must divide first. The type of nuclear division that occurs during growth is mitosis. This makes sure that the cells all have the same number of chromosomes and are genetically identical. See page 204 for more about this.

Implantation

Fertilisation takes place in the oviduct. After this, the fertilised egg or **zygote** begins to divide. It divides once to form a two-celled embryo. Then it continues to divide to give four cells and then eight, but after a while this cycle of divisions becomes less regular. Some cells continue to divide while others stop or slow down the rate at which they divide. After a few hours the embryo is a hollow ball of cells. It moves down the oviduct, pushed along by peristaltic contractions of the oviduct and the beating of the ciliated epithelial cells lining the oviduct.

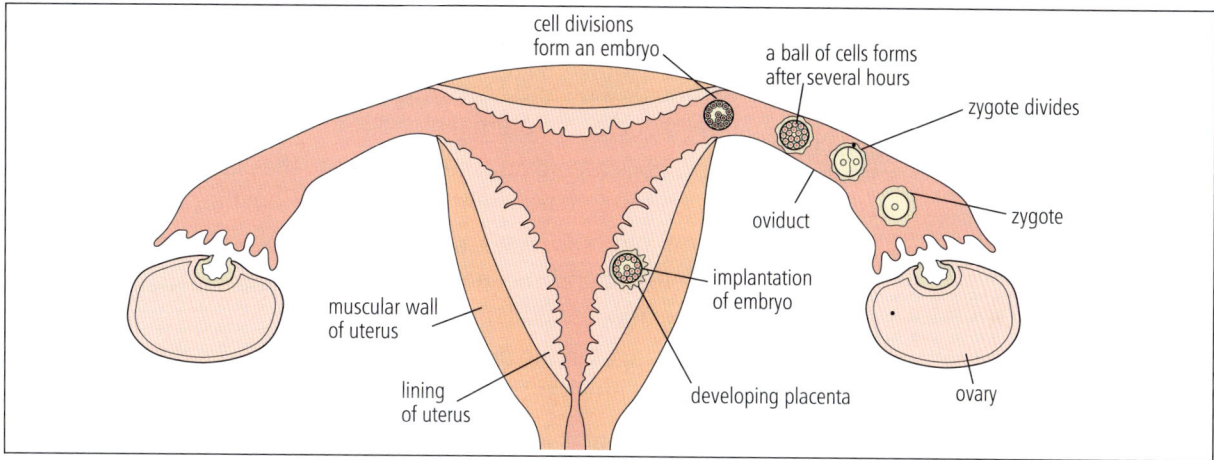

Figure 15.7.3 Implantation of the embryo.

It may take a number of days for the embryo to reach the uterus. The embryo embeds into the soft lining of the uterus. This is called **implantation**. The uterus lining has thickened in preparation and contains numerous blood vessels. The embryo obtains nutrients and oxygen from these blood vessels by diffusion. Carbon dioxide and chemical waste diffuse out in the opposite direction.

SUMMARY QUESTIONS

1 Copy and complete the sentences using these words:

> zygote   sexual   ovaries   implants   uterus
> sperm cells   vagina   fertilisation   weeks   embryo

Sperm cells enter the _____ of the female during _____ intercourse. The sperm cells swim through the cervix and up through the _____ and then enter the oviduct. An egg is made in one of the _____ about every four _____ . The egg passes into the oviduct where _____ may take place if _____ are present. The fertilised egg or _____ divides to form an _____, which passes down to the uterus where it _____ in the lining of the uterus.

2 a How many sperm cells are needed to fertilise an egg?
  b Why does the jelly coat around the egg change after fertilisation?

3 State the number of chromosomes in the nuclei of the following human cells:
  a an egg   b a sperm cell   c a zygote

KEY POINTS

1 Fertilisation occurs when the nucleus of a sperm cell fuses with the nucleus of an egg cell.

2 For fertilisation to occur in the oviduct, ovulation must have taken place and a sperm cell must have reached the oviduct.

3 The zygote divides to form a ball of cells called the embryo, which passes down the oviduct and implants in the lining of the uterus.

15.8 Pregnancy

Pregnancy

Pregnancy is the period of time between fertilisation and birth, which in humans is 9 months. This period of time is called the **gestation period**.

After the embryo has implanted into the lining of the uterus, it grows projections into the soft tissue to gain nutrients and oxygen to support its growth. These projections continue to grow into the **placenta**. An **umbilical cord** grows to attach the fetus to the placenta so it fills the **amniotic sac** which surrounds it.

- The placenta is the site of exchange of oxygen and nutrients for carbon dioxide and other waste products; they diffuse between fetal blood and maternal blood.
- The umbilical cord attaches the fetus to the placenta and contains blood vessels to transport blood to and from the placenta.
- The amnion sac makes **amniotic fluid** to surround and protect the fetus against mechanical damage, for example when the mother makes sudden movements.

The main stages in the development into a baby ready to be born at the end of pregnancy are summarised as follows.

- One month after fertilisation, a human embryo looks a bit like a fish embryo or tadpole. The embryo does not yet have arms or legs, but it is clear where these will develop. The heart has started to beat.
- Two months after fertilisation, the embryo has a face, limbs, fingers and toes and looks human. It is now called a fetus. Most of the organs are formed.
- Three months after fertilisation, the nerves and muscles of the fetus are developing rapidly.
- Five months after fertilisation, although only about 180 mm in length, the fetus has perfectly formed eyebrows, fingernails, fingerprints and body hair. Its movements may have been felt by the mother for the last month.
- Seven months after fertilisation, development is almost complete.

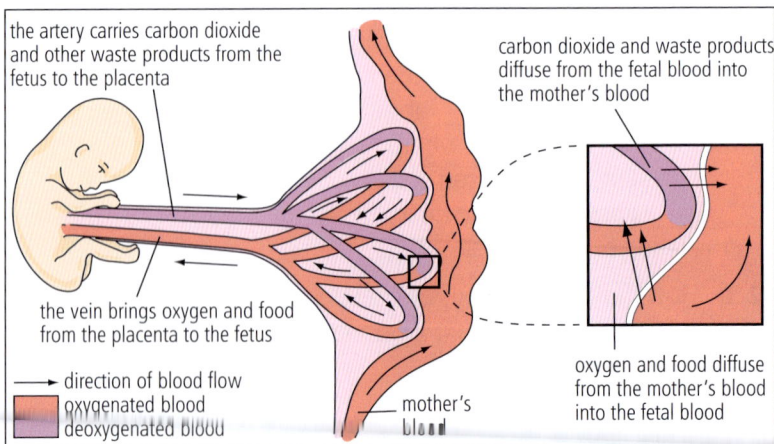

the artery carries carbon dioxide and other waste products from the fetus to the placenta

carbon dioxide and waste products diffuse from the fetal blood into the mother's blood

the vein brings oxygen and food from the placenta to the fetus

→ direction of blood flow
oxygenated blood
deoxygenated blood

mother's blood

oxygen and food diffuse from the mother's blood into the fetal blood

Figure 15.8.1 The role of the placenta.

Supplement

The placenta is the gas exchange surface for the fetus. It is also where dissolved food substances are absorbed. This is why it has the same features as the lung and the small intestine: a large surface area and a short distance for diffusion. Since the exchanges are between fetal and maternal blood there are *two* good blood supplies to maintain concentration gradients (see page 28).

The fetus obtains all its food from its mother, including glucose, amino acids, fats, mineral ions, vitamins and water. The fetus produces some urea, which diffuses into the mother's blood to be excreted through her kidneys.

The umbilical cord joins the fetus to the placenta. The cord contains an artery that transports deoxygenated blood from the fetus to the placenta and a vein that returns oxygenated blood to the fetus.

Amniotic fluid is formed from filtered blood and has a composition similar to plasma. Cells from the fetus' skin are shed into the amnion sac and when the organs have developed the fetus carries out some breathing movements taking the fluid into the lungs. The fetus urinates into the fluid and drinks it. The amniotic fluid does not provide any nutrients for the fetus.

The placenta acts as a barrier to most toxins and pathogens. However, even though the womb is a sterile environment, the fetus is still susceptible to disease. Rubella (German measles) is caused by a virus that can get across the placenta and harm the fetus. This is why all young women should be vaccinated against rubella before they become pregnant. Young men should also be vaccinated since they can act as a reservoir of infection for the disease.

The HIV virus may cross the placenta, so a baby may be born HIV positive if the mother is infected with the virus. However, with careful management during pregnancy a mother who is HIV+ can give birth to a child who does not carry the virus. Nicotine from smoking cigarettes can also pass across the placenta.

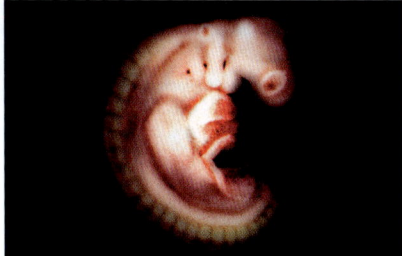

Figure 15.8.2 The fetus at 26 days.

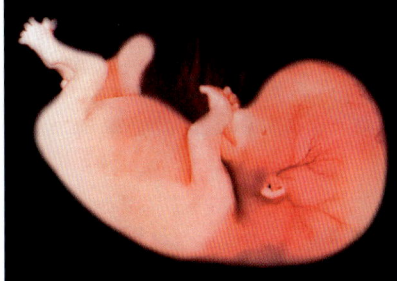

Figure 15.8.3 The fetus at 8 weeks.

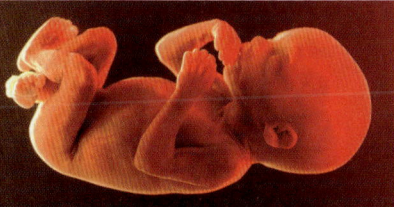

Figure 15.8.4 The fetus at 5 months.

KEY POINTS

1 The fetus grows and develops inside the uterus. It increases in size and develops all the major organs.

2 The umbilical cord attaches the fetus to the placenta, which is the site of exchange of substances between fetus and mother; the amnion makes amniotic fluid to protect the fetus from mechanical damage.

S 3 The placenta has a large surface area for gas exchange and diffusion of nutrients and waste between fetal blood and mother's blood.

4 Some toxins and pathogens cross the placenta into the fetus; the placenta acts as a barrier to many others.

SUMMARY QUESTIONS

1 **a** How is the fetus connected to the placenta?
  b List two substances that pass from the mother's blood to the fetus and two substances that pass in the opposite direction.

2 What is the gestation period? How long does it last in humans?

3 What are the functions of the amnion and amniotic fluid?

S 4 Describe two ways in which the structure of the placenta helps efficient diffusion between fetal blood and mother's blood.

15.9 Sex hormones

Figure 15.9.1 A group of adolescents.

Puberty and adolescence

You are born with a complete set of sex organs. But they become active only later in life. Between the ages of about 10 and 14, the testes start to make sperm and follicles start to develop in the ovaries. This time of development in your life is called **puberty**.

Girls usually develop earlier than boys. The actual age of puberty varies from person to person. The changes that take place are all controlled by **hormones**. At the beginning of puberty the **pituitary gland** at the base of the brain starts to make hormones that stimulate the testes and the ovaries.

These make the sex organs active. The sex organs start to produce sex hormones, which develop our **secondary sexual characteristics**.

Puberty in boys

The testes start making **testosterone**, which stimulates:

- the growth of the male sex organs
- the testes to make sperm cells
- growth of hair on the face
- the deepening of the voice
- development of muscles in the body.

Puberty in girls

The ovaries start making **oestrogen**, which stimulates:

- the growth of female sex organs
- the start of the first menstrual cycle and the first period
- growth of hair on parts of the body
- growth and development of breasts
- widening of the hips.

A person becomes an **adolescent** when puberty starts. Adolescence finishes when you stop growing – at about 18 years of age. Adolescence can be an emotional time. Hormones can bring about mood changes and increased sexual urges.

Menstrual cycles are the changes that occur in the ovary and the uterus. A complete cycle lasts for about 28 days and is controlled by hormones that coordinate the activity of the ovary and the uterus. This coordination is important so that the lining of the uterus is ready to receive an embryo so that it can start its development.

Girls start to have periods between the ages of about 10 to 15 years old. During a period, the lining of the uterus breaks down and blood and cells pass out of the vagina. This is called **menstruation**. When this happens for the first time, it shows that a girl has had her first menstrual cycle.

Stages in the menstrual cycle

Girls are born with a very large number of potential egg cells in their ovaries. They do not produce any more during their lifetime. Each potential egg is surrounded by a small group of cells and together they form a **follicle**. There are thousands of these follicles in each ovary.

Each month after puberty one or a few follicles start to develop.

A follicle starts to develop in the ovary at the beginning of the cycle as menstruation finishes. As the follicle grows it enlarges and fills with fluid and moves towards the edge of the ovary. The egg cell increases in size as the cytoplasm fills with food and the nucleus divides. At this time oestrogen, produced by the ovary, causes the lining of the uterus to thicken. After 2 weeks, the follicle bursts, releasing the egg, some follicle cells and the fluid into the oviduct. This is called **ovulation**. The follicle cells left behind in the ovary form the **yellow body**, which remains for the next 2 weeks secreting hormones to maintain the lining of the uterus so menstruation does not occur. If implantation does not occur, the yellow body then decreases in size. If implantation does occur, it remains active during pregnancy.

While the egg is developing in the ovary, the lining of the uterus starts to thicken. In the week after ovulation the uterus has a thick lining of blood vessels and glands. If fertilisation occurs, the fertilised egg divides to form a ball of cells, which **implants** into the thick lining and the woman becomes pregnant.

If fertilisation does not occur, the egg dies and passes out of the vagina and the yellow body in the ovary breaks down. The thick lining of the uterus breaks down and is lost during menstruation. The cycle now begins again.

If a pregnancy does occur, the embryo releases a hormone that stimulates the yellow body to remain active and to release hormones, which in turn stimulate the lining of the uterus to continue to thicken. This thickened lining supplies the embryo with food and oxygen as it continues its development. The release of hormones also ensures that menstruation will not occur.

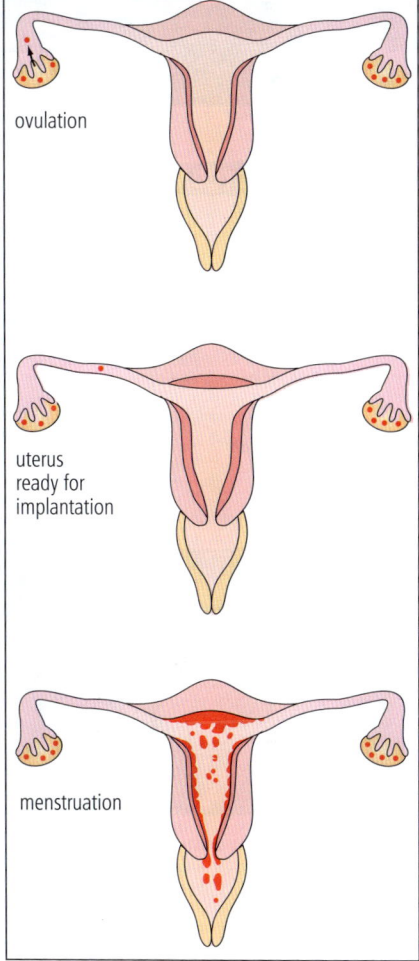

Figure 15.9.2 During the menstrual cycle the activities of the ovaries and the uterus are synchronised under the control of the female sex hormones.

ovulation

uterus ready for implantation

menstruation

SUMMARY QUESTIONS

1 Explain the role of the testes and the ovaries in the development and regulation of secondary sexual characteristics in males and females at puberty.

2 a List the effects that testosterone has on the body of a boy.
   b List the effects that oestrogen has on the body of a girl.

3 Describe the functions of the follicle and the yellow body during the menstrual cycle.

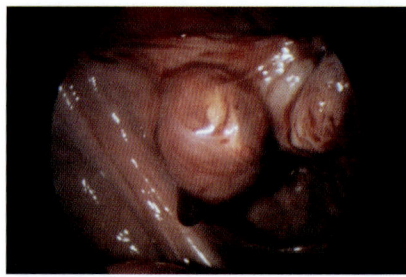

Figure 15.10.1 In the centre is a yellow body, which enlarges in the second half of the menstrual cycle to produce and release progesterone.

There are four hormones that control the menstrual cycle.

The pituitary gland at the base of the brain secretes:

- **follicle-stimulating hormone (FSH)**
- **luteinising hormone (LH)**.

The ovary secretes:

- **oestrogen** and **progesterone**.

Each cycle starts with the secretion into the blood of FSH from the pituitary gland. This hormone is carried in the blood to the ovary where it stimulates the development of one or more follicles. Inside each follicle there is a cell that has the potential to grow into an egg. This cell divides by meiosis and its cytoplasm fills with stores of fat and protein (see page 183).

FSH stimulates other cells in the follicle to secrete oestrogen into the bloodstream.

Oestrogen has three main effects.

- It stimulates the repair and thickening of the lining of the uterus.

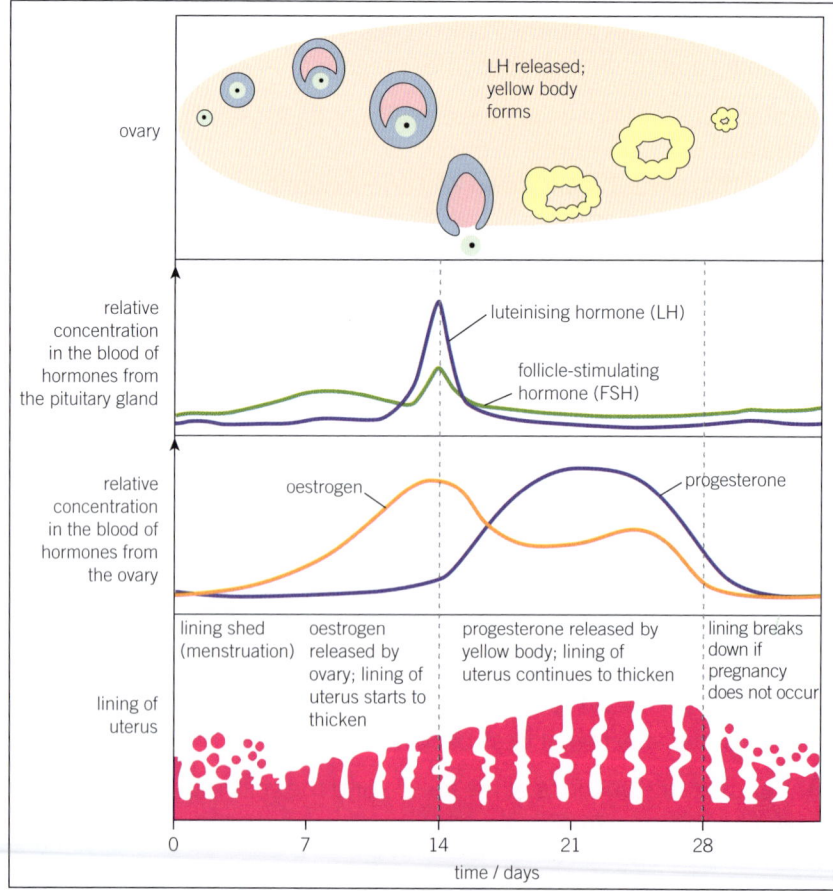

Figure 15.10.2 Control of the menstrual cycle by hormones from the pituitary gland.

- It stops the secretion of FSH from the pituitary gland.
- It stimulates the pituitary gland to secrete LH.

The lining of the uterus thickens in preparation to receive an embryo.

As you can see from the graphs in Figure 15.10.2, FSH and oestrogen are the important hormones during the first half of the menstrual cycle from day 0 to day 14. During this time the egg develops and increases in size within the follicle.

LH stimulates the follicle to burst and release its egg. This happens at around day 14 of the cycle.

The top part of Figure 15.10.2 shows that after ovulation the remains of the follicle become the yellow body. LH stimulates the yellow body to secrete progesterone.

In the second half of the cycle, progesterone together with oestrogen secreted by the ovary have these effects:

- stimulate further growth of the lining of the uterus,
- maintain the lining in its thickened state and prevent it breaking down, and
- reduce the secretion of FSH and LH from the pituitary gland.

The graph shows that the concentration of LH decreases in the second half of the cycle. This reduces the stimulation of the yellow body so eventually it stops secreting progesterone and decreases in size.

As the concentration of progesterone decreases there is less stimulation for the lining of the uterus and menstruation occurs.

If intercourse occurs around the middle of the cycle and the egg is fertilised, then it may implant in the lining of the uterus so that pregnancy begins. If this happens, progesterone continues to be secreted by the yellow body so that the lining of the uterus remains thick. This stops menstruation occurring and prevents the menstrual cycle starting again.

The embryo secretes a hormone that stimulates the yellow body to continue secreting progesterone.

During pregnancy, the yellow body continues to secrete progesterone, but gradually the placenta takes over this role. Progesterone from the placenta continues to maintain the lining of the uterus. This prevents menstruation occurring during pregnancy. This does happen sometimes and is known as a miscarriage.

KEY POINTS

1 The pituitary hormone FSH stimulates the development of a follicle inside the ovary and the secretion of oestrogen.

2 Oestrogen stimulates the thickening of the uterus lining, inhibits FSH production and causes the pituitary to release LH.

3 LH stimulates ovulation and the formation of the yellow body from the remains of the follicle. The yellow body secretes progesterone, which maintains the lining of the uterus.

EXAM TIP

Use the timeline you made in answer to question 3 on page 191 to show when the changes in concentrations of LH, FSH, oestrogen and progesterone occur.

EXAM TIP

Some of the hormone released by the embryo in early pregnancy is excreted in the urine. If a woman misses a period she may be pregnant. She can use a pregnancy testing kit that detects this hormone in her urine to find out.

SUMMARY QUESTIONS

1 Name four hormones involved in coordinating the menstrual cycle and state where each is secreted.

2 Describe the roles of these hormones in controlling the menstrual cycle.

3 Explain why it is important that the lining of the uterus is thick at the time of ovulation.

4 State what happens as a result of the decrease in concentration of oestrogen and progesterone towards the end of the menstrual cycle.

5 Explain why it is important that progesterone is secreted if pregnancy occurs, but FSH is not.

6 Draw a timeline to show the changes that occur in the ovary and in the uterus during a menstrual cycle.

15.11 Sexually transmitted infections (STIs)

EXAM TIP

AIDS is not a disease; it is a collection of diseases that result from a weakening of the immune system. This is what the word 'syndrome' means. Often the term HIV/AIDS is used.

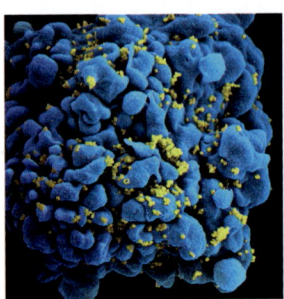

Figure 15.11.1 Lymphocyte infected with HIV (yellow) (×5000).

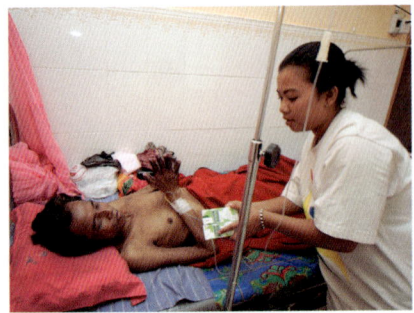

Figure 15.11.2 A health worker treating a patient with AIDS in Cambodia.

A sexually transmitted infection (STI) is a disease that is transmitted via body fluids during sexual contact. The major STIs include HIV, syphilis and gonorrhoea.

Human immunodeficiency virus (HIV)

The human immunodeficiency virus (HIV) is a human pathogen that was first identified in the early 1980s. Like all viruses it is composed of a few genes and a protein coat. It enters human cells and uses each cell to make more viruses, which then enter even more cells. The main type of cell that HIV infects is a type of white blood cell.

Supplement

HIV and lymphocytes

HIV attacks and destroys an important type of lymphocyte, called a T-lymphocyte, that coordinates the immune system. During an infection, T-lymphocytes stimulate other lymphocytes to produce antibodies. During an HIV infection, the number of T-lymphocytes decreases and so fewer antibodies are produced every time there is another infection. By reducing the number of T-lymphocytes, HIV weakens the body's ability to defend itself against pathogens. Eventually the person infected with HIV will be infected with other pathogens because of their weakened immune system. Various types of cancer may also develop.

How is HIV transmitted?

HIV is transmitted in the blood and semen. The virus can pass from one person to another during unprotected sexual intercourse.

Either partner may infect the other. The virus can also be passed via hypodermic needles contaminated with infected blood. In this way, HIV has spread very quickly among drug addicts who share needles.

Babies are also at risk from HIV. This is because the virus can pass from the mother's blood to the baby's blood at birth when the two bloodstreams come into close contact. HIV can also be transmitted in breast milk. Blood used for transfusions is another way in which HIV has been transmitted.

HIV and AIDS

If untreated, HIV infection may lead to the development of **acquired immunodeficiency syndrome** or **AIDS**. If someone has been infected with HIV they are described as being HIV positive (HIV+). The virus weakens the body's immune system. It is easier for a person who is HIV+ to be infected by other diseases, such as tuberculosis (TB) and pneumonia. Various cancers may also develop. People who do not have HIV are described as HIV negative (HIV−).

AIDS is the name given to a collection of diseases brought on by the weakening of the body's immune system. The early symptoms of HIV infection are very much like flu, with swollen glands and a high temperature. If untreated, AIDS may develop after several years. Symptoms may include weight loss, fever, fatigue and diarrhoea. There may be a decrease in brain function. Not all HIV+ people develop AIDS. Some people remain HIV+, but without any symptoms at all.

How can HIV infection be prevented?

There is no cure for HIV/AIDS and as yet no vaccine to give protection from HIV.

Most people across the world who are living with HIV receive antiretroviral treatment (ART), which involves taking several drugs, such as zidovudine that prevents the replication of the virus inside host cells. This treatment is so successful that most people who receive ART are expected to live a full life. The drugs not only prevent progress of HIV infection to AIDS, but also reduce the chance of transmitting the virus.

The success of the treatment depends on people having access to medical facilities that can test for the presence of HIV and provide ART to people found to be living with HIV. It is estimated that nearly 70% of people across the world living with HIV have access to ART. The introduction of home testing kits for HIV may well increase the this percentage and help to reduce transmission of the virus. Between 2010 and 2019 there was a 23% decrease in the number of people who became newly infected with HIV and this is expected to continue to fall.

Other methods to reduce the spread of HIV include:

- the use of condoms to provide a physical barrier to the transmission of the virus during sexual intercourse
- setting up free needle exchange schemes for those people who inject drugs; this reduces the risk of transmission from the use of shared needles and syringes
- screening donated blood for HIV antibodies and eliminating contaminated blood being used for transfusion.
- education programmes to make people aware of the methods of the transmission of the HIV virus and how it can be prevented.

SUMMARY QUESTIONS

1. **a** What is the difference between AIDS and HIV?
   b What exactly is HIV?
   c Why can an HIV infection not be treated by antibiotics?

2. **a** State the ways in which HIV is transmitted.
   b State the measures that can be taken to reduce the spread of HIV infection.

S 3 Describe the effect of HIV on the immune system and explain how HIV infection leads to the development of AIDS.

EXAM TIP

There are many myths about HIV and AIDS. People cannot catch AIDS. They can become infected with HIV but only through the methods described here. For example, HIV is not transmitted by blood-sucking insects, by touching someone who is HIV+ or sharing food, cups or glasses with them.

KEY POINTS

1. A sexually transmitted infection (STI) can be passed on via body fluids during sexual contact.

2. The spread of STIs can be controlled by the use of condoms, preventing the sharing of hypodermic needles, and education programmes.

3. HIV is a virus that infects cells in the immune system. It is transmitted in the semen or in the blood. With time, HIV infection may lead to a collection of diseases that is known as AIDS.

4. The spread of HIV can be reduced by men using condoms during sexual intercourse, by reducing the use of shared needles between drug users and by careful screening of donated blood used in transfusions.

S 5 HIV destroys lymphocytes leading to a weakening of the immune system. This makes people susceptible to many diseases, such as TB.

195

Practice questions

1 The drawing shows a half flower. Which row is the correct identification of the labels **1** to **4**?

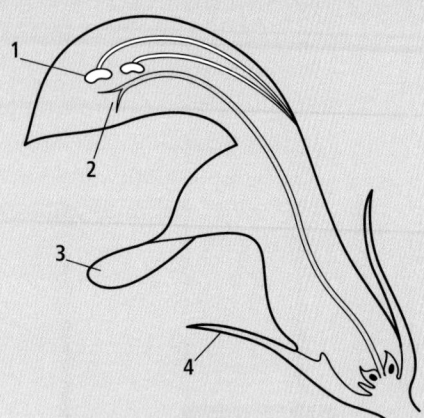

	1	2	3	4
A	anther	stigma	petal	sepal
B	petal	stigma	sepal	anther
C	sepal	petal	anther	stigma
D	stigma	anther	sepal	petal

(Paper 1) [1]

2 Which row shows the correct functions of an insect-pollinated flower?

A	anthers produce pollen grains	carpels contain ovules	sepals provide landing site for insects
B	sepals attract insects	anthers produce pollen grains	carpels contain ovules
C	stigmas receive pollen grains	ovaries contain ovules	anthers produce pollen grains
D	petals protect a flower in the bud	stigmas are site of pollination	anthers produce ovules

(Paper 1) [1]

3 Which is a method of transmission of HIV?

   A injection using sterile needles

   B skin contact with contaminated surfaces

   C breathing in airborne droplets

   D sexual contact with an infected person

(Paper 1) [1]

4 The diagram shows the female reproductive system. Which row identifies the places where the events occur?

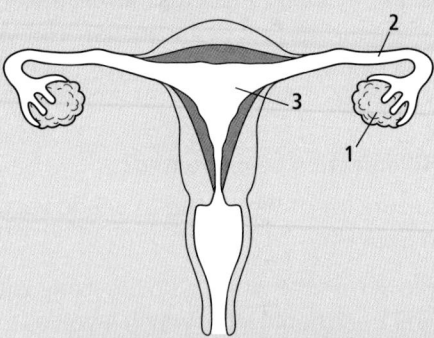

	1	2	3
A	fertilisation	implantation	ovulation
B	implantation	ovulation	fertilisation
C	ovulation	production of eggs	fertilisation
D	production of eggs	fertilisation	implantation

(Paper 1) [1]

5 The following four processes occur during sexual reproduction in flowering plants.

   1 Pollen grains lands on the stigma.

   2 The male gamete fuses with the female gamete.

   3 The male gamete travels down the pollen tube.

   4 The pollen tube grows out from the pollen grain.

In which order do these processes occur?

   A 3, 4, 1, 2

   B 4, 3, 2, 1

   C 1, 4, 3, 2

   D 4, 2, 1, 3

(Paper 2) [1]

6 The symbol n is used to represent haploid nuclei; the symbol $2n$ is used to represent diploid nuclei. Which row is correct for the gametes, zygote and embryo in mammals?

	sperm	egg	zygote	embryo
A	$2n$	$2n$	n	n
B	$2n$	n	$2n$	n
C	n	n	$2n$	$2n$
D	n	$2n$	n	$2n$

(Paper 2) [1]

7 The diagram below shows the changes that occur to the uterus during the menstrual cycle.

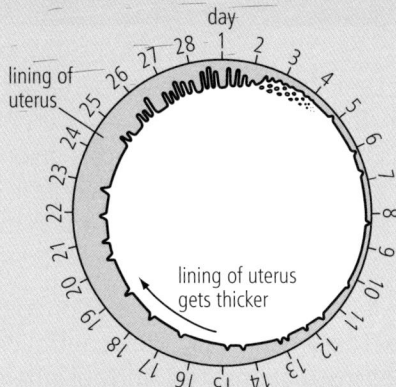

Which row shows the times (in days) when the events during the cycle occur?

	ovulation is likely to occur	menstruation occurs	uterus lining gets thicker
A	10–27	25–28	4–10
B	13–15	1–4	6–25
C	22–25	1–4	7–14
D	10–27	13–15	25–28

(Paper 2)   [1]

8 Potato plants reproduce by means of flowers and stem tubers. Which results in the least variation among offspring in potato plants?

A cross-pollinating plants of different varieties

B cross-pollinating plants of the same variety

C production of stem tubers

D self-pollination

(Paper 2)   [1]

9 (a) Outline the events that occur in sexual reproduction in both flowering plants and mammals.   [4]

(b) Explain the difference between the following pairs of terms: (i) ovule and ovary, and (ii) pollination and fertilisation.   [4]

(c) State three factors required for the germination of seeds.   [3]

(Paper 3)

10 (a) Describe the difference between self-pollination and cross-pollination.   [2]

(b) Hibiscus flowers are insect pollinated.

Describe three features of insect-pollinated flowers that attract pollinators.   [3]

(c) The photograph shows part of an hibiscus flower immediately after pollination.

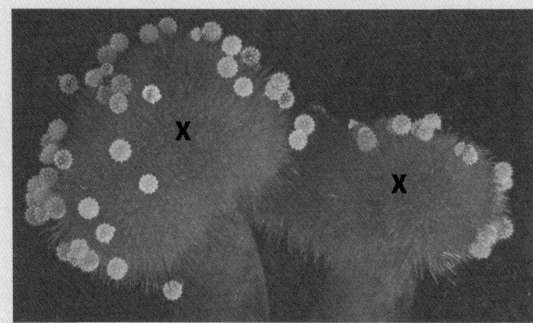

(i) State the name of the structures labelled **X**.   [1]

(ii) Describe how a male gamete in a pollen grain reaches the female gamete inside an ovule.   [4]

(iii) Describe the process of fertilisation in flowering plants, such as hibiscus.   [3]

(d) (i) Explain the advantages of cross-pollination for a population of plants of the same species growing in a habitat that is affected by climate change.   [4]

(ii) Outline the advantages of self-pollination for populations of plants that live in habitats that are not changing. [3]

(Paper 4)

11 (a) Explain how a fetus is supplied with the substances it needs for growth **and** how its excretory products are removed.   [6]

(b) Rubella is a transmissible disease that is caused by a virus. In most people it causes mild symptoms, but can cause a miscarriage or severe defects in a developing fetus if a pregnant woman is infected.

Explain why it is important that all children are vaccinated against the viral disease rubella. [3]

(Paper 4)

16.1 Chromosomes, genes and DNA

Figure 16.1.1 Family members often resemble each other.

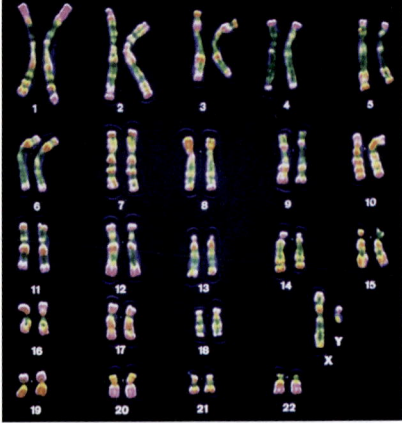

Figure 16.1.2 Chromosomes from a human male.

Table 16.1 Numbers of chromosomes in different species.

species	number of chromosomes
fruit fly	8
broad bean	12
locust	24
yeast	34
potato	48
chimpanzee	48
horse	64

People who belong to the same family often resemble each other – sometimes quite closely; in the case of identical twins, very closely. Characteristics or traits, such as physical appearance, personality and a talent for sport or music, are often shown by people in different generations of a family. These people have inherited the feature from their parents.

The transmission of these features from one generation to the next is called **inheritance**. The characteristics are controlled by genes, which are like coded instructions. The study of inheritance is known as **genetics**.

In Unit 15 we looked at how new generations of flowering plants and humans are produced. Genes from one generation are transmitted to the next in the gametes. At fertilisation, the gametes fuse to form a zygote, which contains the genetic information from both parents. A new individual grows from the zygote – half of its genetic information comes from its male parent and half from its female parent.

Chromosomes

We cannot see genes through a light microscope, but we can see **chromosomes**. Most cells have nuclei containing chromosomes. The first step in visualising the genetic information that we inherit is to look at these chromosomes.

Chromosomes are visible through a microscope during cell division. Figure 16.1.2 shows the chromosomes of a human white blood cell specially prepared so that they can be counted and analysed.

There are 46 chromosomes in the nucleus of a human cell. This is called the **diploid number**. The chromosomes are different shapes and sizes. They can be sorted into 23 pairs based on their size and shape. In each pair, one of the chromosomes has been inherited from the male parent and one from the female.

You can see that each of the chromosomes in Figure 16.1.2 is double stranded. They have two strands joined together usually at the centre, but not always. Figure 16.1.3 shows the two strands of one chromosome in two different colours.

If we could unravel a chromosome, it would form an extremely long thread. That thread is made up of a long chain molecule called **DNA (deoxyribonucleic acid)**. The DNA is wound around molecules of protein. During cell division, the DNA and protein are packed very tightly together and the chromosomes are coiled up. At other times they are uncoiled so the cell can use the information in the DNA.

What is a gene?

Genes are lengths of the DNA that you can see in Figure 16.1.4. Each chromosome has many genes along its length, although the actual number depends on its length and the length of the genes. You can see that some of our chromosomes are much longer than others; these chromosomes have large numbers of genes.

Each gene is a unit of inheritance in that it codes for a specific protein. It is a chemical code that the cell interprets as an instruction to make a protein molecule. You know about several proteins – haemoglobin, amylase and lipase, for example. There is a gene for each of these proteins and many more. Humans have between 20 000 and 25 000 different genes. The proteins produced by cells influence how the body works and what it looks like. Some features that genes control are influenced by the environment. Your height is determined by both genes you inherit and your environment – for example by your diet and how much exercise you take. Other genes are not affected by the environment. Your blood group is determined by a gene and the environment has no affect on this at all.

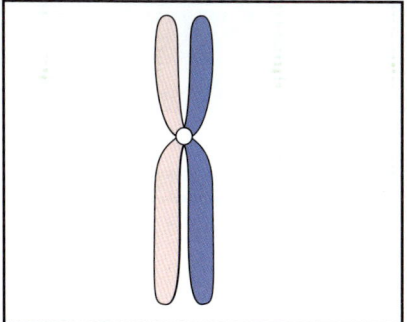

Figure 16.1.3 A double-stranded chromosome.

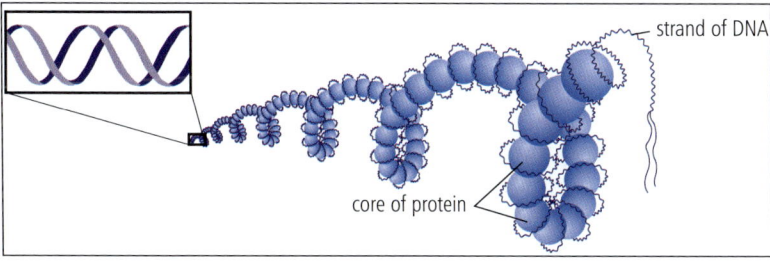

Figure 16.1.4 The arrangement of DNA in a chromosome.

strand of DNA

core of protein

EXAM TIP

'Diploid' means that there are two sets of chromosomes – one set inherited from the mother and the other set inherited from the father. A set consists of one of each type of chromosome.

EXAM TIP

Each chromosome is made of DNA (deoxyribonucleic acid) and protein. DNA is wound around molecules of protein. During cell division, DNA and protein are packed very tightly together. At other times the DNA is packed less tightly so cells can use the information in the genes.

SUMMARY QUESTIONS

1 Copy and complete the sentences using these words:

> **chromosomes   DNA   genes   genetic information
> gametes   nuclei   resemble**

Offspring _____ their parents because of the _____ passed on to them in the _____ (sex cells) from which they developed.

This information is contained in the thread-like structures called _____ that are inside the _____ of most cells. These threads are made of proteins and _____ . The information is found in small units called _____ .

2 a Define the terms *gene* and *chromosome*.

  b Name three human proteins.

  c Write a few sentences to explain the relationship between the following:

  nucleus, chromosome, gene, protein.

3 a What is meant by the *diploid number*?

  b State the diploid numbers of three non-human species.

KEY POINTS

1 Inheritance is the transfer of genetic information from generation to generation.

2 A chromosome is a thread of DNA carrying genetic information in the form of genes.

3 A gene is a length of DNA that codes for a protein.

16.2 Protein synthesis I

Supplement

LEARNING OUTCOMES

- Explain that the sequence of bases in a gene code for the correct order of amino acids in a protein
- Explain how a protein is made

Each chromosome is made up of a long, super-coiled molecule of DNA. Each molecule is divided into thousands of shorter sections called **genes**. A gene is a small part of the DNA strand.

The length of DNA making up a particular gene carries the information needed to make a particular protein. Remember that inside every cell there are thousands of different chemical reactions taking place.

Enzymes control all these chemical reactions. You should remember that all enzymes are proteins. Because DNA codes for proteins, it determines which enzymes are produced in each cell and therefore which chemical reactions take place inside cells.

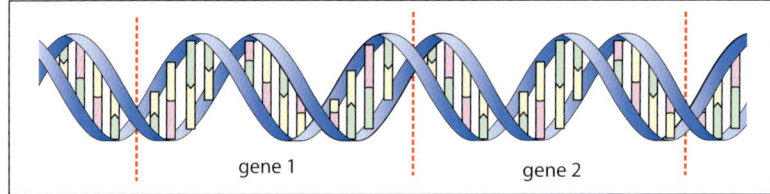

Figure 16.2.1 A gene is a short section of DNA.

The structure of DNA (see Unit 4) contains bases. The information is found in the **sequence of bases** along the length of DNA. These determine the sequence of amino acids in the protein. DNA also controls the production of other important proteins, including antibodies and receptors for neurotransmitters.

DNA carries information to build proteins from amino acids. The information moves from the DNA to the site of protein synthesis in the ribosomes via a kind of 'messenger molecule'. This molecule is RNA and is given the name **messenger RNA** (mRNA). The role of mRNA is to carry a copy of the base sequence on DNA out of the nucleus to the ribosomes in the cytoplasm where protein synthesis occurs.

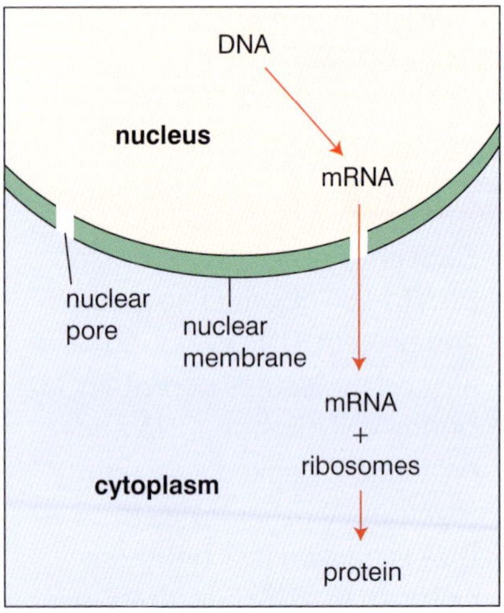

Figure 16.2.2 The vital role of mRNA carrying the code to the ribosomes.

The genetic code

There are **four** different bases found in DNA. We refer to them by their letters A, T, C and G. Each amino acid is coded for, by a sequence of *three* of these bases on DNA. There are about 20 different amino acids and each is coded for by a different base triplet on DNA. When mRNA copies the base sequence on DNA it does so by the 'rule of base pairing', so for instance C will copy G, and G will copy C. However, mRNA does not have the base T – this is replaced by another base U (uracil). Therefore, A on DNA will code for U on mRNA.

EXAM TIP

Do not confuse the genetic code with the sequence of bases in DNA that determine the sequence of amino acids in each protein. It is **not** correct to say that 'DNA contains the genetic code to make a protein'. The different base triplets like those shown in Figure 16.2.3 make up the genetic code.

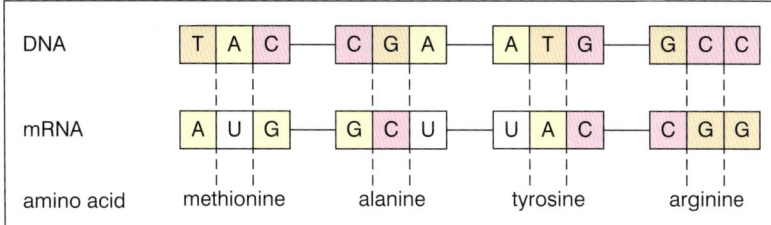

Figure 16.2.3 How DNA base triplets code for amino acids.

The coded information on mRNA is used to assemble amino acids in the correct sequence for each protein.

- The mRNA arrives at the ribosomes from the nucleus.
- The mRNA strand contains the base triplets for each particular amino acid in the protein.
- The mRNA then passes through the ribosomes and each ribosome assembles amino acids into protein molecules.
- The amino acids bond together forming a long chain – the protein.
- The specific order of amino acids is determined by the sequence of bases in the mRNA and each amino acid is joined to the next one by a peptide bond.

All body cells in an organism contain the same genes. However, in a particular cell many genes may not be expressed because the cell makes only the specific proteins it needs.

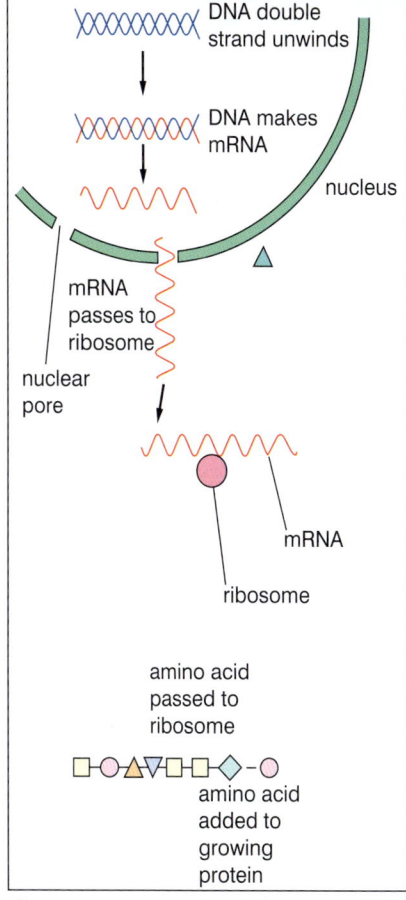

Figure 16.2.4 How DNA controls protein synthesis.

SUMMARY QUESTIONS

1 What is the function of each of the following in protein synthesis?
   a DNA
   b mRNA
   c ribosomes

2 Look at the part of a molecule of messenger RNA here:
   AUGACGCAUGCAGUCCGA.
   a How many base triplets are shown in this section of mRNA?
   b What is the role of each of these base triplets in protein synthesis?
   c Write down the DNA base triplets that coded for this mRNA.

3 Indicate if these statements are true or false.
   a DNA is a protein
   b DNA has a double helix
   c DNA contains four bases: A, G, C and U
   d The molecular shape of DNA is maintained by weak bonds between complementary base pairs.

KEY POINTS

1 The sequence of bases on DNA codes for the correct sequence of amino acids in a protein.

2 The role of mRNA is to copy the DNA code and carry it to the ribosomes in the cytoplasm.

3 The ribosomes assemble amino acids into protein molecules.

16.3 Protein synthesis II

Supplement

LEARNING OUTCOMES

- The order of base triplets in DNA determines the sequence of amino acids in a protein

- DNA can code for a different proteins including enzymes, antibodies, membrane carriers and receptors for neurotransmitters

A **gene** is a sequence of DNA nucleotides. Each nucleotide has one of four bases (A, T, C or G). So a gene will contain a particular sequence of bases. This sequence instructs the cell to make a particular protein.

As you know, proteins are made up of amino acids linked together by peptide bonds.

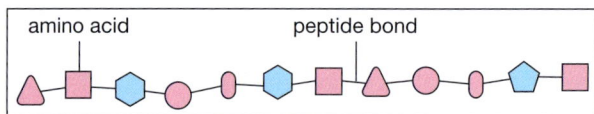

Figure 16.3.1 Part of a protein chain.

It is the type and sequence of the amino acids that are joined together that determines the type of protein formed. For example, one protein molecule may start with the amino acid sequence:

glutamic acid – lysine – arginine – serine – arginine

Another protein may start with the amino acid sequence:

aspartic acid – cysteine – valine – valine – arginine

It is the sequence of bases that determines which amino acids are used and in which sequence they are joined together.

Each group of **_three_** bases (a base triplet) codes for one particular amino acid. For example, the DNA base triplet TTT codes for the amino acid lysine, AGC codes for serine, ACC for tryptophan and AGT for amino acid histidine.

In Figure 16.3.2, the DNA code CGA-CCA-CAT codes for the amino acid sequence: alanine – glycine – valine

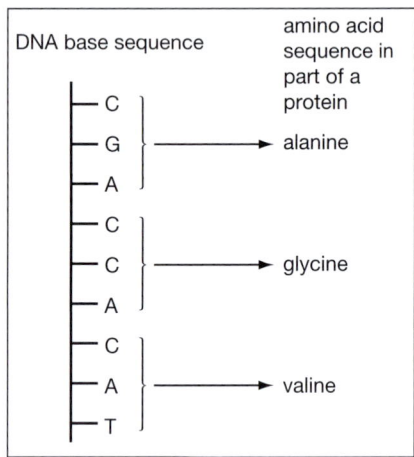

Figure 16.3.2 DNA base sequences determine the sequence of amino acids.

Protein structure

The basic structure of a protein is the sequence of amino acids in the chain. It determines the eventual shape of the protein and hence its function. There are 20 different amino acids and all their different combinations can build the 100 000 known proteins. But proteins are three dimensional molecules. The amino acid chains do not lie flat like a string of beads on a table. They form of helices, which twist and fold around themseleves.

The result is a complex three-dimensional structure held together by a number of different chemical bonds. It is this resulting intricate structure that gives the protein its function (Figure 16.3.3).

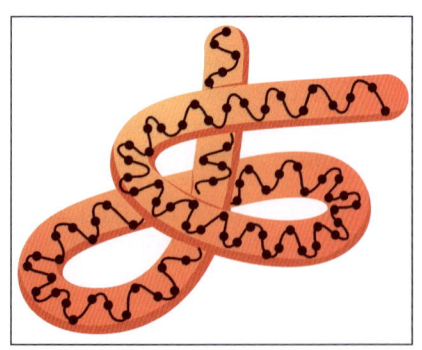

Figure 16.3.3 The complex three-dimensional structure of a protein.

The only genes that are expressed are those that code for the proteins that each cell needs to function correctly. Some genes are expressed in all cells, but many are only expressed in specialised cells.

Important proteins

DNA controls cell functions by controlling the production of proteins including enzymes, antibodies, membrane carriers and receptors for neurotransmitters.

Enzymes control the chemical reactions that take place in a cell. As you know from chapter 5, all enzymes are proteins. So the *genetic code* of DNA, by determining which proteins, particularly enzymes, are made in the cell, will determine the cell's structure and function. The three-dimensional shape of an enzyme is critical to its function.

The active site on the surface of the enzyme has a complementary shape to the enzyme's substrate (Figure 16.3.4). This is important because it gives the enzyme its specificity.

Antibodies are proteins that are made by white blood cells called lymphocytes. Each antibody has a binding site that can attach to pathogens like bacteria. This either directly destroys the pathogen or 'labels' it in a way that can be detected by other white blood cells called phagocytes.

Each pathogen has antigens on its surface that have a particular shape, so only specific antibodies, with a complementary shape can attach to a particular antigen.

Membrane carriers are proteins that control the movement of water-soluble ions and certain molecules across the cell membrane. They are able to maintain different concentrations of ions on either side of the cell membrane (Figure 16.3.5).

Receptors for neurotransmitters are proteins found in the membrane at the synapses between neurones. They have a complementary structure to the transmitter molecules, which attach to them and enable an impulse to pass across the synapse.

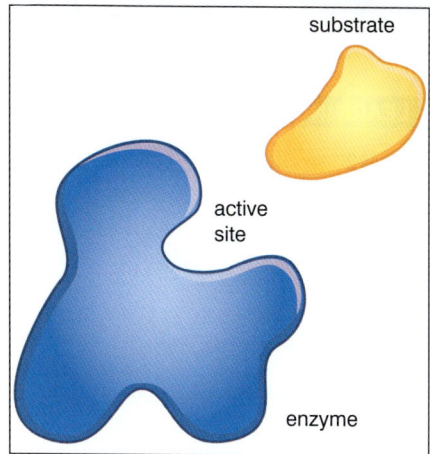

Figure 16.3.4 The active site of the enzyme has a complementary shape to that of the substrate.

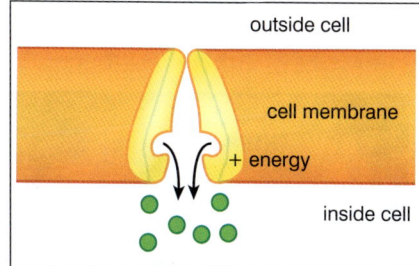

Figure 16.3.5 A carrier protein transports particles across a membrane

KEY POINTS

1. The order of the base triplets in a strand of DNA detrmines the sequence of amino acids in a protein.

2. The complex three-dimensional structure of a protein makes it complementary to particular other molecules.

3. DNA can code for different proteins including enzymes, antibodies, membrande carriers and receptors for neurotransmitters.

SUMMARY QUESTIONS

1. Explain how the order of base triplets in a strand of DNA determines the sequence of amino acids present in a protein.

2. Explain the function of each of the following important proteins.
   a An enzyme
   b A membrane carrier
   c A receptor for a neurotransmitter.

3. The sequence of base triplets in a strand of DNA is:

   ACCAGCTTTACCAGT

   Use information on page 202 to list the sequence of amino acids that this section of DNA codes for.

16.4 Mitosis

Supplement

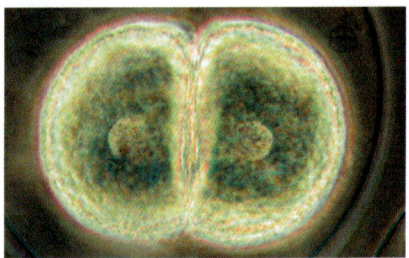

Figure 16.4.1 A zygote of a sea urchin has just divided in two by mitosis. You can see that each cell has a nucleus.

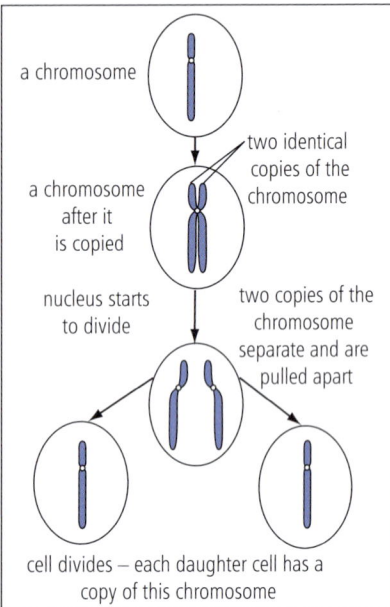

a chromosome

a chromosome after it is copied

two identical copies of the chromosome

nucleus starts to divide

two copies of the chromosome separate and are pulled apart

cell divides – each daughter cell has a copy of this chromosome

Figure 16.4.2 This is what happens to each chromosome when cells divide by mitosis.

New cells are produced by the division of old ones. In Figure 16.4.1, the fertilised egg of a sea urchin is in the process of dividing into two **daughter cells**. Before the cell divides, the nucleus must divide. As the nucleus is finishing its division into two, the cytoplasm divides. In animal cells, a constriction occurs between the two nuclei to divide the cell into two. In plants, a new cell wall is formed between the nuclei.

There are two ways in which nuclei divide.

- **Mitosis** is the type of nuclear division that occurs during growth and asexual reproduction. The daughter cells are genetically identical.
- **Meiosis** occurs in sex organs to form gametes. It is the nuclear division that gives rise to daughter cells that are not genetically identical.

The process of mitosis

Before a cell can divide, new copies of the genetic information in the DNA of the chromosomes must be made. This is a very reliable process in which exact copies are made. This copying process occurs *before* the nucleus divides. While it is going on, the DNA in the chromosomes is uncoiled and arranged very loosely in the nucleus. You cannot see chromosomes in the nuclei when this is going on.

The new copy of DNA of each chromosome is attached to the original copy. As mitosis begins, the DNA coils up so that each chromosome becomes thicker. They become visible with the microscope when a suitable staining technique is used.

During mitosis, the two copies separate so each new cell gets a copy of each chromosome. Figure 16.4.2 shows what happens to one chromosome.

As a result of **mitosis**, each daughter cell has the same chromosome number as the original parent cell. As the chromosomes have been copied by a reliable system, they are genetically identical to each other and to the parent cell. If they were genetically different they would be rejected by the body's immune system.

The significance of mitosis

Mitosis produces cells that are an exact copy of the parent cell. These daughter cells have the same number of chromosomes and are genetically identical to the parent cell. Mitosis is important for the following processes:

- **growth** – in animals this happens all over the body, in plants it happens in special growing areas such as the tips of stems and roots
- **repair** of wounds – your skin cells divide by mitosis to repair damaged tissues and wounds

- **replacement** of cells that wear out and die, such as red blood cells, which only live for a short time
- **asexual reproduction** – this occurs in fungi and in plants, but is rare in the animal kingdom. There are some examples of asexual reproduction on page 174.

Growth: As multicellular organisms grow, the number of cells making up their tissues increases. The new cells must be identical to the existing ones. Growth by mitosis takes place over the whole body in animals. In plants, group is confined to certain areas called **meristems**.

Repair and replacement of cells: Damaged cells must be replaced by identical new cells. Your skin cells and the cells lining your gut are constantly dying and being replaced by identical cells. Red blood cells, which live only for a short time, will wear away and die. They also need to be replaced by mitosis.

Asexual reproduction: Asexual reproduction results in offspring that are identical to the parent. Mitosis occurs when unicellular organisms reproduce – for example, yeast and potato tubers (Topic 15.1).

Stem cells

Some of the cells that form from an embryo remain as unspecialised cells. They do not develop into any type of specialised cell. These cells keep dividing by mitosis to provide the body with a constant supply of new cells. These cells are needed to replace cells that get worn out and die. There is a layer of stem cells just under the outer layer of skin cells. These stem cells continually divide to make cells that become toughened for protection. They get rubbed off and need to be constantly replaced. When skin is wounded, these stem cells produce more cells to repair the damage. There are many stem cells in the bone marrow to produce new red blood cells and phagocytes as replacements.

Each time a stem cell divides, one daughter cell starts to become specialised for some specific function and the other daughter cell remains unspecialised to go through the process of mitosis again. The cells in meristems in plants have the same role as stem cells in animals.

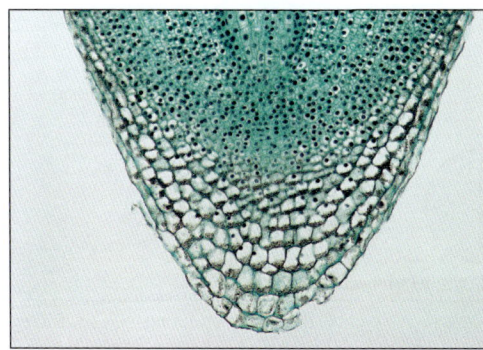

Figure 16.4.3 Mitosis takes place in the meristem behind the root tip.

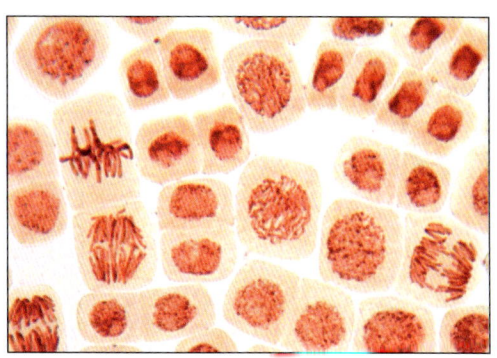

Figure 16.4.4 Cells at various stages of mitosis.

16.5 Meiosis

Supplement

LEARNING OUTCOMES

- Define meiosis as a reduction division in which the chromosome number is halved, resulting in genetically different cells

- Explain how meiosis produces variation by forming new combinations of maternal and paternal chromosomes

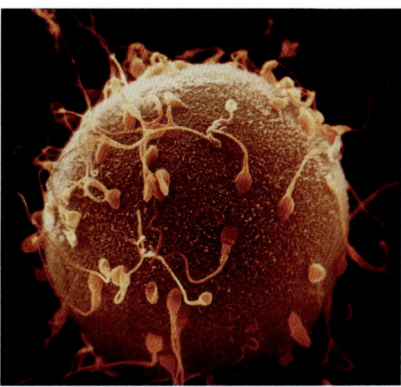

Figure 16.5.1 Human sperms and eggs contain 23 chromosomes as a result of meiosis.

Meiosis

Meiosis is a type of cell division that is essential for sexual reproduction. It takes place in the reproductive organs and results in the formation of gametes with half the normal chromosome number. This is known as the **haploid number**, as opposed to the full chromosome number, which is the **diploid number**.

If human sperm cells and egg cells both had 46 chromosomes the zygote formed at fertilisation would have 92 chromosomes. If this continued generation after generation the number of chromosomes in the nuclei would double every generation. This does not happen. Instead, the number remains at 46. This is because there is a different sort of nuclear division in the life-cycle involved in producing gametes. This type of nuclear division is a reduction division called **meiosis**.

Meiosis halves the number of chromosomes so egg cells and sperm cells have only 23 chromosomes each. The zygote has 46 chromosomes, 23 from the mother, in the egg, and 23 from the father, in the sperm. In sexual reproduction the number of chromosomes stays constant from generation to generation.

Meiosis occurs in sex organs in humans – the ovaries and testes. In flowering plants meiosis occurs in the anthers and in the ovules.

In meiosis the daughter cells are not identical. They are genetically different and this contributes to the genetic variation that appears in the next generation following sexual reproduction. This genetic variation is important in natural selection as you will see in page 226.

Figure 16.5.3 shows what happens to one pair of chromosomes during meiosis. As you can see there are **two divisions** of the cell

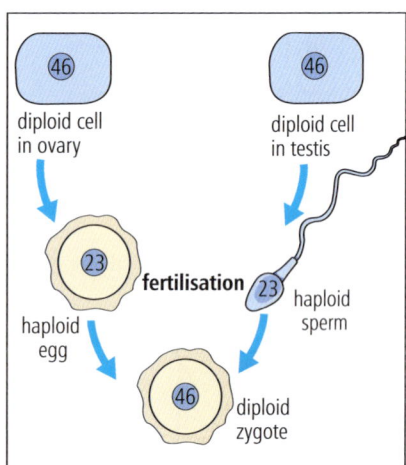

Figure 16.5.2 The number of chromosomes in the gametes is called the haploid number. The number in a zygote is the diploid number.

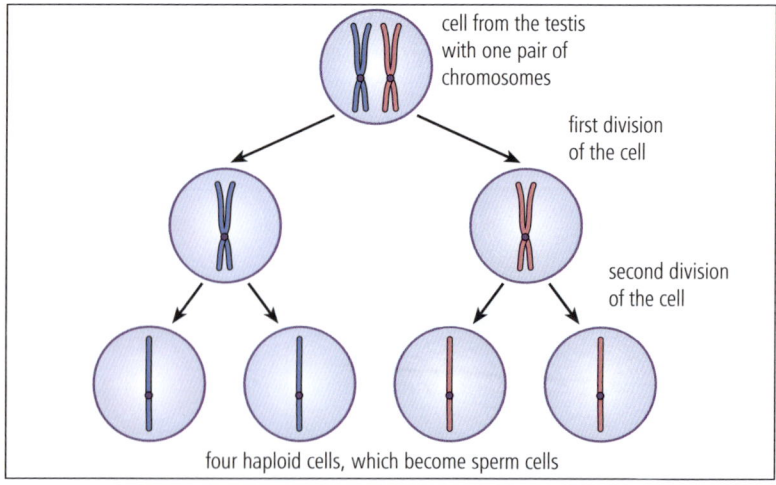

Figure 16.5.3 This shows what happens to one pair of chromosomes during meiosis. There are two divisions of the cell to give four cells. Each of these has one chromosome so the number has been halved.

resulting in four cells. Each of these four cells has one chromosome, so the number has been halved to give a haploid gamete.

The pair of chromosomes shown in the first cell in Figure 16.5.3 are different from each other. One is paternal in origin as it was inherited from the father, and one is maternal as it was inherited from the mother. When the 23 pairs of chromosomes separate in the first division of meiosis, the daughter cells gain different combinations of paternal and maternal chromosomes. This is one way in which the haploid cells produced in meiosis are genetically different from one another.

Differences between mitosis and meiosis

The differences between mitosis and meiosis are summarised in Table 16.5.1 and in Figure 16.5.5

Table 16.5.1 Comparing mitosis and meiosis.

mitosis	meiosis
one division	two divisions
the number of chromosomes remains the same	the number of chromosomes is halved
the daughter cells are genetically identical	the daughter cells are genetically different from the parent cells
two daughter cells are formed	four daughter cells are formed

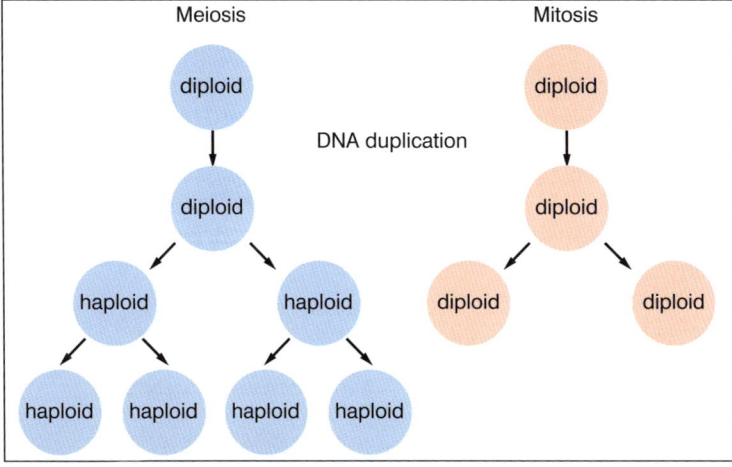

Figure 16.5.5 Meiosis and mitosis compared.

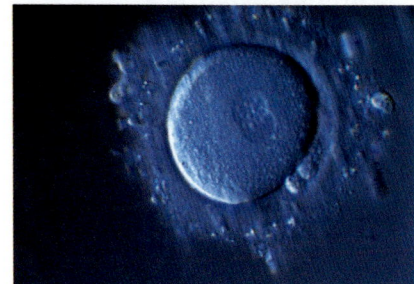

Figure 16.5.4 A fertilised human egg or zygote. The nucleus of the sperm is fusing with the nucleus of the egg to give a cell with 46 chromosomes.

EXAM TIP

Be careful when you write the words <u>mitosis</u> and <u>meiosis</u>. They are easily confused and you should learn how to spell them correctly.

KEY POINTS

1 Meiosis is a reduction division in which the chromosome number is halved from diploid to haploid resulting in cells that are genetically different.

2 Meiosis produces variation by forming new combinations of maternal and paternal chromosomes.

SUMMARY QUESTIONS

1 a How many pairs of chromosomes are there in the nucleus of a human body cell?

  b State the number of chromosomes in these human cells:

    i an egg cell   ii a sperm cell

    iii a zygote   iv a red blood cell

2 a State three ways in which meiosis differs from mitosis.

  b Explain how meiosis produces variation.

  c Explain the advantage of genetic variation among gametes.

16.6 Inheritance and genes

Cat breeders know that if they breed together two cats that both have long hair, all the kittens will have long hair. If two short-haired cats are bred together then there may be some long-haired and some short-haired kittens in the litter, or there may be no long-haired kittens at all. Breeders can make pedigree diagrams to show the features that appear within a family of cats as in Figure 16.6.1. You will see that a pedigree diagram looks very like a family tree, but symbols are used to indicate the gender of individuals.

Symbols for pedigree diagrams

☐ = male; ◯ = female

■ ● = individual with the condition studied

☐ ◯ = individual not showing the condition

individuals may be identified by generation and number, e.g., I – 2, II – 5 and III – 1.

shading = long hair   no shading = short hair

Figure 16.6.1 A pedigree diagram showing the inheritance of hair length in a family of cats.

Genes and alleles

A gene is a length of DNA that is the code for making a protein molecule. Each gene is always located in the same place on one of the chromosomes. Each species has its own genes and every individual from the same species has the same genes. However, genes vary between individuals within each species. The different versions of each gene are known as **alleles**. There may be two alleles of a gene (see below), but usually there are many more.

In Figure 16.6.2, you can see three genes on one pair of chromosomes. You can also see the alleles **A/a**, **B/B** and **d/d**.

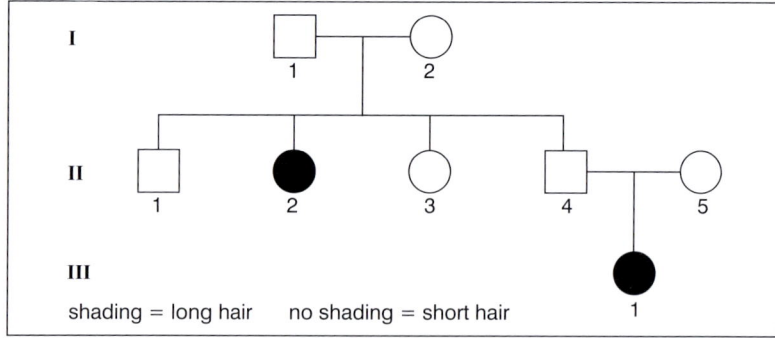

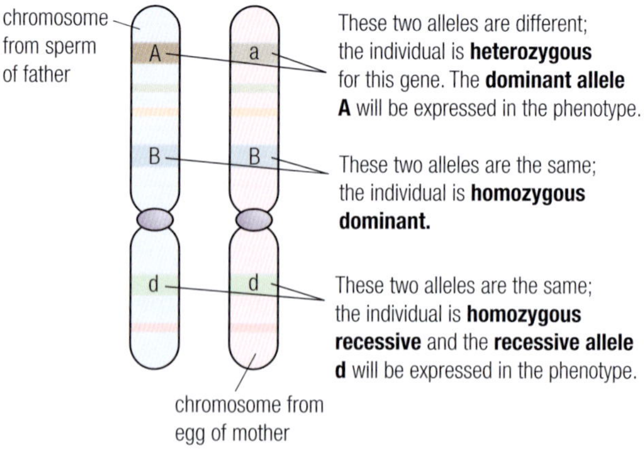

chromosome from sperm of father

These two alleles are different; the individual is **heterozygous** for this gene. The **dominant allele A** will be expressed in the phenotype.

These two alleles are the same; the individual is **homozygous dominant.**

These two alleles are the same; the individual is **homozygous recessive** and the **recessive allele d** will be expressed in the phenotype.

chromosome from egg of mother

Figure 16.6.2 This shows three genes that are on the same chromosome.

Sometimes an organism gains two copies of the same allele so there is only one type of instruction to use during growth, development or metabolism. But many organisms contain two different alleles (Figure 16.6.2). If an allele is expressed and affects the appearance of the individual, it is said to be a dominant allele. If it is not, then it is said to be a recessive allele.

In the 1850s, Gregor Mendel investigated inheritance in pea plants. He noticed that among garden pea plants one variety grew to be tall while another remained short (dwarf). He used pea plants that always bred true – from generation to generation they were either tall plants or dwarf plants.

When he cross-bred these two varieties, all the next generation were tall. None of them were dwarf. But when this new generation of pea plants self-pollinated and produced seeds, the next generation consisted of many tall plants and some dwarf. He reasoned that dwarfness had been passed on although it had 'skipped' a generation. He did not know anything about genes and DNA, but he used his knowledge of algebra to give letters to represent what he called 'factors' that controlled height.

Using modern terminology, we say that there is a gene that controls height in pea plants. This gene has two alleles.

Let **T** represent the allele for tall, and let **t** represent the allele for dwarf. A capital letter is used for the dominant allele and a lower case letter for the recessive allele.

The **genotype** of an organism is the alleles it has. So the genotype of a pea plant could be **TT**, **Tt** or **tt**. If the two alleles are the same, **TT** or **tt**, the plant is homozygous. They can be homozygous dominant (**TT**) or homozygous recessive (**tt**). Individuals are heterozygous if they have two alleles that are different, for example **Tt**.

The **phenotype** is the way the alleles are expressed in an individual. It refers to all the aspects of an organism's biology except its genes. We will apply the term to one feature of an organism, for example height in pea plants. The genotypes **TT** and **Tt** result in plants with the tall phenotype, whereas the genotype **tt** results in dwarf plants.

Figure 16.6.3 Mendel studied the inheritance of seven different features in pea plants including the shape of the seeds – round and wrinkled.

KEY POINTS

1 Pedigree diagrams show how features are inherited in families.

2 An allele is a version of a gene. Most genes have many alleles.

3 The genotype is the genetic makeup of an organism; the phenotype is all the features of an organism other than its genotype.

4 A dominant allele is always expressed in the phenotype.

5 A recessive allele is expressed only when no dominant allele is present.

6 A homozygous individual has two identical alleles of a gene and a heterozygous individual has two different alleles of a gene.

SUMMARY QUESTIONS

1 Chromosomes occur in pairs in all cells except the gametes. Explain why they are not in pairs in gametes.

2 a What are alleles?
  b Explain what is meant by the terms *dominant allele* and *recessive allele*.

3 Explain each of the following genetic terms:
  a genotype   b phenotype
  c homozygous   d heterozygous

4 Explain how you would find out whether the allele for wrinkled seed shape is dominant or recessive.

16.7 Monohybrid inheritance

Figure 16.7.1 Flower colour is often controlled by a single gene.

Monohybrid inheritance concerns the inheritance of a *single* characteristic, such as plant height or flower colour.

It involves the inheritance of the alleles of one gene. We will use the example of Mendel's peas.

There is one gene for height in pea plants. The gene for plant height has two alleles: tall, **T**, and dwarf, **t**.

There are three possible genotypes for plant height:

TT = homozygous tall

Tt = heterozygous tall

tt = homozygous dwarf.

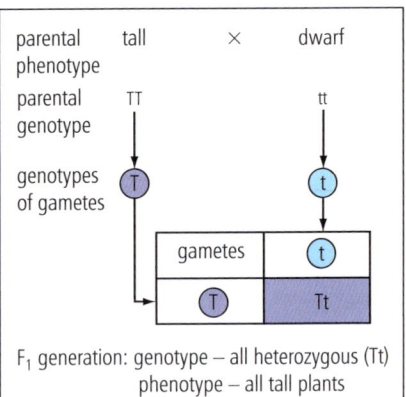

Figure 16.7.2 A genetic diagram for a single characteristic. The box showing the combination(s) of gametes is called a **Punnett square**.

The parental cross

The genetic diagram in Figure 16.7.2 shows a cross between a homozygous tall plant, **TT**, and a homozygous dwarf plant, **tt.** These represent the pure breeding plants that Mendel used in his crosses. This is called the parental cross.

As a result of meiosis, all the gametes from the tall plant contain the dominant allele, **T**, and all the gametes from the dwarf plant contain the recessive allele, **t**. When fertilisation occurs, the new plants receive one dominant allele and one recessive allele so will be heterozygous and all have the genotype, **Tt**. Their phenotype will be tall, because the allele, **T**, is dominant to the recessive allele, **t**.

The first generation is known as the F_1 generation, and in this particular genetic cross, all the plants in the F_1 generation are heterozygous tall (**Tt**).

Crossing the F_1 generation (3:1 ratio)

The F_1 plants are allowed to self-pollinate (Figure 16.7.3). When meiosis occurs, half of the gametes of *each* plant will have the dominant allele, **T**, and half will have the recessive allele, **t**.

This time when fertilisation occurs, there are three possible combinations of alleles in the F_2, or second generation:

TT (homozygous tall)   **Tt** (heterozygous tall)

tt (homozygous dwarf).

Among this generation ¼ (or 25%) will be homozygous tall; ½ (or 50%) will be heterozygous and therefore tall; ¼ (or 25%) will be homozygous dwarf. Since the phenotypes of **TT** and **Tt** are the same, ¾ of the F_2 plants will be tall and ¼ will be dwarf. This can also be written as 3 tall : 1 dwarf.

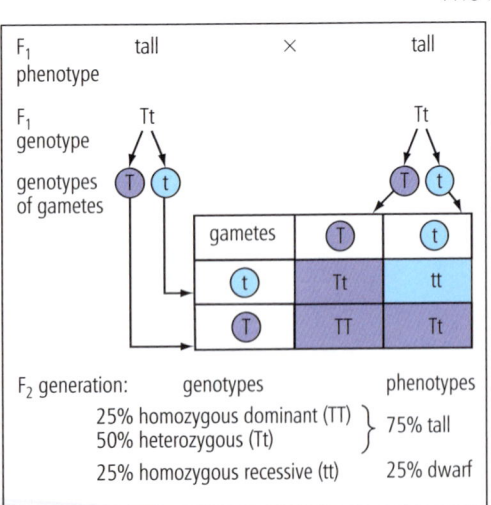

Figure 16.7.3 75% : 25% is a 3:1 ratio.

Supplement

A test cross

If you look at the last two crosses, you will see that the recessive allele is not expressed in the F$_1$ generation but is expressed in the F$_2$ generation. This means that we do not know the genotype of a tall plant just by looking at it because plants with genotypes **TT** and **Tt** have the same phenotype. We do know the genotype of a dwarf plant because it can only be **tt**.

A test cross is used to determine the genotype of an individual to find out whether it is homozygous or heterozygous (Figure 16.7.4).

If our tall plant is homozygous (**TT**), then crossing it with a dwarf plant (**tt**) gives all tall plants (**Tt**). If, however, our tall plant is heterozygous (**Tt**), then crossing it with a dwarf plant (**tt**) will give half tall plants (**Tt**) and half dwarf plants (**tt**) in a ratio of 1:1.

EXAM TIP

Genetic diagrams should always include a Punnett square to show all the possible fusions between the gametes.

EXAM TIP

Heterozygous individuals are not pure breeding as they pass on different alleles of the gene.

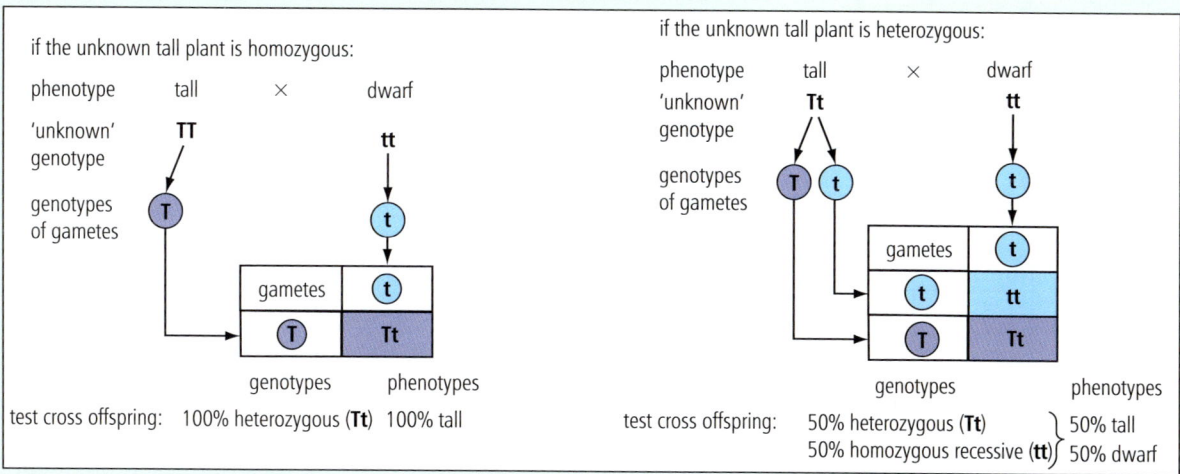

Figure 16.7.4 Genetic diagrams like these show the ratios expected in the offspring, not the actual numbers of offspring.

SUMMARY QUESTIONS

1 Complete these sentences by choosing the correct word from inside the brackets.

   a A pea plant with a tall (genotype/phenotype) could have the (genotype/phenotype) **TT** or **Tt**.

   b A tall pea plant with genotype **TT** is (homozygous/heterozygous) dominant.

   c A tall pea plant with genotype **Tt** is (homozygous/heterozygous).

   d A dwarf pea plant with genotype **tt** is (homozygous/heterozygous) recessive.

2 Fruit flies have a gene that determines the length of their wings.

   a Using the symbols **W** = long wing and **w** = short wing:

      i State the genotype of a fruit fly that is heterozygous for this characteristic.

      ii Give the genotypes of its gametes.

   b A fruit fly heterozygous for wing length is mated with one that is homozygous for short wing. Use a genetic diagram to predict the proportions of each wing type in their offspring.

KEY POINTS

1 Monohybrid inheritance involves the inheritance of a single characteristic such as plant height or flower colour.

2 Crossing a homozygous tall plant with a homozygous dwarf plant produces all heterozygous tall plants.

3 Crossing two heterozygous tall plants gives a ratio of 3 tall plants to 1 dwarf plant (3:1 ratio).

4 Crossing a heterozygous tall plant with a homozygous dwarf plant gives a ratio of 1 tall plant to 1 dwarf plant (1:1 ratio).

16.8 Codominance

Supplement

Codominance

Up to now the examples we have looked at involve alleles that are either dominant or recessive. The recessive allele is not expressed in the phenotype of individuals with the heterozygous genotype. Sometimes, however, *both* alleles are expressed and neither is dominant.

So the phenotype is a mixture of the effects of each allele.

This condition is known as **codominance** and the alleles are called codominant alleles. The four o'clock plant can have red flowers or white flowers. Flower colour is determined by one gene. However, if you cross homozygous red-flowered plants with homozygous white-flowered plants, the offspring are all pink. The two parents produce heterozygous offspring (the F_1 generation) in which both alleles are expressed to give pink – an intermediate colour.

You can see in Figure 16.8.1 what happens when the pink plants are pollinated among themselves. There are red-, pink- and white-flowered plants in the F_2 generation in a ratio of 25% : 50% : 25% or 1:2:1. Notice that the ratio of the phenotypes is the same as the ratio of the genotypes when the alleles are codominant.

Notice also that each allele is represented as a capital letter, so red flower = C^R and white flower = C^W. This shows that neither allele is recessive and that they both exert an equal effect on the phenotype.

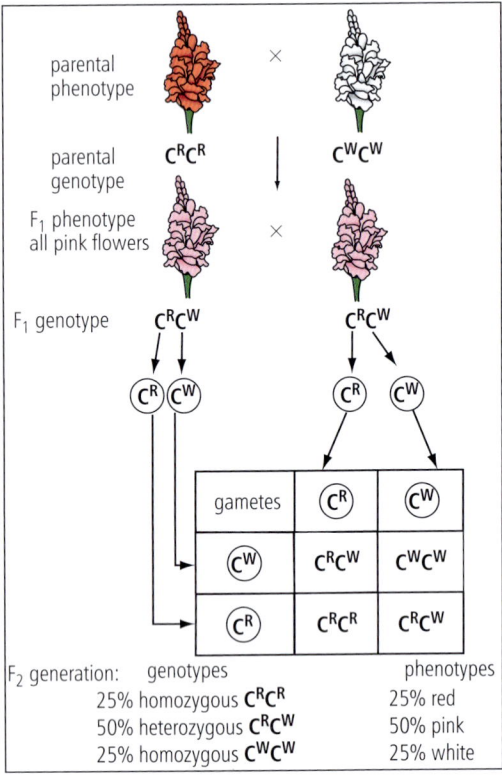

Figure 16.8.1 Codominance in flower colour.

F₂ generation: genotypes
25% homozygous $C^R C^R$
50% heterozygous $C^R C^W$
25% homozygous $C^W C^W$

phenotypes
25% red
50% pink
25% white

Multiple alleles

So far, we have looked at examples where a gene has only two alternative alleles. There are examples where a gene may have three or more alleles.

The four groups of the human ABO blood group system are determined by one gene, **I**, with three different alleles: I^A, I^B and I^O (Figure 16.8.2).

The alleles I^A and I^B are codominant and code for slightly different molecules on the surface of red blood cells. The allele I^O is recessive to both I^A and I^B and does not code for one of these molecules.

Genotype	Blood group
$I^A I^A$	A
$I^A I^O$	A
$I^B I^B$	B
$I^B I^O$	B
$I^A I^B$	AB
$I^O I^O$	O

Look at the table. You can see the possible genotypes for each blood group. Remember any person can only have two alleles as we are all diploid.

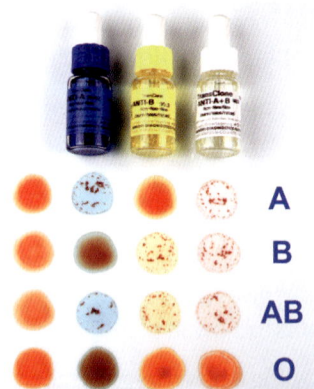

Figure 16.8.2 Blood of the four different groups tested with anti-sera A (blue), anti-sera B (yellow) and anti-sera A and B (white). Red blood cells clump together if the anti-sera reacts with the A or B molecules on their surfaces.

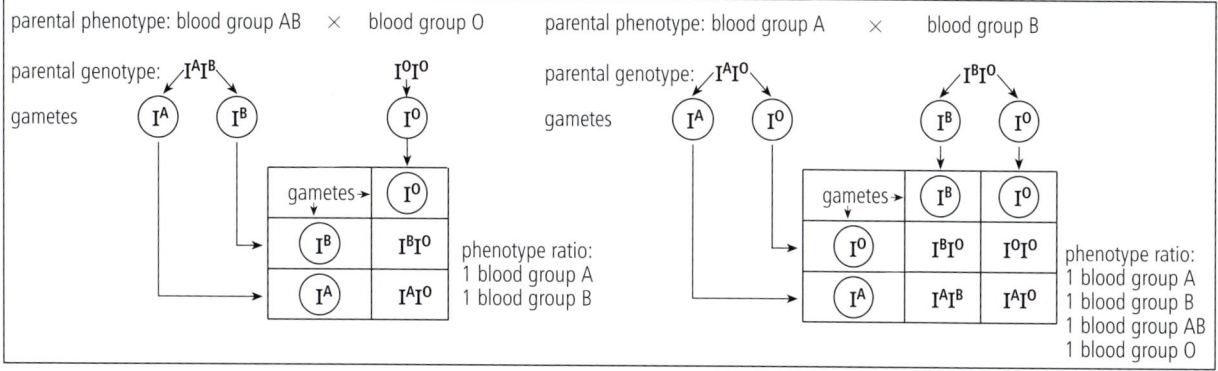

Figure 16.8.3 Inheritance of blood groups

Look at the genetic diagrams in Figure 16.8.3. Two parents have blood groups AB and O. The children will have blood groups A or B. None of them will have the same blood group as their parents.

Another two parents have blood groups A and B. If they are both heterozygous, they could have children with any of the four blood groups – A, B, AB and O.

SUMMARY QUESTIONS

1 There are red-flowered, white-flowered and pink-flowered varieties of the four o'clock plant (Figure 16.8.4). The results of a number of crosses between the varieties are shown in the table. Explain the results of each cross.

crosses	number of plants of each colour		
	red	pink	white
red × pink	126	131	
white × pink		92	88
red × white		115	
pink × pink	39	83	43

2 Use a genetic diagram to work out the possible blood groups of the offspring where the father is heterozygous for blood group A and the mother is blood group O.

3 The inheritance of coat colour in some cattle is an example of codominance. If a homozygous red cow is crossed with a homozygous white bull, all the calves have coats with a colour known as 'roan'.

The allele R^R causes red hairs to be produced and the allele R^W causes white hairs. Heterozygous cattle have coats that are a mixture of red and white hairs giving roan.

Use genetic diagrams to show the genotypes and phenotypes of the offspring in the following crosses:

a a roan bull and a roan cow

b a roan bull and a red cow

c a roan bull and a white cow

Figure 16.8.4 Flowers of the four o'clock plant.

16.9 Sex linkage

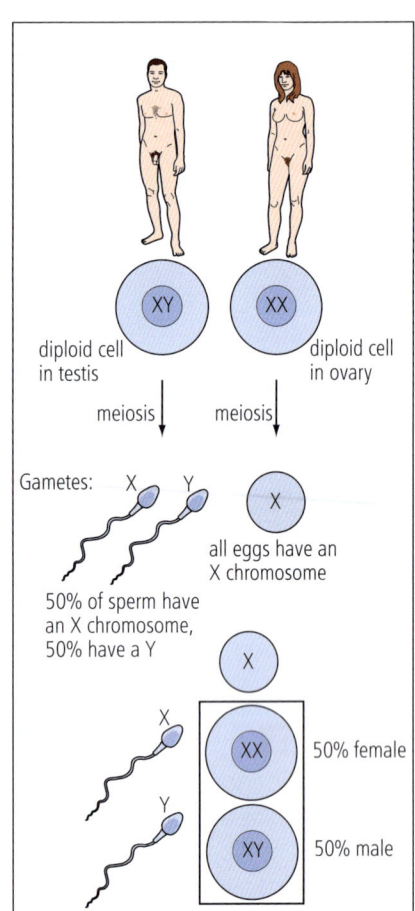

Figure 16.9.1 The diagram shows the inheritance only of the sex chromosomes, X and Y. Do not forget that each of the gametes (egg and sperm) contain 22 other chromosomes to give the haploid number of 23.

Sex determination

Your chromosomes determine which sex you are. In humans, there are 46 chromosomes and these occur in 23 pairs. In females, all 46 chromosomes can be matched together into pairs. This is done by taking images of the chromosomes and matching them for size and shape. In males, there are two chromosomes that are not alike. These are known as the X and Y chromosomes. Females have two X chromosomes and males have an X chromosome and a Y chromosome.

In Figure 16.9.1, you can see how sex chromosomes in the gametes determine the sex of individuals. Remember that the number of chromosomes is halved in meiosis so each gamete can have only one of the two sex chromosomes.

All egg cells contain an X chromosome. Half of the sperm contain an X chromosome and half of the sperm contain a Y chromosome. At fertilisation, the egg may fuse with either a sperm with an X chromosome or a sperm with a Y chromosome. Since there are equal numbers of sperm with the X chromosome and sperm with the Y chromosome, there is an equal chance of the zygote being XX or XY and the child being female or male.

Sex inheritance

As you can see in Figure 16.9.2, the Y chromosome is very small and much smaller than the X chromosome. Therefore, the Y chromosome has far fewer genes than the X chromosome or any other chromosome. One gene on the Y chromosome stimulates the development of testes in the embryo. If it is not present (or if it has mutated) then the body that develops is female.

Supplement

Sex linkage

The genes that are located on the X chromosome are described as sex-linked even though they have nothing to do with determining gender or controlling sexual characteristics. Among the genes on the X chromosome are genes involved with controlling vision and blood clotting.

Males only have one copy of the genes that are on X chromosomes. This means that if any of them are recessive, the effect will be seen. Because women have two X chromosomes they are less affected by sex-linked recessive alleles and this is why sex-linked conditions are more common in boys than in girls.

Supplement

Colour blindness

One of the genes on the X chromosome controls the ability to see red and green colours. The gene works in the receptors, known as cones, in the retina of the eye. There is an allele of this gene that does not produce a protein necessary for colour vision. The allele is recessive, so any girl or woman who is heterozygous, **Rr**, has normal colour vision. Males only have one X chromosome so if they have inherited the allele **r** they will have red-green colour blindness.

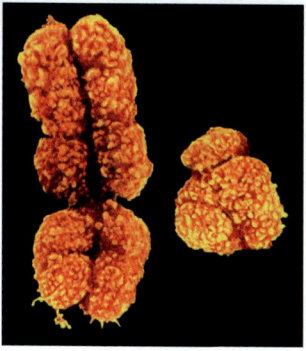

Figure 16.9.2 An X chromosome on the left and a Y chromosome on the right, photographed in an electron microscope.

Figure 16.9.3 shows what to expect if the mother is homozygous dominant and the father is colour blind. Notice that none of the children can expect to be colour blind, but the girls will inherit the allele for colour

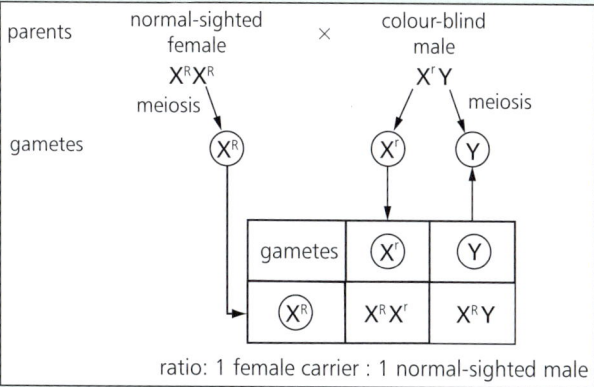

Figure 16.9.3 Inheritance of colour blindness from a colour-blind father and normal-sighted mother.

blindness on the X chromosome from their father. They have this allele, but it does not affect their phenotype. Any females who are heterozygous are **carriers** of red-green colour blindness.

Figure 16.9.4 shows what to expect if the mother is a carrier and the father has normal colour vision. In this case, there is a 1 in 4 chance that one of the children will be colour blind. Note that there is a 1 in 2 chance that any boy will be colour blind.

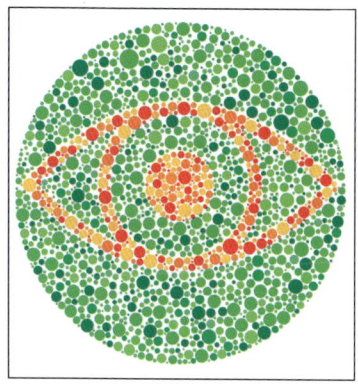

Figure 16.9.5 What can you see? If you cannot see something familiar then you may have red-green colour blindness.

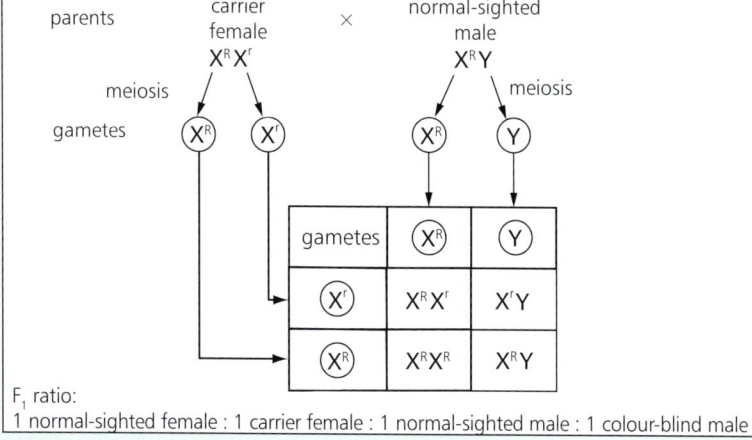

F$_1$ ratio:
1 normal-sighted female : 1 carrier female : 1 normal-sighted male : 1 colour-blind male

Figure 16.9.4 Inheritance of colour blindness from a carrier mother and normal-sighted father.

KEY POINTS

1. Sex is determined by the sex chromosomes, X and Y. Males are XY and females XX.

S 2. A sex-linked characteristic is one in which the gene responsible is located on a sex chromosome, usually the X chromosome.

3. Males have only one X chromosome so recessive alleles are more likely to be expressed in males than in females.

4. Females are carriers of sex-linked disorders, such as red–green colour blindness.

SUMMARY QUESTIONS

1. Explain how sex is determined in humans.

S 2. Explain the following terms:
   a sex linkage **b** carrier **c** red-green colour blindness

Practice questions

1 A male fruit fly with red eyes was crossed with a female fruit fly with orange eyes (parental generation). All the offspring had red eyes (first generation). The fruit flies in the first generation were bred together. Of the offspring, 75% had red eyes and 25% had orange eyes (second generation).

Which row shows the correct description of the genotypes in the first and second generations?

	genotypes in first generation	genotypes in second generation
A	all homozygous	all heterozygous
B	all homozygous	50% heterozygous 50% homozygous
C	all heterozygous	all homozygous
D	all heterozygous	50% heterozygous 50% homozygous

(Paper 1)   [1]

2 A couple discover that they are both carriers of the condition albinism in which skin pigment is not made. What is the probability that their first child will have this condition?

A   0%

B   25%

C   50%

D   100%

(Paper 1)   [1]

3 A recessive allele:

A   causes a harmful feature

B   is never expressed in the phenotype

C   is not expressed when in a genotype with a dominant allele

D   produces the same phenotype when homozygous as when heterozygous

(Paper 1)   [1]

4 Which is meiosis involved with?

A   asexual reproduction

B   growth of stems and roots

C   production of egg cells

D   repair of tissues

(Paper 2)   [1]

5 Which explains the advantage of meiosis occurring in a life cycle?

A   meiosis allows sexual reproduction to occur

B   meiosis produces four times as many cells as mitosis

C   meiosis produces haploid gametes

D   meiosis produces sperm and eggs

(Paper 2)   [1]

6 A man has normal colour vision. His wife does not have colour blindness but her father has red-green colour blindness. What is the probability that their sons will be colour blind?

A   25%

B   50%

C   75%

D   100%

(Paper 2)   [1]

7 The Indian muntjac deer has a diploid number of 6. How many chromosomes are there in a skin cell and in a sperm cell of this species of deer?

	skin cell	sperm cell
A	6	6
B	12	6
C	6	3
D	3	6

(Paper 2)   [1]

8 Which feature is controlled only by a gene and *not* influenced by the environment?

A   blood group

B   concentration of glucose in the blood

C   heart disease

D   blood pressure

(Paper 2)   [1]

9 Which describes the genetic code?

A   the four different bases that are found in DNA

B   the mRNA molecules that leave the nucleus in a cell that is making proteins

C   the sequence of bases in a gene

D the triplet of bases in DNA and RNA that code for the different amino acids in proteins

(Paper 2) [1]

10 (a) Define the term *inheritance*. [2]

(b) Explain the difference between the following pairs of terms:

(i) genotype and phenotype [2]

(ii) gene and allele [2]

(Paper 3)

11 Brachydactyly is a rare genetic condition in which people with the condition have short fingers and toes. A pedigree diagram was drawn for this rare condition:

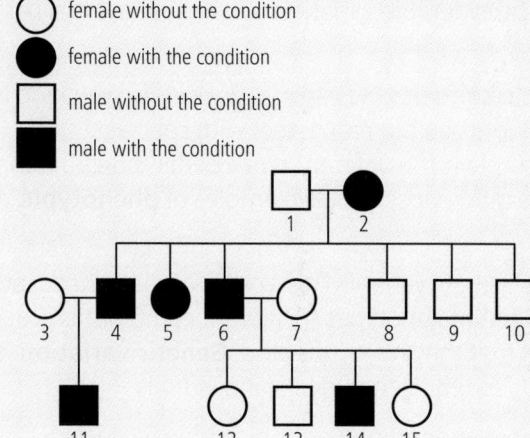

Key

○ female without the condition

● female with the condition

□ male without the condition

■ male with the condition

(a) (i) State whether brachydactyly is caused by a dominant or a recessive allele. [1]

(ii) Give evidence from the family tree that supports your answer to part **(i)**. Refer to the people shown in the tree by numbers. [3]

(b) What are the chances of parents 3 and 4 having another child with this condition? Draw a genetic diagram to explain your answer. [3]

(Paper 3)

12 (a) Use the letters **A** and **a** to explain the difference between a homozygous genotype and a heterozygous genotype. [3]

(b) In tabby cats the pattern of stripes in the coat is controlled by a gene. The allele for parallel stripes (**T**) is dominant to the allele (**t**) for a blotched pattern.

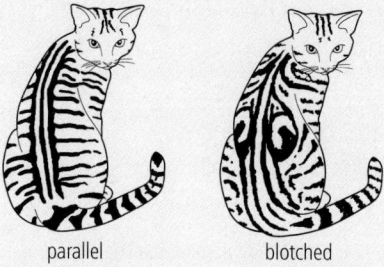

parallel   blotched

Use genetic diagrams to predict the proportions of cats with different coat patterns in the next generation in the following crosses:

(i) homozygous dominant × homozygous recessive, [4]

(ii) heterozygous × homozygous recessive. [4]

(c) Explain why it is possible to state the genotype of a tabby cat with blotched stripes, but not one with parallel stripes. [3]

(d) Explain why male and female tabby cats with blotched stripes cannot produce a kitten with parallel stripes. [2]

(e) Use the example of inheritance in this question to explain the terms *gene* and *allele*. [3]

(Paper 3)

13 (a) Explain how you would use a test cross to find the genotype of an organism that could either be homozygous dominant or heterozygous. [3]

(b) The four o'clock plant, *Mirabilis jalapa*, has a red-flowered variety and a white-flowered variety. When these are cross-bred the offspring have pink flowers.

(i) Explain why this is so. [3]

(ii) Show, by means of a genetic diagram, the result of cross breeding the pink-flowered plants among themselves. [5]

(Paper 4)

17.1 Variation

Figure 17.1.1 Although we show a lot of variation, we are all one species.

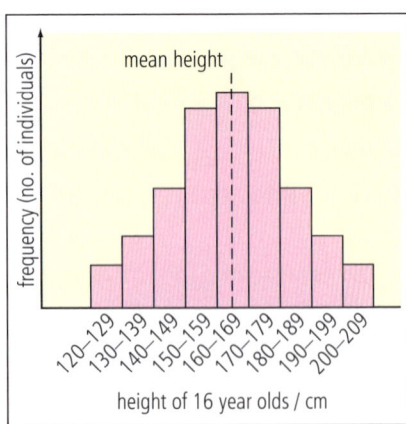

Figure 17.1.2 Frequency histograms are drawn to present data on continuous variation.

Types of variation

All humans have many features in common as we are all members of the same **species**. Now think about the similarities and differences between ourselves and monkeys and apes. We have many features in common, but there are significant differences that we use to categorise animals into different species. In biology, the term *variation* means:

- differences between species – we use these when constructing keys
- differences within a species.

It is differences *within* a species that we are concerned with in this unit. Although we all belong to the same species, there are many differences between us – even between identical twins. Some people are taller or heavier, others have different coloured hair, skin and eyes. There is considerable variation between individuals of the same species.

Besides variation in features that we can see, there is also variation in features that we cannot see but can detect with specialist equipment. An example is the different types of haemoglobin that people have. These differences are all examples of **phenotypic variation**.

As we have seen in Unit 16, variation also exists at the genetic level. Individuals have different genotypes – different combinations of alleles of the genes that they have inherited. **Genetic variation** is the differences between the genotypes of individuals.

There are two types of phenotypic variation within a species:

- continuous variation
- discontinuous variation.

Continuous variation

If you measured the heights of all the students in your year group, you would find a range of heights from the shortest to the tallest. If you divide the class into groups (170 cm to 179 cm, etc.) you can plot a frequency histogram (Figure 17.1.2). It is likely that the mean height will be in the middle of the range and correspond to the group with the largest number of people. This type of distribution is called a normal distribution.

This type of variation is called **continuous variation** because there is a continuous range of heights from shortest to tallest. There are many genes that contribute to your overall height. But environmental factors such as the quantity and quality of the food you eat and the amount of exercise you take influence your height. Also, people go through their growth spurts at different ages and this will influence the results you get from your class.

Other features that you can measure also show this type of variation. Try measuring the length of your index finger or your hand span. You can collect data from the whole class and plot as frequency histograms.

Continuous variation also occurs in plants. You can measure the lengths of leaves or stems and you will get the same sort of results – a range between two extremes with many intermediates in between. There are no clear divisions between the intermediates.

Discontinuous variation

Some people have attached ear lobes. Other people's ear lobes are free and not attached. This is an example of **discontinuous variation**. In this type of variation, there is usually a small number of phenotypes (e.g., attached and not attached ear lobes) and no intermediates. If you collect data about a feature showing this type of variation, you plot it as a bar chart (Figure 17.1.3).

Discontinuous variation is caused by genes alone and the environment has no effect. The ABO blood group system is another example. Everyone has an ABO blood group – it is either A, B, AB or O. There are no intermediates. Some flowering plant species have flowers of distinctly different colours. This is another example.

Inheritance versus environment

Identical twins develop from the same embryo and are genetically identical. However, they are different in many ways. These differences will have come about because they developed slightly differently in the womb and have been exposed to different environmental influences. Identical twins separated at birth or shortly afterwards and brought up apart often show striking similarities. These are often to do with personality, which shows the influence that genes have on us.

The effect of environmental factors can be investigated with plants. Figure 17.1.4 shows the plant sometimes called the Mexican hat plant. It reproduces asexually to make lots of tiny plantlets all around its leaves. You can detach these and grow them in pots and keep them under different conditions. Any variation shown will be because of **environmental** factors.

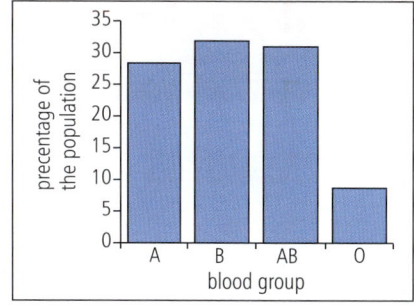

Figure 17.1.3 This bar chart shows the percentage of people with the four blood groups (A, B, AB and O) in Maharashtra, India. The variable 'Type of blood group' is called a 'categoric variable'. Unlike continuous variables, categoric variables are not measured.

Figure 17.1.4 These plantlets grow on the leaves of the Mexican hat plant. They are genetically identical.

EXAM TIP

When the words 'distinguish between' are used in a question, you need to give the differences between two things. Here you could be asked to distinguish between the two types of variation.

KEY POINTS

1 Variation is the difference between individuals of the same species.

2 Genetic variation is the differences in the genotypes of individuals; phenotypic variation is the differences in external and internal appearance.

3 Continuous variation is influenced by genes and the environment resulting in a range of phenotypes (e.g., height in humans).

4 Discontinuous variation results in a small number of phenotypes with no intermediates.

5 Discontinuous variation is cause by genes alone (e.g., human blood groups).

SUMMARY QUESTIONS

1 Distinguish between continuous and discontinuous variation and give three examples of each.

2 Describe how you would investigate the effect of an environmental factor on the growth of plants that all share the same genotype.

17.2 Investigations into variation

LEARNING OUTCOMES

- Describe an experiment to investigate an example of continuous variation.

- Describe an experiment to investigate an example of discontinuous variation

Figure 17.2.1 A quadrat in use.

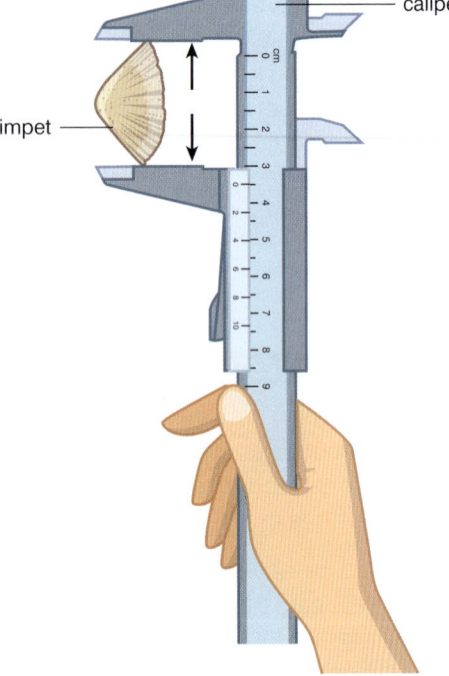

Figure 17.2.3 Measuring the length of a limpet with a caliper.

As we have seen, there are two main types of variation, continuous and discontinuous. There is an example of each of these types of variation in the following investigations.

Investigation into continuous variation: the length of limpets on a rocky shore

Limpets are aquatic snails with a distinctive conical shell and a strong muscular foot. They are found between the high and low watermarks on rocky shores. They are truly international, being found on rocky shores of the Caribbean, South Africa, Mediterranean and North Sea.

You cannot measure all the limpets on a particular rocky shore. This would be far too laborious and time-consuming. The best strategy is to take a representative sample of the population. **Quadrats** are often used when sampling in the rocky shore studies. A frame quadrat is a metal square and the most common size is 0.25 m². About 10 to 15 quadrats can be laid down at random sample points on the shore and the length of any limpet shell found within each quadrat can be measured with a **caliper**.

In our example the data was divided into six equal size groups and the number of limpets in each group was recorded in a tally chart.

shell length / mm	tally (number of limpets)	frequency
10–14	III	3
15–19	⊞⊞ ⊞⊞	10
20–24	⊞⊞ ⊞⊞ II	12
25–29	⊞⊞ ⊞⊞ ⊞⊞ ⊞⊞ ⊞⊞ III	28
30–34	⊞⊞ ⊞⊞ ⊞⊞ ⊞⊞	20
35–40	⊞⊞ ⊞⊞ ⊞⊞	15

Figure 17.2.2 A tally chart for limpet shell length.

This was used to draw a histogram – which is used to show frequency distributions, in this case the number of limpets of different lengths found on the rocky shore. As you can see, the graph displays a normal distribution reflecting continuous variation.

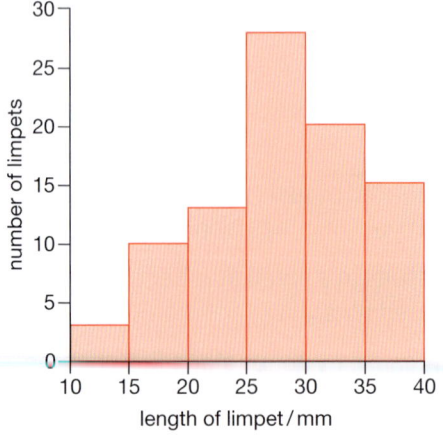

Figure 17.2.4 Histogram of continuous variation in limpet shell length.

Investigation into discontinuous variation: the different coloured flowers of sweet pea plants

With discontinuous variation, the characteristics can be divided into distinct categories into which individuals can be placed. There is no overlapping between categories since categories are distinctive, e.g., ABO blood groups. The characteristics tend to be one thing or another with no 'in-betweens'.

Gregor Mendel (see pages 207–208) tended to choose characteristics for his experiments that were controlled by a single gene displaying discontinuous variation, e.g., shape and colour of pea seeds.

In this investigation, a group of students bought a packet of sweet pea seeds with a variety of coloured flowers.

The seeds were sown in pots of soil, watered and placed in sunlight so that they could germinate.

Eventually they grew into mature plants and produced flowers.

The flowers were placed into four groups depending on their colour.

Numbers were recorded in a tally chart.

This data was used to draw a bar chart since the independent variable (flower colour) was categoric and the dependent variable (number of plants) was continuous.

Figure 17.2.5 Sweet pea flowers are found in a range of different colours.

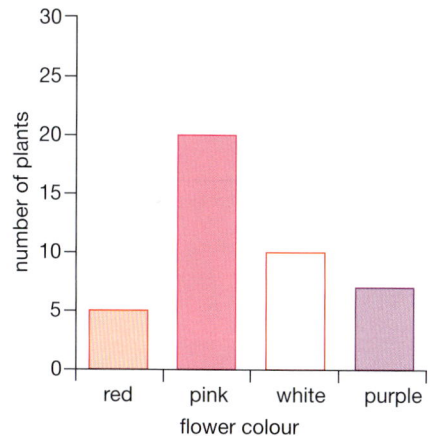

Figure 17.2.6 Bar chart of discontinuous variation in sweet pea colours.

SUMMARY QUESTIONS

1 Carry out one of these two investigations into variation:

   a Do the upper leaves on a bush differ in size from the lower leaves?

   b Does size of the leaves from plants grown on sunny habitats differ from the size of leaves of plants grown in shady habitats?

   Whichever investigation you choose to carry out, make a plan before you start.

2 Some students did a survey on the size of fruits produced by foxglove plants. Here are results:

length of fruit / mm	20	21	22	23	24	25	26	27	28	29
number collected	4	6	14	22	30	26	18	12	12	3

   a Draw a histogram of their results.

   b What sort of variation do you think this shows?

KEY POINTS

1 Data on examples of continuous variation can be measured, recorded and presented as histograms.

2 Data on examples of discontinuous variation can be collected, recorded and presented as bar charts.

17.3 Mutations

LEARNING OUTCOMES

- Define the term *mutation*
- State that mutation is the way in which new alleles are formed
- Describe the possible effects of ionising radiation and certain chemicals on the rate of mutations
- **S** Describe how changes in the base sequences of genes can alter the structure of a protein.

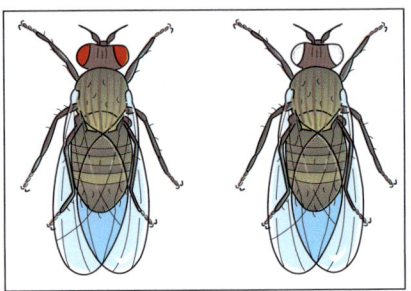

Figure 17.3.1 Red (normal) eye colour versus white (mutant).

A **mutation** is a change in a gene or a chromosome that may cause a change in a phenotypic characteristic. A gene mutation is a change in DNA molecules. Often this change can be harmful, but some mutations are beneficial and some have no effect at all.

Gene mutations

One of the first mutations studied was in the fruit fly. The normal eye colour of fruit flies is red, but a **mutant** form with white eyes was discovered (Figure 17.3.1). A gene controlling eye colour had changed. This was due to a change in the DNA so that the gene no longer coded for the production of the red pigment.

A similar example occurs in **albinos**. A gene controls production of the skin, hair and eye pigment **melanin**, which protects the skin and eyes from ultra-violet light. The gene can mutate to give an allele that does not produce melanin. It is a recessive allele and homozygous recessive individuals are albino.

Causes of mutations

Gene mutations are the only way in which new alleles are formed. They are caused by damage to DNA or by a failure in the copying process that occurs before nuclear division. They occur naturally at random, but the rate at which they occur is increased by exposure to ionising radiation and some chemicals. Ultra-violet radiation, X-rays and gamma rays are the most damaging. The greater the dose of radiation, the greater the chance of mutation. Benzopyrene in cigarette smoke is a chemical cause of mutations.

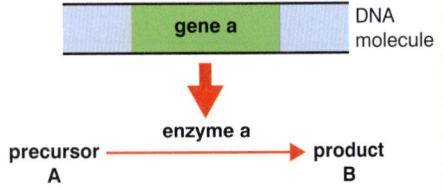

Figure 17.3.2 If gene a mutates it no longer codes for enzyme a and the pathway will be blocked

Supplement

Changes in base sequences

Gene mutations result from a random change in the base sequence of the DNA of a gene. For example, one strand of DNA has the following 'normal' base sequence:

CCT AGT ATT CGC TGA GGC TAA TG ...

A **substitution** has occurred in the following strand, can you spot it?

CCT AGA ATT CGC TGA GGC TAA TG

Can you see that substitutions only alter one or two DNA triplets? The result is that only one or two amino acids in the protein are changed. In some cases the new amino acid does not alter the structure of the protein a great deal, so the mutation may not matter at all.

Now see if you can spot the **deletion** in this DNA strand:

CCT AGT TTC GCT GAG GC AAT G

Only one base has been removed but this causes a shift in the whole sequence of bases and all the base triplets are changed after that point. this is known as a **frame shift**. Such mutations can result in a completely different protein structure from the original. The three-dimensional shape of the protein also becomes altered and the protein is no longer able to do its job. So deletions are far more damaging than substitutions.

ABO blood groups

All red blood cells may look the same, but they have different antigens on their cell membranes.

People can be put into one of four blood groups under the **ABO system**. These blood groups are called **A**, **B**, **AB** and **O**.

They are determined by which antigens are present on the red blood cells.

There are two main antigens: **A** and **B**.

- People in blood group **A** have only **A** antigens on their red blood cells.
- People in blood group **B** have only **B** antigens.
- People in blood group **AB** have both **A** and **B** antigens.
- People in blood group **O** have neither **A** nor **B** antigens.

The plasma will not contain antibodies that will attack its own antigens. So depending upon the blood group, the plasma may contain just **anti-A** antibiodies or just **anti-B** antibiodies, both of them or neither of them.

Blood type is determined genetically

The A and B antigen molecules on the surface of red blood cells are made by two different enzymes. These two enzymes are coded by different versions or alleles of the same gene.

The A allele codes for an enzyme that makes the A antigen and B allele codes for an enzyme that makes the B antigen.

The enzymes responsible for A and B antigens differ by 4 out of 354 amino acids. A third version of this gene, the O allele, codes for a protein that is not functional, so it makes no surface molecules at all.

In the case of the O allele there is a deletion that results in a loss of enzymic activity. The O allele differs slightly from the A allele by a deletion of a single nucleotide in its base sequence.

This deletion causes a frame shift and results in a translation of an almost entirely different protein that lacks enzymic activity so no A or B antigen can be made.

EXAM TIP

Remember that mutations are not the only source of genetic variation in populations. Exchange of chromosome material in meiosis, random mating and random fertilisation all contribute to genetic variation.

KEY POINTS

1 Mutation is a change in a gene caused by a change in the base sequence of DNA.

2 Ionising radiation and some chemicals increase the mutation rate.

S 3 Changes in the base sequences of a gene can alter the structure of a protein.

father	mother		
	A	**B**	**O**
A	AA	AB	AO
B	BA	BB	BO
O	OA	OB	OO

alleles		blood type
A + A	=	A
A + O	=	A
A + B	=	AB
B + B	=	B
B + O	=	B
O + O	=	O

Figure 17.3.3

SUMMARY QUESTIONS

1 a Explain what is meant by a *gene mutation*.

  b Explain how the rate of mutation can be increased.

  c Find the names of some human genetic diseases that are caused by gene mutation.

S 2 a Explain how changes in the base sequences of a gene can alter the structure of a protein.

  b Explain the following terms:
   i A substitution
   ii A deletion
   iii A frame shift

17.4 Adaptive features

LEARNING OUTCOMES

- Describe an *adaptive feature* as an inherited feature that helps in the survival and reproduction of an organism
- Interpret images or other information about a species to describe its adaptive features
- **S** Explain the adaptive features of the leaves, stems and roots of plants to aquatic and desert environments

An **adaptive feature** is an inherited feature that helps an organism to survive and reproduce in its environment. The photographs in Figure 17.4.1 show examples of adaptive features. The thistle (a) has tufts on its seeds so that they are carried by the wind to disperse them widely. The mayfly nymph (b) has a flattened, streamlined body and clings to the underside of rocks so it is not washed away. The Arctic fox (c) grows a thick white coat in winter for insulation and camouflage. Its body has a relatively low surface area to volume ratio because it has short legs and small ears, so there is less surface area to lose heat from. It can shelter in burrows in the snow and has a very keen sense of hearing.

Adaptive features like those in Figure 17.4.1 are likely to increase an individual's chances of surviving in its environment, and its chances of reproducing and having offspring, which are likely to show the same features.

The feathery tufts of thistle seeds are dispersed by the wind.

A mayfly larva has a flattened body so it does not get washed away in fast-flowing streams.

The Arctic fox has many features to enable it to survive extreme cold.

Figure 17.4.1 Adaptive features for different environments

Supplement

Figure 17.4.2 Water lilies, *Nymphaea alba*, are adapted to live in water.

Adaptations to aquatic environments

Hydrophytes are plants that grow submerged or partially submerged in water (Figure 17.4.2). Living in water has both costs and benefits.

Buoyed up by water and with no need for water transport, floating plants save energy because they produce little or no xylem tissue.

Roots, if present, are for anchorage and because there is no need for the roots to absorb water or mineral ions, there are no root hairs. The leaves and stems of hydrophytes have little or no cuticle because there is no need to conserve water.

The problem for hydrophytes is that carbon dioxide, which is needed for photosynthesis, diffuses through water much more slowly than it does through air. The same applies to oxygen (needed for plant respiration) because it is not very soluble in water.

Therefore, many hydrophytes have an extensive system of air spaces in their stems and leaves through which gases diffuse quickly. These air spaces provide buoyancy to keep the plants close to the light and are a reservoir of oxygen and carbon dioxide.

Desert adaptations

All plants have to balance water uptake with water loss. It is important that they maintain the turgor in their cells, or they will wilt. Very high rates of transpiration can kill a plant if it cannot absorb enough water to prevent long-term wilting.

Xerophytes are plants are that are able to exist in conditions where water is scarce. Cacti are xerophytes that survive in hot, dry (arid) desert regions (Figure 17.4.3). Cacti reduce water loss and conserve water in the following ways.

- Their leaves are reduced to spines. This reduces the surface area of the leaf over which water can be lost.
- A thick, waxy cuticle covers the plant's surfaces and reduces transpiration.
- They have swollen stems containing water-storage tissue.
- They have a shallow, spreading root system to quickly absorb any water from rain and overnight condensation.
- Many cacti have a round, compact shape that reduces their surface area so there is less surface through which water can be lost.
- They have shiny surfaces that reflect heat and light.
- Their stomata are closed during the day to reduce water loss. They open their stomata at night to absorb carbon dioxide, which they store for use in photosynthesis during the day. Photosynthesis occurs in the outer layers of cells in their stems.

(a) Swollen stems store water.

(b) Spines provide some shade and protection from herbivores.

Figure 17.4.3 Cacti have adaptions to survive in hot, dry conditions.

SUMMARY QUESTIONS

1 Research how the following animals are adapted to extreme conditions:
   a a camel to desert conditions
   b an Arctic fox to arctic conditions.

S 2 Make a list of four ways in which desert plants are able to reduce water loss and conserve water.

3 Explain each of the following statements about hydrophytes.
   a They need little xylem in their stems and leaves.
   b Their roots do not have root hairs.
   c They have an extensive system of air spaces inside their stems and roots.

KEY POINTS

1 An adaptive feature is inherited and helps an organism to survive and reproduce.

S 2 Pond plants do not need much transport tissue since they are surrounded by water. They have extensive air spaces to store carbon dioxide and oxygen, which diffuse very slowly in water.

3 Xerophytes such as cacti have thick cuticles, leaves reduced to spines, swollen stems to store water and extensive root systems. Often their stomata close during the day and absorb carbon dioxide at night.

17.5 Natural selection

LEARNING OUTCOMES

- Describe natural selection
- Describe evolution as the change in adaptive features over time
- **S** Describe how selection is responsible for antibiotic resistance in bacteria

Figure 17.5.1 Charles Darwin as a young man.

Figure 17.5.2 Sockeye salmon.

Figure 17.5.3 Vultures and a hyaena compete for food from a dead gnu.

Natural selection is the process that gives organisms that are well adapted to their environment a greater chance to breed and pass on their alleles to the next generation than those that are less well adapted.

Variation within populations

Variation is the term used to describe all the differences that exist within populations of organisms. Gene mutation is the only way in which completely new genetic material is produced. Some mutations may give an advantage to the individual that expresses them. An example is having good camouflage to avoid being seen by a predator. But variation is also produced by sexual reproduction between two individuals.

During meiosis, the alleles of different genes are 'shuffled' to give new combinations in the gametes. At fertilisation, when the gametes fuse, alleles from the two different individuals are combined within the same nucleus. Gene mutation, meiosis, mating between individuals with difference genotypes and fertilisation give rise to variation between individuals in every generation.

Over-population of offspring

Sockeye salmon (Figure 17.5.1) return from the Pacific Ocean to breed in the rivers in British Colombia in which they hatched. The females produce millions of eggs, many of which are fertilised. We see the same in other species, for example a poppy plant releases over 15 000 seeds a year. Yet, in spite of this huge over-production of eggs and seeds, populations of animal and plant species remain fairly stable.

Populations remain stable because, in the case of sockeye salmon, most of the eggs , and many of the newly hatched young fish, are eaten by predators. Many young salmon die from disease and starvation. Similarly, many poppy seeds are eaten by birds; others may land in places where they cannot grow because there is no soil or water. Production of large numbers of eggs and seeds ensures that enough offspring make it into the early stages of life to maintain the population from year to year.

Competition for resources

There is competition for resources among the organisms that survive the early stages of life. Plants compete for space, light, water and nutrients. Animals compete for food, water, space (territories) and mates. This is referred to by biologists as the '**struggle for existence**' or '**struggle for survival**'.

Competition is fiercest between individuals of the same species as they have the same adaptive features to obtain their resources from the environment. Individuals of different species also compete, but often the competition is not as fierce. For example, one species may feed at night and the other during the day, or they may eat slightly different foods.

Reproduction

The individuals with features that best adapt them to the conditions in their environment are those most likely to survive and reproduce. Individuals that are not so well adapted are likely to lose out in the competition for resources. They may die before they have a chance to reproduce or, if they do reproduce, have few offspring. The better adapted individuals have a greater chance to pass on their alleles to the next generation.

Natural selection

If the environment does not change, then natural selection maintains populations of organisms so they do not change much, as many are already well adapted to their environment. Any that are not well adapted are unlikely to survive to breed. When the environment changes, individuals with features that help them to survive in the new conditions are at an advantage over others. These individuals that are now better adapted, compete successfully, survive, breed and pass on their alleles. Natural selection will bring about a change to a species over time. Thus selection is the mechanism by which evolution occurs – an idea first proposed by Charles Darwin in 1859. The process of adaptation results from natural selection whereby populations become more suited to their environment over generations.

Supplement

Antibiotic resistance

Antibiotics are chemicals that kill bacteria or inhibit their growth. Soon after the introduction of antibiotics in the 1940s, some bacteria developed a resistance to their effects.

When bacteria are exposed to an antibiotic, such as penicillin, most are killed. However, there may be some individual bacteria that have a mutation giving them resistance to the antibiotic. The bacteria may be able to produce an enzyme that breaks down the antibiotic. These individual bacteria survive and now have more resources available because all their competitors – the non-resistant bacteria – have died. The resistant bacteria survive to reproduce and pass on the gene for resistance to their offspring. This is an example of natural selection. An example of this is MRSA – the strain of the bacterium *Staphylococcus aureus* that is resistant to the antibiotic methicillin.

Humans have been responsible for the change in the environment by introducing antibiotics, but we have not consciously chosen the bacteria that are resistant to antibiotics. There is a constant search for new antibiotics as bacteria develop resistance to existing ones.

SUMMARY QUESTIONS

1 Explain the following terms in the context of natural selection: *variation*, *over-production*, *competition*, *struggle for survival*, *adaptation to environment*.

S 2 a Antibiotics are prescribed for bacterial diseases. Explain why people are told to follow the advice given by doctors about taking antibiotics.

b Why do drug companies invest in finding new antibiotics?

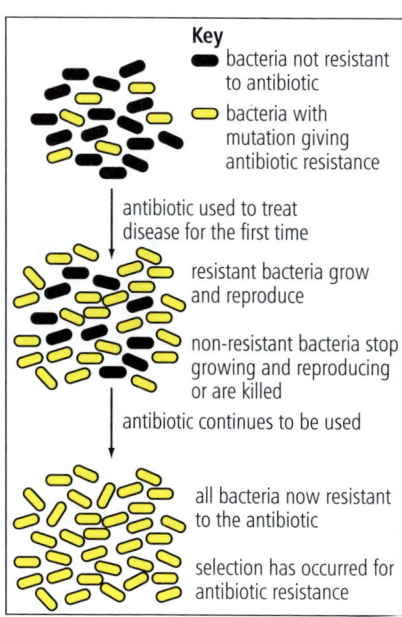

Figure 17.5.4 Selection for antibiotic resistance.

17.6 Natural selection in action

- Describe adaptation as the process, resulting from natural selection, by which populations become more suited to their environment over many generations.

We often get the impression that evolution by natural selection, takes place over vast periods of geological time. But given the right circumstances, there are examples of rapid evolution taking place well within a person's lifetime.

PCB resistance in tomcod fish

PCBs are chemicals that can kill fish and sea birds and have been linked to cancer and other health problems in humans. Between 1947 and 1970 huge amounts of PCBs were discharged into the Hudson River in New York. Despite being banned in 1979, these toxins have remained in the sediment on the river bed since they do not break down.

Scientists have now discovered that after a period of over 60 years, one bottom-feeding species of fish, the Atlantic tomcod, has evolved resistance to PCBs.

It is thought that this resistance has developed in the same way that antibiotic resistance came about in bacteria, namely natural selection.

Mutation results in random genetic variation. In this case individuals with a mutation that gives protection from PCBs to the tomcod appear in the population. These individuals carrying the gene have a better chance to survive and reproduce and so even more offspring are left behind with the resistant gene. Over generations the gene that codes for resistance to PCBs becomes more and more common in the population. The tomcod that carry the mutant gene are able to store high levels of PCBs in their body fat so PCBs are passed up the food chain. Their predators with no genetic defence eventually die.

Figure 17.6.1 The Atlantic tomcod, *Microgadus tomcodi*, has developed resistance to PCBs by natural selection.

Bird pollination in South Africa

Plants have a great number of strategies to attract animals to pollinate them. They are able to attract insects, birds and small mammals with colourful, shapely and sweet-smelling flowers.

Researchers suspected that the vertical, branchless stem of the South African plant, known locally as the Rat's Tail, had evolved to encourage pollinating birds to visit its flowers.

Plants reliant on bird-pollination tend to keep their flowers higher up because most birds avoid feeding on or close to the ground to keep clear of ground dwelling predators. The team observed birds hanging upside down from the branchless stem, which acted like a perch enabling them to push their beaks into the red flowers to drink nectar and at the same time pollinate the plant. The research team wondered if this perch-like structure had evolved giving the plants a selective advantage by providing pollinating birds a foothold from which to feed.

Figure 17.6.2 A malachite sunbird feeding on the nectar of a Rat's tail plant, *Babiana ringens*. See Exam Question 2 on page 307 for more about this example.

Weed seeds in the concrete jungle

Increased urbanisation is causing plants to evolved quickly and they have become more likely to drop their seeds onto their home patch of ground.

Crepsis sancta is a daisy-like annual weed that possesses two types of seed. One is big, heavy and tends to drop straight down to the ground. The other is light, floaty and can be caught on the wind and spread farther afield.

In the French city of Montpellier, there has been much urbanisation and few open areas of soil remain available for these seeds to germinate.

Over the past 12 years, scientists have noticed that weeds growing in patches of soil around trees have evolved to produce more of the big seeds than the lighter, far-reaching ones. At the time of dispersal, these heavier seeds tend to fall vertically and so the land in an area of soil that is occupied by the parent plant. This gives them a greater chance of successful germination and growth.

In cities, areas available for plant growth are becoming more fragmented, these plants can only grow in certain isolated areas. Research on *Crepsis sancta* found that the small, light seeds had a 55% lower chance of settling and growing than the heavy seeds that fall straight down into the same patch of ground as the parent. Light seeds were more likely to land on concrete than a patch of soil.

These results showed that evolution favouring dispersal of the heavier seeds in these plants had taken place in a remarkably quick period.

Figure 17.6.3 This weed's relatives won't travel far from home.

Figure 17.6.4 Some seeds are built to drop; others to fly.

SUMMARY QUESTION

Cane toads were introduced into Australia about 70 years ago in an attempt to control beetles. But this biocontrol agent began eating other insect species and became a pest itself. The cane toad is a fairly slow plodder. But by strapping tiny radio transmitters to the toads' waists scientists have found that they can now crawl up to 1.8 km per night. So how are they making the distance so easily? Researchers have found that the toads back legs have lengthened by 25% making them five times faster than when they were first introduced into the country.

1 Explain why the researchers concluded that this was an example of **natural selection** taking place over a relatively short time span.

2 What selective advantage has this evolution given to the cane toad?

3 What are the possible consequences of these changes to the environment?

KEY POINTS

1 Evolution by natural selection can take place over a relatively short period of time.

2 Adaptation is a process that results from natural selection. Populations become more suited to their environment from generation to generation.

17.7 Selective breeding

Figure 17.7.1 Ankole longhorn **(a)** and **(b)** Zebu are breeds of cattle well adapted to conditions in Africa such as shortages of water.

Figure 17.7.2 Merino sheep.

Selective breeding

People have grown crops and kept animals for at least 10 000 years. Over that time they have kept seed from one year to another and have bred their livestock. Domesticated plants and animals have changed considerably over that time due to this **artificial selection**. Here is how these changes have happened.

1 Humans choose a desirable feature or features of an animal or plant to improve – for example, fast growth, short stems (so less straw), docile (not fierce) nature, high yield, disease resistance, resistance to drought.

2 Animals or plants showing these features are bred to produce the next generation.

3 The offspring are checked to find those that show an improvement in the desired feature or features. These are kept for breeding the next generation.

4 This process of **selective breeding** continues for many generations.

Commercial farmers want to increase the yields of their crops and animals. They want to improve features of economic importance to maximise their profits. Plant breeders have increased the yield of grain in cereal crops such as rice, wheat and maize. Animal breeders have increased the milk yield in dairy cattle, and meat quantity and quality in beef cattle. Breeders of dogs and growers of ornamental plants are looking to improve other non-economic features such as appearance.

This is called *artificial* selection because it is humans who are the selective agent, not the environment as it is with *natural* selection. Humans decide which feature to improve and which individuals survive to breed and pass on their genes.

Merino sheep have been selectively bred for their wool. They have been particularly successful since being introduced to Australia and New Zealand from Europe.

The danger of selective breeding is that there may be too much inbreeding between closely related individuals. This may result in harmful recessive alleles being passed on to the descendants and a reduction in variation. Many breeds of dog suffer from the effects of inbreeding such as the failure of the hip joint to develop properly leading to lameness.

(a) These ewes from a flock of sheep have thick wool. They are mated with this ram who also has a thick coat...

(b) to give these offspring – some of whom have thick wool...

(c) but these sheep have thicker wool and are selected to breed the next generation

Figure 17.7.3 Artificial selection in sheep.

Supplement

Differences between artificial selection (selective breeding) and natural selection

There are two main methods of carrying out selective breeding.

Outbreeding involves the breeding of unrelated animals or plants. This may be used to combine the good characteristics of separate individuals, for example crossing a crop plant with high yield, with another crop plant that is resistant to disease, can produce offspring with both a high yield and is resistant to disease. Outbreeding often results in tougher individuals with a better chance of survival. This is called **hybrid vigour**.

Inbreeding involves breeding close relatives in an attempt to retain desirable characteristics. However, there can be harmful effects as a result of inbreeding. These can include a loss of vigour, with the population weakened by a lack of gene diversity and reduced fertility. There is also a greater susceptibility to disease as a result.

Some differences between artificial selection and natural selection (the evolutionary process) are shown in the table.

Table 17.7.1 Comparing artificial and natural selection.

artificial selection	natural selection
selection due to human influences	selection due to environmental factors
produces varieties of organisms very different from native generations	produces greater biodiversity
does not result in new species	may lead to new species
inbreeding is common, leading to loss of vigour in the offspring	outbreeding is common, leading to hybrid vigour
a relatively fast process	a slow process, taking many years
proportion of heterozygous individuals in the population is reduced	proportion of heterozygous individuals in the population remains high

Figure 17.7.4 Hybrid maize plants display hybrid vigour.

SUMMARY QUESTIONS

1 State the useful features that are improved by breeders in:
   a race horses
   b sheep
   c cereal crops (e.g., wheat, rice or maize)

2 Describe the procedure that a farmer would take to improve the milk yield of a herd of cows.

S 3 **a** List four ways in which artificial selection differs from natural selection.
   b What is meant by **i** outbreeding and **ii** inbreeding?
   c What is meant by the term *hybrid vigour?*

KEY POINTS

1 Selective breeding (artificial selection) involves humans finding organisms with desirable features, crossing them and selecting the best from the next generation.

2 Selective breeding has produced new varieties of animals and plants with increased economic importance (e.g., high-yielding crops, cattle that produce more milk or better meat and sheep that produce more wool).

S 3 Differences between artificial and natural selection are summarised in the table.

Practice questions

1 Merino sheep have much thicker wool than wild sheep. Merino sheep are the result of:

   A   mutation of a gene

   B   natural selection

   C   phenotypic variation

   D   selective breeding

   (Paper 1)   [1]

2 Which is an example of discontinuous variation in humans?

   A   blood group

   B   body mass

   C   foot length

   D   hand span

   (Paper 1)   [1]

3 On very rare occasions a white mouse may appear in a wild population of mice that all have brown hair. This is the result of:

   A   adaptation

   B   artificial selection

   C   competition

   D   mutation

   (Paper 1)   [1]

4 An island was invaded by a species of bird that preyed on butterflies. In the population of butterflies, only those individuals that produced a toxic substance as a protection against predation survived. This is an example of:

   A   artificial selection

   B   competition

   C   immunity

   D   natural selection

   (Paper 1)   [1]

5 Individuals of a species of flightless sand-burrowing beetle range in colour from pale brown to almost black. The beetles have the same colour as the sand in which they live. The reason for this is likely to be:

   A   Beetles are well camouflaged so that predators cannot see them.

   B   Beetles change colour to match their surroundings.

   C   Beetles are subject to artificial selection.

   D   Dark beetles absorb more heat.

   (Paper 2)   [1]

6 Which row correctly identifies the causes of continuous and discontinuous variation?

	continuous variation	discontinuous variation
A	environmental factors only	genes only
B	genes and environmental factors	genes only
C	genes only	environmental factors only
D	genes only	genes and environmental factors

   (Paper 2)   [1]

7 Why is it necessary for drug companies to find and develop new types of antibiotics?

   A   all viruses have become resistant to antibiotics

   B   antibiotics have become drugs of misuse

   C   antibiotics have been in use for over 60 years

   D   many bacteria are resistant to some antibiotics

   (Paper 2)   [1]

8 Which correctly defines the term *gene mutation*?

   A   a change in the position of a gene on a chromosome

   B   a change in the DNA base sequence of a gene

   C   a change in the way a gene is inherited

   D   a change in the amino acid sequence of a gene

   (Paper 2)   [1]

9 Match each of the following terms with the correct definition on the right:

   mutation   outward appearance of an organism
   variation   genetic constitution of an organism
   selection   range of forms found in a species
   phenotype   change in DNA
   genotype   organisms best adapted to their environment survive and breed

   (Paper 3)   [5]

10 The distance between the outstretched thumb and little finger is the handspan. A group of students measured their handspans. The data is shown in the table.

handspan / mm	number of students	frequency / %
160–69	7	4.7
170–79	12	8.0
180–89	25	16.7
190–99	45	30.0
200–09	34	
210–19	19	12.7
220–29	8	5.3
Total	150	100.0

(a) (i) Calculate the percentage of students who had a handspan measurement between 200 and 209 mm. [2]

(ii) State the range in the handspan measurements. [1]

(b) Draw a frequency histogram of the results shown in the table. [5]

(c) (i) State the type of variation shown in your graph. [1]

(ii) Explain your answer to part (i). [2]

(Paper 6)

11 Surveys were carried out in Pakistan and New Zealand to find the percentage of the population in the different blood groups of the ABO blood group system.
The results are shown in the table.

country	percentage of people in the sample in each blood group			
	A	B	AB	O
Pakistan	31.0	36.2	7.7	25.1
New Zealand	35.5	8.8	2.6	53.1

(a) Draw a bar chart to show the data. [5]

(b) (i) What type of variation is shown by the data? [1]

(ii) Explain your answer to part (i). [2]

(c) A couple are blood group B. They have a child who is blood group O. Use a genetic diagram to explain how this is possible. [5]

(Paper 4)

12 Xerophytes are plants that grow in dry places. They have many adaptive features.

(a) Define the term *adaptive feature*. [2]

(b) Describe three features of xerophytes that are adaptations for growing in dry habitats. [3]

(c) An area becomes very dry as a result of climate change. Explain how a species of plant may change over time to become adapted to this environmental change. [6]

(Paper 4)

13 Gonorrhoea is a bacterial disease. Antibiotic sensitivity tests are carried out on samples of bacteria taken from people with gonorrhoea to ensure that they are treated with an appropriate antibiotic.

(a) Describe how bacteria become resistant to antibiotics. [3]

(b) Explain how antibiotic resistance is transmitted from one bacterial cell to others. [2]

(c) Explain why it is important to carry out antibiotic sensitivity tests before prescribing a course of antibiotics. [2]

(d) Explain why antibiotics should not be used to treat viral diseases. [3]

(e) Antibiotic resistance is a serious medical problem worldwide. Suggest ways in which this problem may be overcome. [3]

(Paper 4)

14 (a) Explain how gene mutation, meiosis and fertilisation lead to genetic variation in populations of organisms that reproduce sexually. [6]

(b) A couple have a son who has red-green colour blindness. There is no history of colour blindness in the families of the husband or wife. Explain how the son inherited colour blindness. [5]

(c) Hydrophytes are plants that are adapted to grow in very wet places.

Explain how hydrophytes are adapted to grow in very wet places. [5]

(Paper 4)

- State that the Sun is the main source of energy for biological systems
- State that a food chain shows the transfer of energy from one organism to another
- Define the terms *producers*, *consumers* and *decomposers*
- Construct simple food chains
- Describe a trophic level as a feeding level in a food chain or food web

Figure 18.1.1 Sunlight is the source of energy for all living things in this ecosystem.

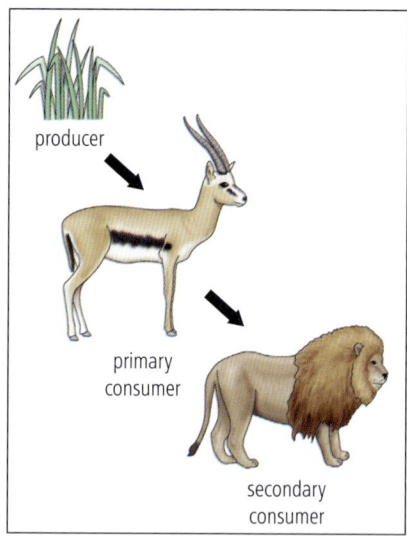

Figure 18.1.2 A food chain.

In this unit we are looking at the flow of energy through living organisms and the recycling of nutrients to organisms in biological systems.

There is an important difference between these two processes. The Sun is a principal source of energy input into biological systems. Light energy is absorbed by photosynthetic organisms and made available to all other organisms as chemical energy. **Energy transfer** occurs when animals feed on plants and when animals feed on other animals. Eventually the energy is transferred to the environment, heating it up, and is wasted. Nutrients, on the other hand, consist of the chemical elements that make up living organisms. They are taken up by plants, passed on to animals and eventually recycled back into biological systems. This is known as **nutrient cycling**, because they are constantly reused.

Communities

Living organisms can be placed into groups based on similar features. However, when looking at a community of living things, it is often more useful to group them according to the way that they feed. Living organisms can be categorised as those that can make their own food and those that cannot.

Producers make their own organic nutrients from simple raw materials, such as carbon dioxide and water. Green plants use light as a source of energy to make sugars from carbon dioxide and water by photosynthesis. Some bacteria are also producers and they are either photosynthetic or obtain their energy from simple chemical reactions. Some of the bacteria involved in recycling nitrogen obtain their energy from reactions involving compounds of nitrogen (see page 243).

Producers make energy available for all the other members of the community in the form of energy-rich carbon compounds, such as carbohydrates, fats and proteins. Sunlight is the ultimate source of energy for almost all food chains. The major exceptions are the vent communities in the deep ocean that rely on chemical energy rather than light energy.

Consumers obtain energy by eating other organisms, either plants or animals or both. All animals are consumers. They cannot make their own food, so they have to eat (consume) it. Herbivores are primary consumers – they eat the producers. Carnivores are secondary consumers and they eat herbivores. Tertiary consumers are carnivores that eat other carnivores, and are sometimes called top carnivores.

Decomposers are fungi and bacteria that gain their energy from waste organic material (see pages 6 and 7 for details of how they feed).

Each of these feeding groups is a **trophic level**. (The word *trophic* comes from a Greek word meaning 'to feed'.) You can put each organism in an ecosystem into a trophic level as you will see in these food chains.

Food chains

Food chains show what living organisms feed on in a community. They show the flow of food and energy from one organism to the next. Look at the food chain in Figure 18.1.2 from an East African ecosystem. The arrows show the direction in which energy in food is transferred from one organism to the next. This food chain tells us that the gazelle ingests grass and that the lion ingests the gazelle. Notice that the food chain always begins with a producer, often a green plant. This can include parts of a plant, such as seeds, fruits or even dead leaves. Food chains not only show the flow of food, they also show the flow of energy. Producers gain their energy from the Sun, so an *energy chain* would look like this:

Sun $\longrightarrow$ producers $\longrightarrow$ primary $\longrightarrow$ secondary
                                      consumers   consumers

But we do not include the Sun in food chains, as it is not a 'food'. Food chains start with producers.

Energy flows to decomposers, but they are rarely included in these simple food chains.

Food webs

In most communities, animals will eat more than one type of living organism. A food web gives a more complete picture of the feeding relationships in an ecosystem (Figure 18.1.3).

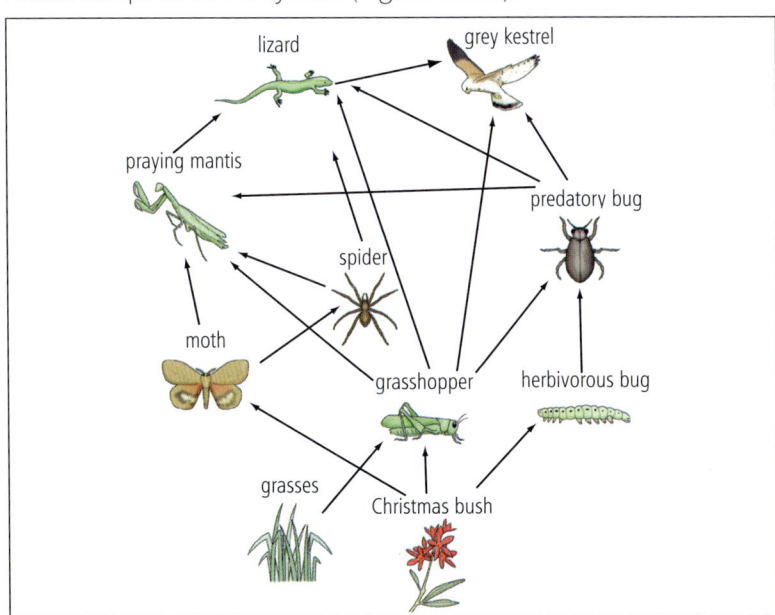

Figure 18.1.3 A food web for a West African forest ecosystem.

The over-harvesting of food species by humans in an unsustainable manner can result in a decrease in their numbers and extinction. For example, overfishing results in a decrease of fish stocks in the ocean. The introduction of foreign (alien) species can displace native organisms due to predation and competition for resources such as food. Both of these actions result in disruption of food chains and webs, due to the removal of a species or a decrease in their numbers.

SUMMARY QUESTIONS

1 Look at the food web for the forest ecosystem here:

   a Give one example from the food web of each of the following:

      i a producer

      ii a primary consumer

      iii a secondary consumer

   b Name the carnivores in the food web.

   c Draw a food chain with four organisms. Indicate the trophic levels on the food chain.

2 Define the term *trophic level* and explain how energy is transferred between trophic levels.

18.2 Pyramids of numbers and biomass

Pyramids of numbers

Food chains and food webs show feeding relationships in a community, but they do not show *how many* living organisms are involved. For instance, grasshoppers feed on many grasses and other plants and many grasshoppers and lizards are eaten by grey kestrels. Food chains and food webs do not show this.

Look at Figure 18.2.1. There are far more plants than caterpillars because each caterpillar eats leaves from many plants. There are far fewer owls than shrews because each owl eats many shrews.

Look at the numbers in this food chain:

plants 600 $\longrightarrow$ caterpillars 100 $\longrightarrow$ mice 10 $\longrightarrow$ owl 1

This information can be shown in a **pyramid of numbers**. The area of each box in the pyramid shows roughly how many living organisms there are at each trophic level (Figure 18.2.2).

The producers are the first level, herbivores are the second level and carnivores are the third. Notice that there are fewer primary consumers than producers. There are also fewer secondary consumers than primary consumers. But secondary consumers are bigger than primary consumers as they have to eat them!

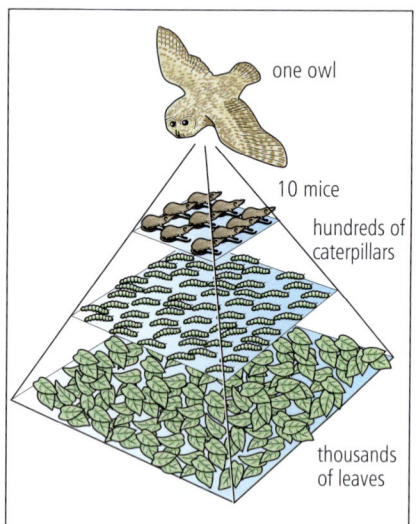

Figure 18.2.1 This shows the producers and consumers that support an owl for several days.

one owl

10 mice

hundreds of caterpillars

thousands of leaves

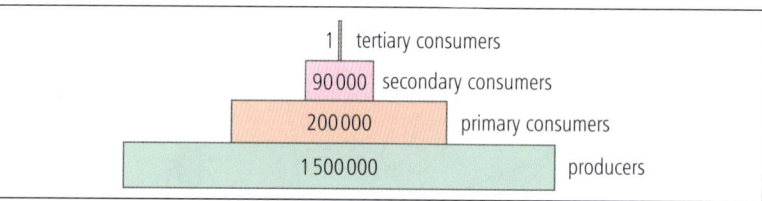

1	tertiary consumers
90 000	secondary consumers
200 000	primary consumers
1 500 000	producers

Figure 18.2.2 Pyramid of numbers for a grassland community in 0.1 hectare.

A problem with pyramids of numbers is that they do not take into account the **size** of organisms at each trophic level.

For instance, a mahogany tree and a grass plant each count as one organism. But one mahogany tree can support many more herbivores than one grass plant. As a result some pyramids of numbers can have unusual shapes (Figure 18.2.3).

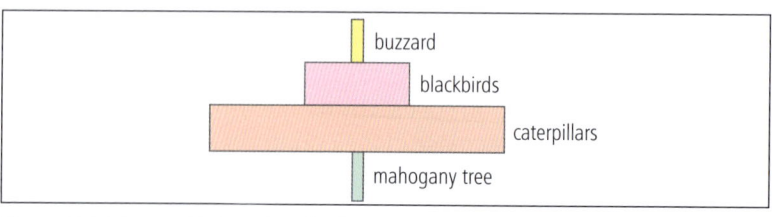

buzzard

blackbirds

caterpillars

mahogany tree

Figure 18.2.3 A pyramid of numbers for the organisms living in a mahogany tree.

In Figure 18.2.4 the tertiary consumers are parasites. Many of them feed on a single ladybird so this pyramid is inverted and it looks top heavy.

parasites

ladybirds

greenfly

rose bush

Figure 18.2.4 An inverted pyramid of numbers.

Pyramids of biomass

One way to overcome the problem of size is to measure **biomass** instead of numbers. Biomass is the mass of living material. So a biomass pyramid shows the actual weight or mass of living things at each trophic level.

To draw a biomass pyramid, the data first needs to be collected. A sample of the organisms from each trophic level is taken and weighed. The average mass for the sample is calculated. The average mass is then multiplied by the estimated number of organisms present in the community.

The dry mass is determined for the plant matter by drying it in an oven until it reaches a constant mass. It is usually possible to estimate the dry mass of animals: approximately 65% of their body mass is water and this does not change as it does in plants. Dry mass tells us how much useful biological material is present as carbohydrates, proteins and fats as food for the organisms in the community. The data is usually expressed in grams per square metre.

The dry mass recorded is taken at one instant in time. Biomass pyramids do not take into account how fast an organism grows. For instance, grass often grows at a fast rate, but because it is grazed by animals, such as grasshoppers and cattle, its biomass at any instant in time will be low. This means it is better to calculate the biomass produced over a period of time such as a year or a growing season.

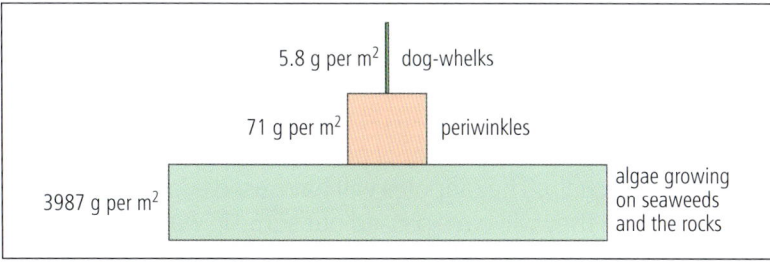

Figure 18.2.6 A pyramid of biomass for a rocky seashore community (over a long period of time).

Phytoplankton consists of tiny photosynthetic organisms that float in water – both freshwater (lakes and ponds) and in the sea. The plankton grow quickly at times of the year when there are nutrients and the sea is warm with plenty of light. They live for only a few days as they are grazed by animal plankton. So their biomass at any one time is small. But over say a year, their biomass is huge. This Figure 18.2.7 records only a few days' growth and so is inverted.

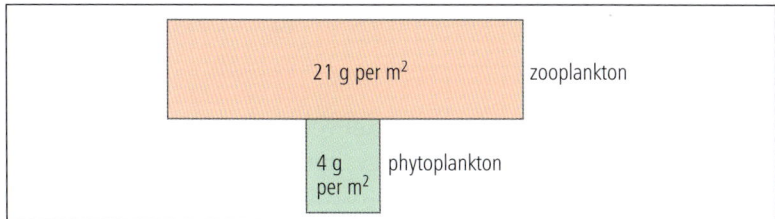

Figure 18.2.7 Biomass pyramid for the English Channel (over a few days).

Biomass can vary with the seasons. In temperate latitudes, deciduous trees shed their leaves in winter. Their biomass is greater in summer than it is in winter as they will have leaves, flowers, fruits and seeds.

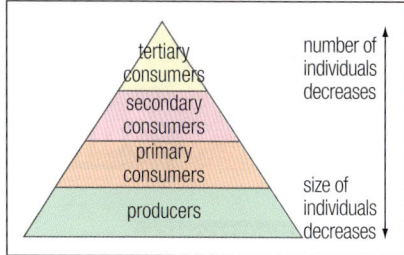

Figure 18.2.5 Pyramids of numbers and biomass show these trends.

KEY POINTS

1 Pyramids of numbers show how many individuals there are at each trophic level but give no indication of their size.

2 Pyramids of biomass indicate the mass of living material at each trophic level, but give no indication of the rate of growth.

SUMMARY QUESTIONS

1 Draw two pyramids of numbers with different shapes. Explain their shapes.

2 Draw a pyramid of biomass for an ecosystem and explain its shape. Make sure that you include the unit of measurement for biomass on your pyramid.

3 What criticisms have been made about pyramids of numbers and pyramids of biomass?

The efficiency of energy transfer

The efficiency of energy transfer between trophic levels can be seen by using **pyramids of energy**. These show the energy transferred from one trophic level to the next. The energy pyramid in Figure 18.3.1 shows that 87 000 kJ per m² per year is passed to the tadpoles from the water plants. The tadpoles pass on 14 000 kJ per m² per year to the small fish, and so on.

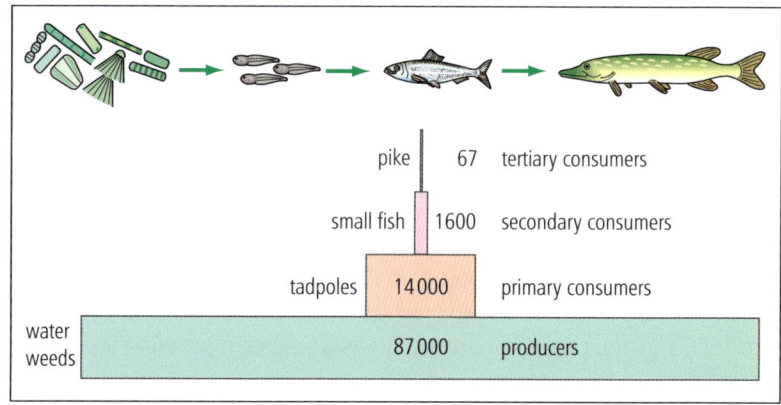

Figure 18.3.1 An energy pyramid for a lake. Figures are in kJ per m² per year.

The tadpoles obtain 87 000 kJ per m² per year from the water plants but only pass on 14 000 kJ per m² per year of this to the small fish. The remaining 73 000 kJ per m² per year has been lost from the food chain. The tadpoles will have used a lot of energy in swimming and they will have passed out some in waste – urine and faeces. The only energy that they pass on to the small fish is the energy in their biomass – their muscles and other organs that the fish eats every time it swallows a tadpole.

Energy is always wasted in this way as it passes from one trophic level to the next. Of the 87 000 kJ per m² per year of energy in the producers, only 67 kJ per m² per year becomes energy in the flesh of the top carnivore, the pike. Since only some of the energy is passed from one trophic level to the next, an energy pyramid is never inverted. Its shape is not affected by either the size of the organisms or how many of them there are because it simply looks at the energy that is passed on.

Wasted energy

Energy flow through ecosystems is relatively inefficient. At best, plants absorb only 2 to 5% of the light energy that strikes their leaves. The rest this energy is not trapped in photosynthesis because it is reflected from the surface of the leaves, passes straight through them or is green light, which plants cannot absorb.

Plants use much of the energy they trap in photosynthesis in their own respiration. As an energy transfer process, respiration is about

40% efficient so about 60% of the energy is wasted heating the plant and then heating the atmosphere.

Much of the plant's biomass is passes to decomposers in the form of dead leaves, twigs, fruits, etc. About 10% of the energy trapped by the plant is available to primary consumers. Much of what they eat is not very edible or easy to digest. So primary consumers make available only about 10% of what they have eaten to secondary consumers.

However, the animal flesh in primary consumers, for example grasshoppers and gazelles, is easier to digest and carnivores may obtain slightly more than 10% of the energy.

At each trophic level, energy is wasted heating the environment and consumers waste energy in their faeces and urine, which pass to decomposers. This all means that there is less energy for the next trophic level. Food chains on land rarely have more than four trophic levels and those in the sea rarely more than five.

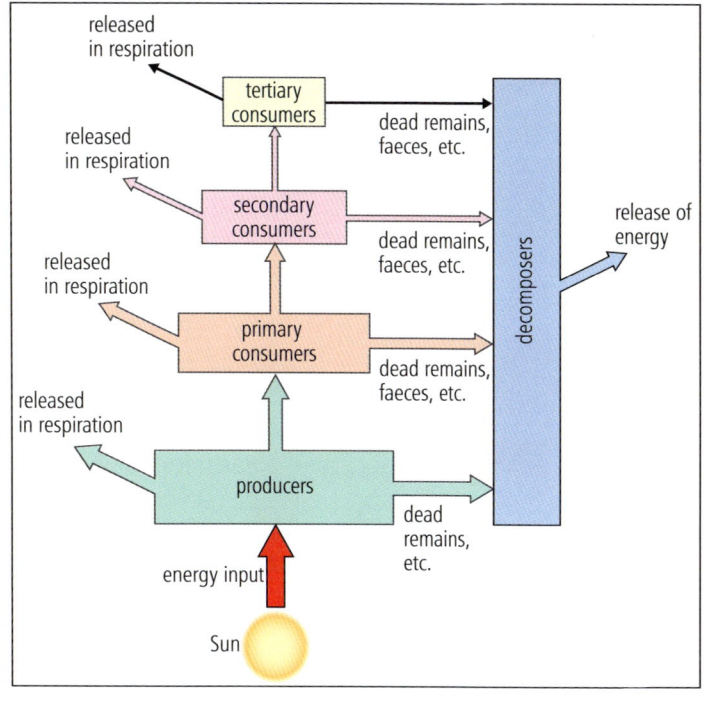

Figure 18.3.2 Wasted energy in an ecosystem.

When farmers grow crops such as rice for human consumption, all of the energy in the grain is available to us as primary consumers. Livestock farmers grow or buy plant food to feed their animals. They may buy cereals such as barley or maize or they may grow fodder crops such as grass, alfalfa and clover for grazing. Animals such as pigs, cattle and chickens use up much of the energy in their food. This means that there is not as much energy available to us.

A vegetarian diet can support far more people. If we cut down the number of links in the food chain, more individuals at the end of the food chain can be fed. This is because we are cutting down the 90% 'wastage' of energy that occurs between each

trophic level; that is, the energy that is uneaten, undigested or used in respiration at each level.

People in many countries tend to have diets that mainly consist of food of plant origin with some meat or fish. In Western developed countries many people have a varied diet that includes a higher proportion of meat from poultry, fish, lamb, beef and pork.

The human population is increasing. If farmers are to provide enough food for everyone, our diet may have to change to one that includes more plant foods and less food of animal origin.

KEY POINTS

1 Energy is wasted between trophic levels as a result of respiration, and in waste materials produced.

2 More energy is available in foods for human consumption if we feed as primary consumers rather than as secondary consumers.

3 A vegetarian diet can support far more people than one that includes meat products.

SUMMARY QUESTIONS

1 Explain the advantages of using pyramids of energy instead of pyramids of numbers or biomass for a food chain or web.

2 Explain why food chains usually have fewer than five trophic levels.

3 a Explain why it is efficient, in terms of energy transfer, to supply green plants as human food.

  b Explain why a diet rich in meat products is relatively inefficient in terms of wasted energy.

18.4 Nutrient cycles

LEARNING OUTCOMES

- Describe the carbon cycle
- Describe the role of decomposers (fungi and bacteria) in recycling carbon

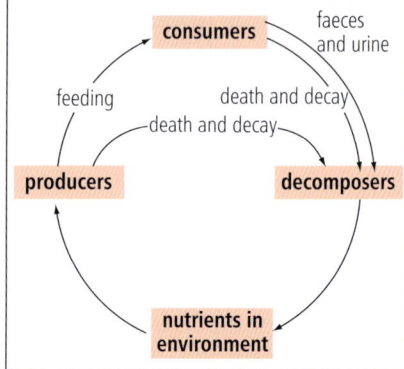

Figure 18.4.1 The role of decomposers in ecosystems.

Nutrient cycling

So far we have been concerned with food and energy. There is a constant flow of energy into ecosystems from the Sun; energy flows through the different trophic levels and is wasted warming the atmosphere. This topic and the next are about some of the elements that living organisms are made of and how they are cycled in nature.

Most living matter (95%) is made up of six elements: carbon, hydrogen, oxygen, nitrogen, phosphorus and sulfur. Living things must have a constant supply of these elements if they are to make proteins, carbohydrates, fats and other organic materials. Unlike energy, there is a finite source of these nutrients. Life has existed on Earth for millions of years. For it to continue, these elements must be recycled or they would become locked up in dead bodies and life would become extinct.

Bacteria and **fungi** feed on dead and decaying matter from plants and animals (Figure 18.4.1). They also feed on the waste matter of animals (urine and faeces). They feed by breaking down carbohydrates, fats and proteins into small molecules that they absorb. Many of these are respired. To do this, these organisms require water and a warm temperature because the processes of digestion and respiration are catalysed by enzymes. They also need oxygen for aerobic respiration.

Decomposers release carbon as carbon dioxide when they respire. They also break down amino acids releasing ammonia into their surroundings. In this way, they recycle carbon and nitrogen – two of the most important elements for organisms. Simple compounds are absorbed by plants and converted into complex compounds (Figure 18.4.2).

These nutrients are then passed on to animals when they eat the plants. We say that **nutrients** are **cycled**.

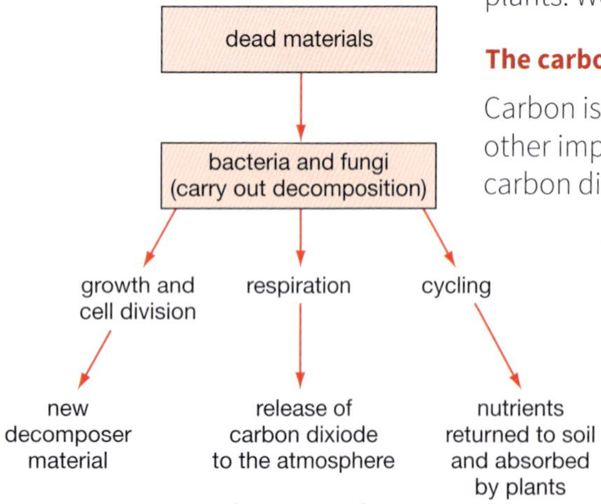

The carbon cycle

Carbon is used to make carbohydrates, proteins, fats, DNA and other important biological molecules. The carbon comes from carbon dioxide in the air. Plants absorb carbon dioxide from the atmosphere and use it in photosynthesis to make food. Animals get the carbon compounds by eating plants.

Carbon dioxide gets into the air in the following ways.

- Plants and animals use some of their food for respiration, releasing carbon dioxide (as well as water and energy).
- Decomposers use dead plants and animals for food. They also use some of the decaying material for respiration, releasing carbon dioxide (plus water and energy).

Figure 18.4.2 The roles of bacteria and fungi in decomposition.

- Fossil fuels like oil, peat, coal, and natural gas contain carbon. Carbon dioxide is one of the gases released into the air during combustion when these fuels are burnt.

These processes all release carbon dioxide into the air and this balances the uptake by of carbon dioxide photosynthesis. Thus, the concentration of carbon dioxide in the atmosphere remains constant. Sometimes, dead plants and animals do not decompose. Their dead bodies become fossilised to form coal, oil and gas – fossil fuels.

The effects of combustion of fossil fuels and deforestation on the concentration of carbon dioxide in the atmosphere are discussed in Topics 19.2 and 19.5.

EXAM TIP

During photosynthesis carbon is fixed as simple sugars. This means a simple compound of carbon is converted into a complex compound. The same idea – fixation – occurs in the nitrogen cycle, but is not as common because carbon dioxide is much more reactive than nitrogen gas (N_2).

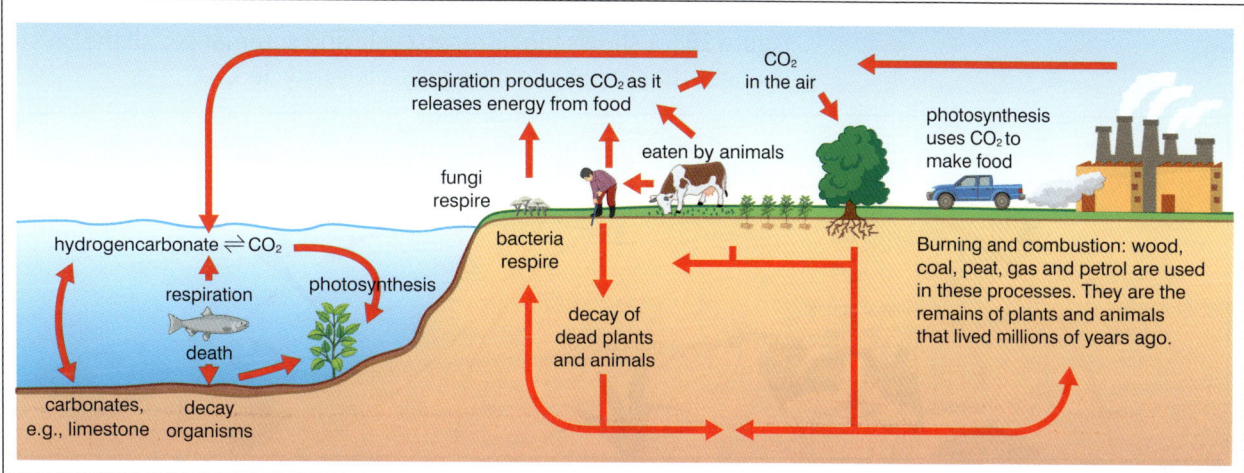

Figure 18.4.3 The carbon cycle.

SUMMARY QUESTIONS

1 **a** Describe what is meant by the term *cycle*, in the context of nutrients and living organisms.
   b State two substances that are absorbed and recycled by decomposers.
   c Name the six elements that make up 95% of living matter.

2 State the roles of the following in nutrient cycles:
   a decomposers   **b** green plants   **c** animals

3 Look at the diagram of the carbon cycle (Figure 18.4.3). Carbon exists in different forms. Much of the carbon on Earth is in the atmosphere and in rocks. The rest forms organic compounds in living organisms and their wastes.
   a State the only process that converts inorganic carbon into organic compounds.
   b State the ways in which carbon in organic compounds is converted into carbon dioxide.

4 How are fuels like peat and coal formed? What environmental problem do they cause?

KEY POINTS

1 Carbon is cycled through ecosystems by the processes of photosynthesis, respiration, feeding, decomposition, fossilisation and combustion.

2 Decomposers such as bacteria and fungi are important in the recycling of nutrients.

18.5 The nitrogen cycle

Supplement

Nitrogen

Nitrogen is an element required for many biologically important molecules. Amino acids, proteins, DNA and chlorophyll all contain nitrogen. Approximately 80% of the air is nitrogen gas (N_2) but it is inert (unreactive) and very few organisms can make use of it. Plants, animals and most microorganisms cannot use it in this form. It has to be available to them as fixed nitrogen in the form of compounds such as amino acids and proteins, and as ions such as nitrate and ammonium ions.

Figure 18.5.1 shows the nitrogen cycle and summarises all the changes that happen to nitrogen as it is recycled.

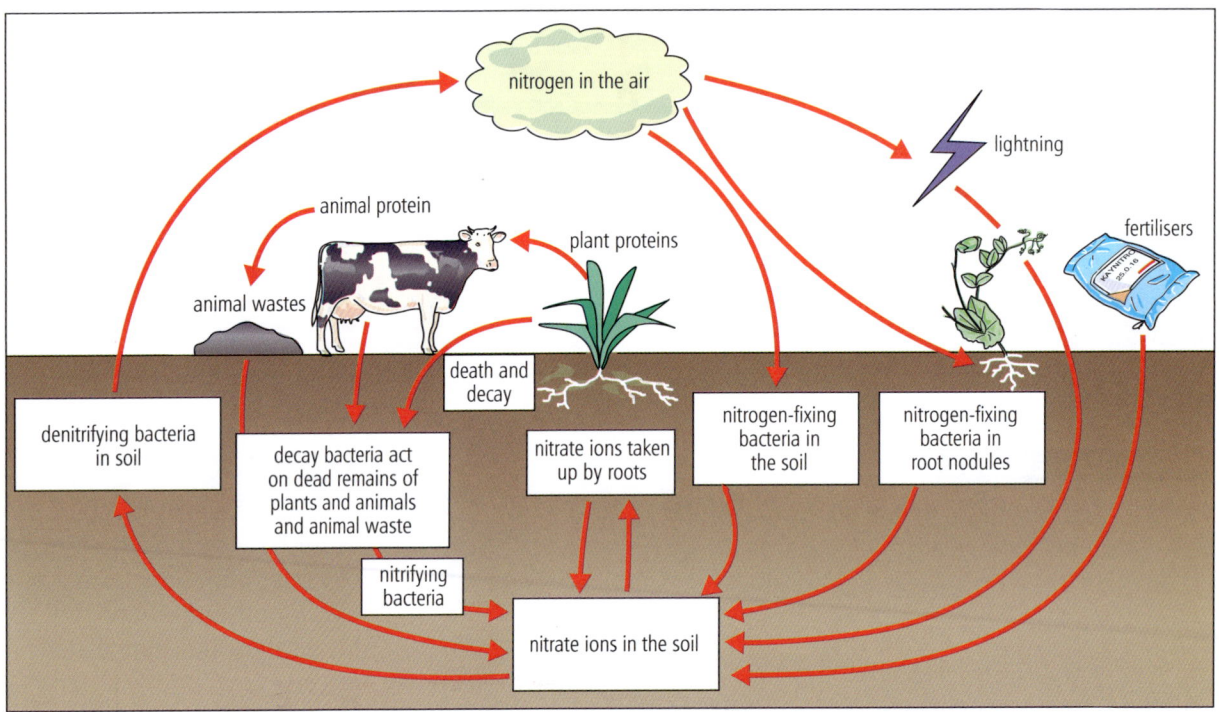

Figure 18.5.1 The nitrogen cycle.

Most of the nitrogen that plants obtain they absorb in the form of nitrate ions (NO_3^-). They use this to make amino acids in their leaves. The amino acids are used to form proteins such as enzymes.

Animals eat plants and so obtain their nitrogen in the form of plant protein. Any excess amino acids are broken down by animals to ammonia in the process of deamination. Animals that live on land (for example, cows and humans) convert ammonia to urea and excrete it in their urine. You can follow this on Figure 18.5.1. Deamination thus results in the production of the excretory product urea, which passes out in animal urine (Topic 13.1).

Decomposers (bacteria and fungi) break down dead remains and animal waste releasing ammonium ions (NH_4^+) into the soil. Bacteria also break down urea in urine to ammonium ions. This process is sometimes called ammonification.

Nitrifying bacteria in the soil change ammonium ions into nitrate ions. These bacteria gain their energy from this oxidation reaction instead of absorbing light or feeding as a decomposer.

If you are tracing the cycle in Figure 18.5.1 you will realise we are back where we started with absorption of nitrate ions by plants.

Two other groups of microorganisms help to recycle nitrogen.

Nitrogen-fixing bacteria are found in the soil. These can convert nitrogen gas from the air into compounds of nitrogen that they use themselves. When they die and are decomposed, this fixed nitrogen is available to plants.

Nitrogen-fixing bacteria are also found in the roots of legume plants such as peas, beans, alfalfa and clover. They are also found in the roots of many tropical trees, such as flame trees. The bacteria are inside swellings on the roots called **root nodules**. These bacteria change nitrogen gas into ammonia that the legume plants can use to make amino acids. In return the legume plants provide a suitable environment for the bacteria (in the nodules) and also provide all the sugars that they need. Nitrogen fixation requires a lot of energy.

Denitrifying bacteria live in water-logged soil. They change nitrate ions to nitrogen gas. It is thought that they balance the uptake of nitrogen gas by nitrogen-fixing bacteria.

Lightning causes nitrogen and oxygen to react together at high temperatures. Gaseous nitrogen oxides are formed in the reactions. These are washed into the soil by rain where they form nitrate ions.

Nitrate ions can be lost from the soil before plants absorb them because they can be washed out of the soil by rainwater. This is called **leaching**.

Farmers add fertilisers containing nitrate ions to the soil. An example is ammonium nitrate, which is manufactured from ammonia made in the Haber process. Such fertilisers are extra sources of nitrate ions for crop plants. The fertilisers replace the nitrate absorbed by the plants and removed at harvest. Some farmers do not use inorganic fertilisers, but instead use natural fertilisers such as manure. This provides dead material that enters the nitrogen cycle at the point where the decomposers act.

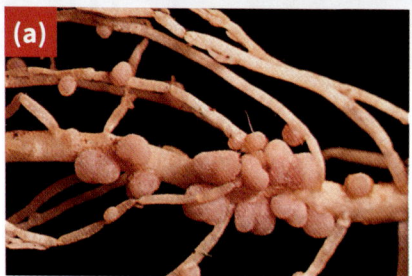

Figure 18.5.2 (a) Roots with some nitrogen-fixing root nodules. **(b)** the cells of a root nodule broken open to reveal the thousands of nitrogen-fixing bacteria that live inside.

SUMMARY QUESTIONS

1 List four substances present in living organisms that contain the element nitrogen.

2 Describe the roles of microorganisms (bacteria and fungi) in the following processes:
   a decomposition   b nitrification
   c nitrogen fixation   d denitrification

3 Explain how nitrogen in the protein of dead animals and plants is made available to plants in the form of nitrate ions.

4 Explain how farmers can increase the quantity of nitrate ions available to crop plants.

KEY POINTS

1 Nitrogen is in the atmosphere as a gas (N_2) but is not available to most organisms as it is not reactive.

2 Some bacteria fix nitrogen (N_2) converting it to ammonia. Some of these live inside root nodules of legumes. They are called nitrogen-fixing bacteria.

3 Bacteria decompose protein and urea to ammonia. Nitrifying bacteria convert ammonia to nitrate ions.

4 Denitrifying bacteria convert nitrate ions to nitrogen gas (N_2).

5 Plants absorb nitrate ions and convert them to amino acids, which they use to make proteins. Animals obtain the amino acids they need by eating and digesting plant or animal protein.

Figure 18.6.1 Wildebeest in East Africa.

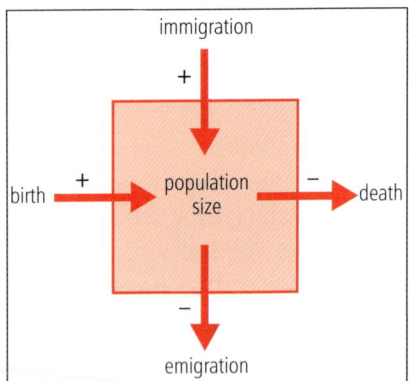

Figure 18.6.2 Factors influencing population size.

Population growth

A **population** is a group of individuals of the same species living in the same habitat at the same time. All the individuals in a population may interbreed. Sometimes it is difficult to tell the geographical limits of a population. This is easiest where the organisms live in a restricted area. For example, Komodo dragons on islands in Indonesia, *Protea kilimanjaro* lives on a mountain. Some populations are easy to identify because they are highly visible, like wildebeest in the game parks of East Africa (Figure 18.6.1).

Some species are very widely distributed and it is difficult to divide them into separate populations. However, there are usually barriers to reproduction, such as rivers and mountains, that divide them into separate populations. Some individuals migrate across these barriers, so there is interbreeding between populations.

Table 18.6.1 The habitat and location of some animals and plants.

population	habitat	place
creole wrasse	coral reef	reefs around Tobago in the Caribbean
wildebeest	grassland	East Africa
Komodo dragon	dry grassland and forest	Komodo Island in Indonesia
Nepenthes rajah – carnivorous plant	mountain forest	Mount Kinabalu in Malaysia
Protea kilimanjaro	shrubland	Mount Kilimanjaro in Kenya

A **community** includes all the populations of different species in an ecosystem. An **ecosystem** is a unit containing the community of organisms and their environment, interacting together, for example in a decomposing log or a lake. An ecosystem is made up of the **biotic** component (the community) and the **abiotic** component (physical factors, such as light, water, pH and temperature).

Population size

Figure 18.6.2 shows the factors that determine the size of a population. The number of births adds to the population and the number of deaths decreases it. Individuals may enter or leave a population from neighbouring populations.

Animals migrate so there is mixing between populations. Flowering plants are fixed so you would think that migration is not possible. However, plants use seed dispersal to colonise new areas and in this way individuals leave one population of plants to join another. The main factors that affect the rate of population growth are food supply, predation and disease. Figure 18.6.3 shows the different stages of population growth. An organism that enters a new habitat may show the lag, exponential and stationary phases if there is plenty of food, no predators and no disease.

- **Competition for resources**. Plants compete for light, space, water and soil nutrients. Animals compete for food, space (territory) and mates. Competition for food is a common limiting factor for species of herbivores, such as wildebeest in East Africa.

- **Predation**. Predators often take young, sick individuals or less well-adapted individuals. Predators may limit the growth of a population of prey animals, but it is more often the case that the numbers of prey animals limit the population size of the predator.

- **Disease**. Disease is an important control factor when populations increase. Pathogens are transmitted between individuals more easily when organisms live close together.

The growth of a population: bacteria

The growth of a bacterial population is a good way to show how the size of a population changes. A small number of bacteria are placed into a flask with a warm nutrient solution, which is aerated so the bacteria can respire aerobically. Bacteria grow and divide under these conditions. Some species of bacteria divide every 20 minutes if conditions are ideal. One cell divides to give two cells, two cells divide to give four, and so on. The four phases of the population growth curve are shown in Figure 18.6.3.

Supplement

Look at the the four phases of the population growth curve, as shown in Figure 18.6.3, which are explained below:

- The **lag phase**, when doubling of the numbers has little effect as the numbers are so small. Bacteria take up water and nutrients, and make new cytoplasm, DNA and enzymes.

- The **exponential** (or **log**) **phase** when the population is increasing rapidly. The population increases by doubling and there are no limiting factors, such as food or water. During this phase of rapid growth there are no factors to limit population growth. But after a while, limiting factors such as lack of food and build-up of waste restrict population growth, which slows down and stops.

- The **stationary phase**, when bacterial cells are dying at the same rate at which they are being produced. This may be because of shortage of food or because waste products are building up.

- The **death phase**, when more cells are dying than are being produced, so the population declines. Causes of death may be lack of food, shortage of oxygen or a build-up of toxic waste products. The curve shown in the graph is known as the **sigmoid growth curve**.

Some species show this population growth in the wild. When resources become available, species that can reproduce rapidly, like algae, grow exponentially. When the resources are used up, most of them die. Most species do not do this – their populations remain fairly stable over time.

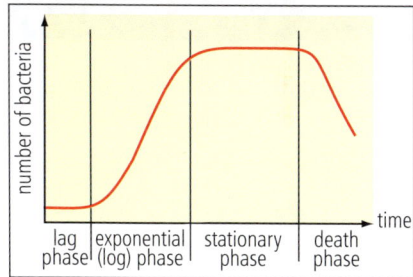

Figure 18.6.3 Growth of a bacterial population.

SUMMARY QUESTIONS

1 a Define *population*.

 b Give three examples of populations in natural ecosystems.

 c Describe the reasons why populations may increase and decrease.

2 Explain how each of the following can act to limit the growth of a population, giving an example in each case:

 a food supply

 b predation

 c disease

S 3 With reference to Figure 18.6.3, describe and explain each of the following phases:

 a lag   b exponential (log)

 c stationary   d death

Practice questions

1 The diagram shows a pyramid of numbers for an ecosystem. Which level indicates the tertiary consumers?

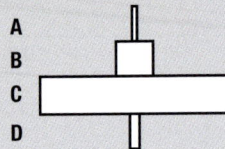

(Paper 1) [1]

2 Fungi and bacteria are important in recycling carbon and nitrogen in ecosystems because they are:

A   decomposers

B   herbivores

C   parasites

D   producers

(Paper 1) [1]

3 In the carbon cycle, which two processes add carbon dioxide to the atmosphere?

A   combustion and respiration

B   decomposition and fossilisation

C   feeding and fossilisation

D   photosynthesis and respiration

(Paper 1) [1]

4 Here is a simple food chain:

grass → grasshopper → bird

The arrows in a food chain represent:

A   the decrease in numbers of the organisms

B   the flow of energy between the organisms

C   the increase in complexity of the organisms

D   the increase in size of the organisms

(Paper 1) [1]

5 From which food chain is *most* energy lost?

A   maize → beef cattle → humans

B   soya beans → humans

C   grass → dairy cattle → humans

D   phytoplankton → mollusc larvae → small fish → tuna → humans

(Paper 2) [1]

6 Some species of bacteria live inside the root nodules of legumes. Which of the following do these bacteria provide to the legumes?

A   compounds of nitrogen

B   protection from infection by fungi

C   sugars, such as sucrose

D   water and ions from the soil

(Paper 2) [1]

7 What is the best explanation for the change in population size shown in the graph?

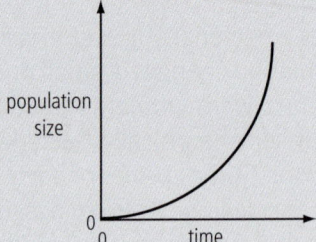

A   birth rate and death rate are the same

B   emigration

C   an outbreak of disease

D   no limiting factors

(Paper 2) [1]

8 The diagram shows a pyramid of biomass.

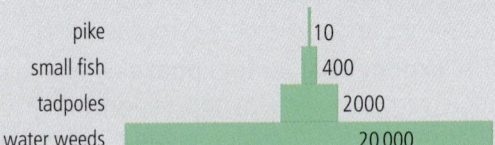

Which statement about the pyramid shown in the diagram is correct?

A   Biomass decreases less between primary consumers and secondary consumers than between producers and primary consumers.

B   The biomass of each individual producer is very large.

C   The biomass of the primary consumers is more than 20% of the biomass of the producers.

D   There are very many producers.

(Paper 2) [1]

9 (a) Use this information to draw a food chain:

Flower beetles eat the pollen and nectar of black cherry trees. Kookaburras are birds that eat flower beetles. [2]

(b) State the role of each organism in the food chain you have drawn. [3]

(c) Define the term *population*. [3]

(d) State three factors that influence the rate of growth of a population. [3]

(e) An ecosystem may contain many different communities. Explain the difference between a *community* and an *ecosystem*. [4]

(Paper 3)

10 The table shows the total biomass formed in one square metre of grassland during one year.

The figures were obtained by measuring the dry mass and not the fresh mass of the living material.

organisms	dry mass / g m^{-2} year^{-1}
green plants	480
herbivores	0.9
carnivores	0.1

(a) Calculate the percentage decrease in the biomass between green plants and herbivores. Show your working. [2]

(b) Explain why the biomass formed during the year decreases between trophic levels. [3]

(c) Explain the advantage of measuring dry mass rather than fresh mass when investigating biomass production in different ecosystems. [3]

(Paper 4)

11 The diagram shows the passage of energy along a food chain in a grassland ecosystem. The light energy that strikes the leaves of the grass each year is 400 000 kJ m^{-2}.

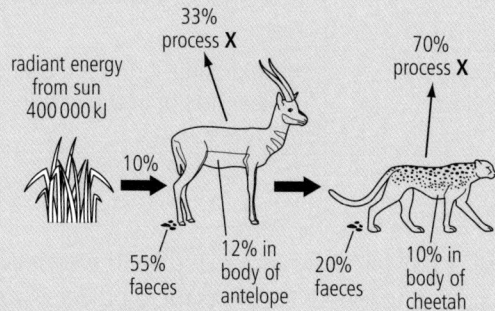

(a) The plant absorbs only about 1% of the energy that strikes the plant and uses it in photosynthesis to make simple sugars.

   (i) State two compounds that are produced in plants from the simple sugars. [2]

   (ii) State three ways in which the leaves of plants are adapted to trap as much light as possible. [3]

(b) Only about 10% of the energy trapped by the plant in photosynthesis is passed to herbivores like antelopes.

   (i) State two ways in which the plant uses the energy that is not available to the herbivores. [2]

   (ii) Calculate how much energy reaches the antelopes each year. Express your answer as kJ m^{-2}. Show your working. [3]

(c) Name process **X**. [1]

(d) Suggest why the antelope loses more of its energy intake as faeces compared with the cheetah. [2]

(e) This food chain has three trophic levels in it. Explain why there are no more. [4]

(Paper 4)

12 (a) The following are biologically important substances that contain the element nitrogen: urea, ammonia (NH$_3$), amino acids, nitrite ions, nitrate ions

Choose one of the substances from the list to match each of the statements **(i)** to **(v)**. You may use each substance once, more than once or not at all.

   (i) the main nitrogen-containing excretory product of animals

   (ii) absorbed by plants

   (iii) a product of deamination

   (iv) synthesised into proteins

   (v) the end product of nitrification [5]

(b) Nitrogen gas (N$_2$) in the atmosphere is inert. Unlike carbon dioxide and oxygen, very few organisms are able to use it. Describe how nitrogen gas (N$_2$) present in the atmosphere is made available to be used by organisms. [4]

(Paper 4)

19.1 Food supply

LEARNING OUTCOMES

- Describe ways in which modern technology has resulted in increased food production
- Describe the advantages and disadvantages of large-scale monocultures of crop plants and intensive livestock production

Figure 19.1.1 Machinery like this combine harvester allows farmers to grow crops in very large fields.

Figure 19.1.2 Plantations of oil palm cover huge areas of land in South-East Asia. Here a lorry from a plantation in Indonesia carries away the fruits of these trees to be processed into palm oil.

Increased food production

Modern technology has resulted in advances in agriculture over the past 60 years and improved food supplies in many parts of the world.

Arable farms use large **agricultural machines** to work very large fields. Examples are tractors and ploughs for preparing land for sowing seeds and combine harvesters for harvesting crops. **Chemical fertilisers** encourage the growth of crop plants, increasing the yield of the crop. **Pesticides** kill pests like insects that feed on crops. **Herbicides** kill weeds that compete with crop plants for water, light and nutrients. **Selective breeding** has increased yield and made crops more resistant to drought and diseases. **Genetic modification** has transferred features, such as herbicide resistance, to crop plants from unrelated species.

A **monoculture** is growing the same crop on large areas, year after year. This has the advantage of allowing farmers to concentrate on growing large quantities of specific crops, such as wheat, barley, maize, soya and rice.

There are negative effects of monoculture on the environment.

Herbicides and pesticides kill plant and insect species that are harmless and even some that may help the crop, such as parasitic wasps that lay their eggs inside pest species. Continuous use of these chemicals act as selective agents, resulting in the evolution of resistance among weeds, pests and plant pathogens (Topic 17.5).

Table 19.1.1 The advantages and disadvantages of monocultures.

advantages	disadvantages
increased machinery means continuous cropping	destroying areas of natural vegetation between fields reduces biodiversity of plants and animals
decrease in labour costs	excess fertilisers can leach into rivers causing eutrophication killing fish and invertebrates
only one crop is grown to it can be selectively bred to increase yield	pesticides can enter the food chain and accumulate to toxic levels
selective use of fertilisers and pesticides increases yield	unlike organic fertilisers, chemical fertiliser do not improve soil structure
soil ph can be regulated to meet the needs of the crop	

The large scale use of chemical fertilisers can reduce the structure of the soil so that over time it no longer supports the biodiversity of soil organisms that supports good crop growth. The addition of organic matter, such as manure, helps to maintain good soil structure but is not always carried out.

Intensive farming is using modern technology to achieve high yields of crop plants and livestock. This involves growing crop plants over large areas at high densities and keeping livestock in large numbers, supplementing their food supply and often restricting their movement by keeping them indoors.

In many intensive systems animals are reared indoors in large numbers.

Intensive methods have negative effects on the environment. Animals, especially cattle, generate lots of methane, which is a greenhouse gas (Topic 19.5). Urine and faeces, often known as slurry, pollutes lakes, waterways and the sea, where it can cause eutrophication (Topic 19.4).

Fish, such as salmon, trout, sea bass and tilapia, are kept in large cages and have their food and growing conditions carefully controlled. The waste food from fish farms can have serious effects on the surrounding waters. High densities of fish mean that parasites and pathogens can spread easily. This means using pesticides and antibiotics to treat the fish, and with this come problems of resistance to these chemicals.

Figure 19.1.3 Intensive turkey farming.

Figure 19.1.4 The waste from this Mongolian farmer's cows pollutes the surroundings of his farm.

Figure 19.1.5 Intensive fish farming in Thailand.

Table 19.1.2 The advantages and disadvantages of intensive livestock production.

advantages	disadvantages
animals reared indoors can be given high-protein diets and additives so they grow as fast as possible	many people do not like to see animals reared in these conditions; they think it is more humane for them to be outside
temperature of housing kept constant because animals grow well in warmth	keeping animals in caged or penned overcrowded conditions leads to them becoming obese, bored and frustrated
antibiotics can be used to control disease	rearing animals in crowded conditions can lead to a rapid spread of disease; free-range products are becoming more popular as people realise the suffering caused by factory farming
keeping animals in cages or pens restricts their exercise so they put on weight faster	fish farming can cause pollution because a high concentration of fish in one place means that animal waste and uneaten food can cause eutrophication

SUMMARY QUESTIONS

1 How have the following increased food production?
   a improved agricultural machinery   **b** artificial selection
   c use of fertilisers   **d** use of pesticides
2 Describe the negative impacts on the environment of intensive livestock production and monoculture of crops.

KEY POINTS

1 The use of machinery, selective breeding and agricultural chemicals has resulted in an increase in food production.

2 Intensive crop and livestock production have advantages, but also many negative impacts on the environment and on people.

19.2 Habitat destruction

LEARNING OUTCOMES

- Describe biodiversity as the number of different species that live in an area
- Describe the reasons for habitat destruction
- State that humans have a negative impact on habitats by altering food chains and food webs
- List the negative effects that deforestation has on habitats
- Explain the effects of deforestation on the environment

Figure 19.2.1 Open cast coal mining destroys large areas of land.

The pressure of an increasing world population has resulted in significant losses of natural habitats. Loss of habitat leads to an inevitable decrease in **biodiversity**. **Species diversity** is one aspect of biodiversity and can be defined as: the number of different species and the number of individuals of each species found within a habitat.

There are many reasons why the natural environment has been and continues to be destroyed. Land is deliberately cleared for a variety of reasons, but we have many indirect effects on natural ecosystems, such as pollution. There is no ecosystem on Earth that is unaffected by human activity.

Reasons for habitat destruction

The natural habitats of plants, animals and microorganisms are destroyed for a number of different reasons:

- The clearance of land for crop production (Topic 19.1) and for the production of biofuels (Topic 20.1).
- Space to build units for intensive livestock production and clearance of land for cattle rearing.
- Digging mines and quarries for the extraction of coal and mineral ores, such as iron ore and bauxite (from which aluminium is produced).
- More people require more housing and the associated infrastructure: roads, industries, shops, etc.
- The human population produces huge quantities of waste. Space is needed for long-term storage of this waste and for recycling it.
- Marine pollution. Major pollutants of the sea include fertilisers, which are thought to cause algal blooms in the sea, oil from oil wells and ships, and industrial chemicals.

At high risk of destruction by pollution are coral reefs, which have very high biodiversity. In the Caribbean it is estimated that over 23% of coral reefs have been destroyed and many are considered at high risk.

Human influences on food chains and webs

If we haven't cleared areas of the natural environment completely or destroyed them by pollution, we have removed organisms either directly or indirectly:

- Killing of large predators – humans have killed many of these because they were dangerous (see also Topic 19.7).
- Killing of large herbivores for food – there were many different species of these in huge numbers.
- Overfishing has had significant effects.

Figure 19.2.2 The population of American bison decreased from an estimated 60 million in 1492 to 750 in 1890 due to hunting.

Removing species results in unbalanced food webs and unbalanced ecosystems that are further at risk of damage by other factors.

Top predators control the populations of primary consumers. Their removal means that populations of herbivorous animals increases and so there is increased competition for the plant food that they eat. The effect can be overgrazing and habitat destruction.

The Pacific sea otter, *Enhydra lutris*, is a secondary consumer that feeds on invertebrates, such as sea urchins and crabs. In the 19th century, sea otters were hunted for their fur and there was a striking change to the whole of the food web. Sea urchins exploded in numbers and ate many of the giant seaweeds that provided a habitat and food for many other species. The loss of one species, the sea otter, led to the catastrophic loss of many other animal species such as small invertebrates, fish, octopus and scallops from the food web.

Deforestation

Humans have been clearing forests for over 10 000 years to grow food and provide land for settlements and provide transport links. There are now very few forests in temperate regions that have not been cleared at some time in the past. Tropical forests in South-East Asia and South America have been cut down over the past 100 years. Many of these forests are rainforests rich in biodiversity.

Deforestation has resulted in a number of environmental problems:

- the extinction of species
- the loss of soil
- an increase in flooding
- the increase of carbon dioxide in the atmosphere.

Soils in tropical rainforests are very thin and when the vegetation is removed the soil is easily washed away. This causes soil erosion, formation of gullies and loss of plant nutrients. The land is rapidly degraded after all the trees are cut down.

Local weather patterns change with more frequent and severe storms. Flooding happens more frequently as water runs off the land more quickly and is not absorbed by plants and transpired into the atmosphere. Forests act as 'stores' of water; their leaves slow down the rate of evaporation from the soil and also decrease the rate at which water reaches the soil.

Carbon dioxide is added to the atmosphere because vegetation is burned. There is increased decomposition that also releases more carbon dioxide that is not absorbed by plants.

Rainforests have no effect on maintaining the correct balance of carbon dioxide and oxygen in our atmosphere. They produce about as much carbon dioxide as they use in photosynthesis and there is such a huge reserve of oxygen in the atmosphere that the quantity produced by all plants on Earth makes little difference to the overall oxygen concentration of the atmosphere. There is no evidence that the rainforests are the 'lungs' of the Earth. There are, however, plenty of other reasons to stop cutting down forests. Destruction of rainforests means the loss of many habitats and the extinction of many species. There are many species in rainforests that have yet to be identified, studied and classified.

Figure 19.2.3 A sea otter eating a crab. Sea otters are protected and their numbers increased in the latter half of the 20th century, but now they are being preyed upon by killer whales that may have less prey to hunt because of overfishing.

KEY POINTS

1 Habitats have been destroyed by clearance of land for farming, extraction of resources, such as fossil fuels and metal ores, housing and industry.

2 Pollution has destroyed many marine habitats, such as coral reefs.

3 Humans have negative impacts on habitats by removing species from food chains and webs.

4 Deforestation leads to extinction of species, loss of soil, flooding and increase of carbon dioxide in the atmosphere.

SUMMARY QUESTIONS

1 Outline some of the effects that an increasing human population has had on the environment.

2 It is estimated that between 1880 and 1980 about 40% of all tropical rainforest was destroyed.

   a Give three reasons for this large-scale deforestation.

   b Outline some of the effects that deforestation has on the environment.

3 Explain the negative effects of deforestation.

19.3 Pollution

LEARNING OUTCOMES

- State the sources and effects of pollution of land and water
- Discuss the effects of non-biodegradable plastics in the environment, in both aquatic and terrestrial ecosystems

Anything released into the environment as a result of human activity that has the potential to cause harm is a **pollutant**. Pollutants are chemical substances, such as fertilisers, oil and carbon monoxide; biological material, such as animal and human waste; heat released from power stations; and noise from industrial and domestic sources. **Pollution** is the release of substances from human activities that are harmful to the environment. We expect natural ecosystems to absorb and break down these substances. In Topics 19.3 to 19.5 we will look at some examples of pollution of land, sea and air.

Figure 19.3.1 We throw away mountains of rubbish every day. In many places this rubbish is taken to rubbish dumps, becoming a visual pollutant as well aas a danger to the environment.

Table 19.3.1 The sources and effects of some pollutants.

pollutant	sources	undesirable effects of pollutant on the environment
carbon dioxide	burning fossil fuels	enhances greenhouse effect (see page 256)
methane	cattle; paddy fields for growing rice; coal and oil extraction	enhances greenhouse effect (see page 256)
sewage	human and livestock waste contains urea, ammonia, protein, carbohydrates, fats and pathogens	reduces oxygen concentration in rivers; destruction of freshwater communities (see page 254)
fertilisers (mostly N and P)	arable agriculture	eutrophication in fresh water (rivers, lakes) (see page 255)
solid waste (biodegradable and non-biodegradable rubbish)	domestic and industrial waste	buried in ground (landfill) or left on rubbish tips; health hazard; leakage of toxic liquids; release of methane

EXAM TIP

Bacteria-like organisms respire in anaerobic conditions, such as soils in flooded fields and the stomachs of cattle to produce methane instead of carbon dioxide.

Figure 19.3.2 Methane gas is a pollutant given off by anaerobic conditions in rice fields.

Figure 19.3.3 Cattle also belch out methane from their stomachs during digestion.

Land pollution

Rubbish is not only a visual pollutant, but it contains many harmful chemicals that leach from rubbish dumps into the ground. Examples are heavy metals that enter water courses and the drinking water supply.

Agricultural chemicals (such as herbicides and pesticides) can affect areas locally and also far away, if not applied correctly. Insecticides have been used to control crop pests and vectors of disease such as mosquitoes. Such insecticides have been detected in the body fat of of Antarctic mammals and birds many thousands of miles away from where they were applied. Herbicides, especially when used in forests, destroy the habitats of animals.

Waste plastics

Some materials we throw away are broken down in the environment by decomposers – these materials are described as **biodegradable**. Some materials are not broken down by decomposers and are described as **non-biodegradable**. Plastics are used for packaging because they last a long time and do not decay easily. But this becomes a problem when it comes to disposing of plastics. Most plastic waste goes into landfill sites or rubbish dumps where it takes up a lot of space. Biodegradable plastics are designed to break down more quickly than conventional non-biodegradable plastics once they are dumped. For example, some plastics now have starch incorporated into their structures, which is digested by bacteria in the soil.

We can recycle many used thermoplastics by melting them and remoulding them into new shapes. Burning plastics in incinerators reduces the volume of waste, but many plastics produce toxic gases as they burn.

Non-biodegradable plastics also pose threats to aquatic life. Sea turtles are threatened if they mistake plastic bags for jellyfish and swallow them. Fish and other aquatic animals can get entangled in discarded plastic nets and other forms of plastic waste. In 2014, the United Arab Emirates banned non-biodegradable plastic products to try to reduce pollution.

Figure 19.3.4 Most plastic bottles have a recycling symbol with a category number. These bottles are in category 1 as they are made of polyethylene terephthalate (PET or PETE), which can be recycled to make products such as packaging materials and carpets.

SUMMARY QUESTIONS

1 List three pollutants of each of the following:
   a the atmosphere
   b aquatic ecosystems
   c the land
2 Discuss the reasons for the large scale use of agricultural chemicals and the harmful effects that they have on the environment.
3 Explain the problems caused to the environment by non-biodegradable plastics.

KEY POINTS

1 Pollution is the harm done to the environment by the release of substances produced by human activities.
2 Pollutants affect the air, ecosystems on land and aquatic ecosystems like rivers, lakes and the sea.
3 Some plastics are biodegradable; others are non-biodegradable but some can be recycled.

19.4 Water pollution

Supplement

LEARNING OUTCOMES

- Describe water pollution by sewage and toxic chemicals
- Describe the pollution of rivers and lakes due to sewage and the over-use of fertilisers
- Explain the process of eutrophication of freshwater ecosystems

Figure 19.4.1 In many parts of the world raw sewage drains into rivers.

Pollution of rivers and seas

Domestic and industrial pollutants are often discharged straight into rivers and into the sea.

Rivers empty toxic waste into the sea, which results in the following effects:

- Fertilisers and sewage encourage the growth of algae that release toxins.
- Pesticides are concentrated in the tissues of some molluscs (shellfish).
- Radioactive chemicals are found in higher concentrations around coastal nuclear power stations.
- Toxic metals such as mercury, copper and lead are found in tissues of marine organisms.

Sewage is the biggest single pollutant. It encourages the growth of algae and bacteria, which use up lots of oxygen, killing fish and small invertebrates. If raw untreated sewage is dumped in a river, it encourages the growth of bacteria that feed on the organic matter. The shortage of oxygen means that many freshwater organisms cannot survive so die or migrate away. However, pollution-tolerant animals, such as midge larvae and sludge worms, which feed on the sewage, increase in number (Figure 19.4.2).

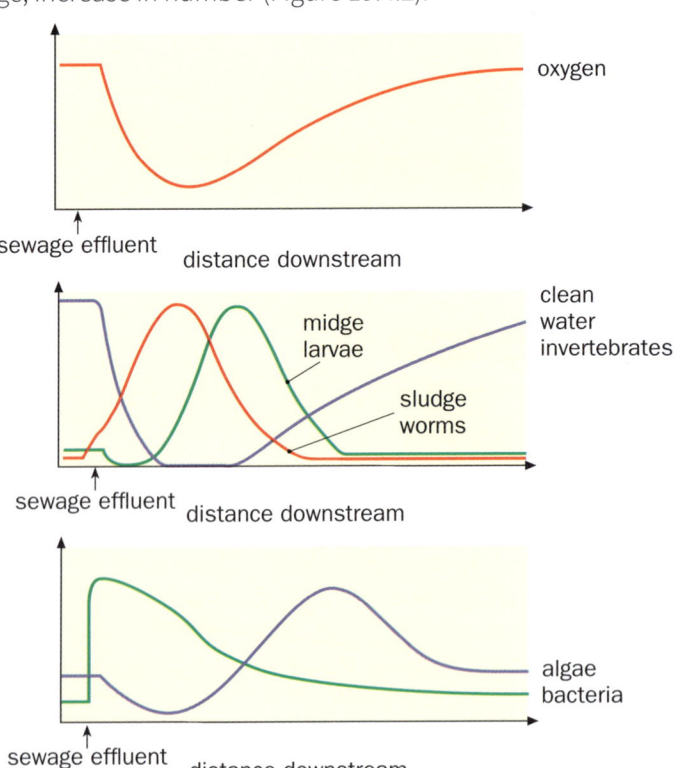

Figure 19.4.3 Effect of sewage effluent on discharge into a clean stream.

Figure 19.4.2 Sludge worms are a pollution-tolerant species.

As the water travels downstream, it gradually improves in quality as suspended waste settles out and is decomposed by bacteria. If no more sewage is dumped in the river, the community recovers as the oxygen concentration of the water increases (Figure 19.4.3). In many countries sewage is treated so that raw sewage is not deposited into rivers.

Fertilisers drain from the land into rivers and lakes. We have seen how using fertilisers can increase the yield of crops. But farmers need to know the best type of fertiliser to use for their particular soil and crop and how much of the fertiliser to add. Problems can occur if a farmer uses too much fertiliser or if the fertiliser is added at the wrong time, e.g., before a period of heavy rain.

The result of either of these is that fertilisers can cause water pollution resulting in **eutrophication**, which means that the water is enriched with plant nutrients. This is the sequence of events:

- Fertiliser can be washed through the soil into rivers and streams – this is called **leaching**. The rivers may flow into a lake.
- Once in the water, this stimulates population growth of algae. Nutrients in the fertilisers are usually in low concentration in the water and so limit the growth of the algae (animals that eat the algae do not multiply fast enough to control their growth).
- Algae cover the surface layers of water, reducing the light reaching plants at the bottom of the lake. These plants eventually die and rot on the river bed. Algae also die as there is competition for resources and many are shaded by the algae on the surface.
- Decomposers, such as bacteria, feed on dead plants and algae. Bacteria respire aerobically, multiply rapidly and use up a lot of dissolved oxygen. The concentration of oxygen decreases and this kills fish and invertebrates that cannot respire properly.

The same chain of events can happen if sewage gets into waterways. Bacteria multiply quickly, use up oxygen in respiration, which can result in the death of fish and pollution-intolerant invertebrates like stonefly larvae.

The problems of eutrophication are caused by nitrate and phosphate. These ions are in short supply in most natural aquatic ecosystems, so they act as a limiting factor to the growth of plants. When they are added in the form of fertilisers, plant growth increases.

Farmers are encouraged to reduce the fertilisers they apply and to make sure that they are applied at a time when crops will take them up. Phosphate tends to remain in soils and much of the phosphate in water comes from domestic sources. Removing phosphate from detergents has helped to reduce this type of pollution. Similar events happen in the sea – nitrate in the sea causes growth of algae, many of which produce toxins and kill other marine life.

Figure 19.4.4 Fertilisers caused algae to grow and turn this water green.

KEY POINTS

1 Water pollutants include sewage, fertilisers, pesticides and toxic metals.

2 If excess fertilisers are leached into rivers and lakes they provide nutrients causing eutrophication by supplying nutrients for algae.

SUMMARY QUESTIONS

1 a Define the terms *pesticide*, *insecticide* and *herbicide*.

 b Explain why they are used.

 c Explain how fertilisers can damage the environment.

2 Look carefully at Figure 19.4.3.

 a Explain the graph for oxygen.

 b Describe and explain the graphs for
   i clean water invertebrates
   ii midge larvae and sludge worms.

 c Explain the graph for bacteria and algae.

19.5 The greenhouse effect

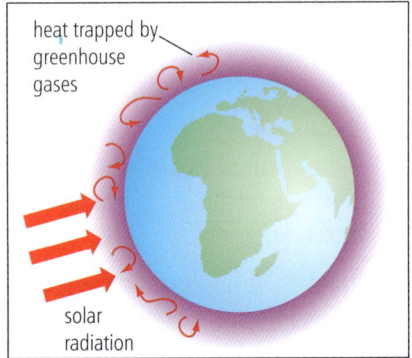

Figure 19.5.1 The greenhouse effect.

Figure 19.5.2 Factories, oil refineries and power stations release huge quantities of carbon dioxide into the atmosphere when they burn fossil fuels.

Greenhouse gases

The atmosphere is like a blanket surrounding the Earth keeping it warm. Some of the gases in the atmosphere act like a greenhouse to keep in heat (thermal energy) that otherwise would be radiated into space. These are the **greenhouse gases** – carbon dioxide, water vapour and methane. Some human made air pollutants, such as CFCs (chlorofluorocarbons), are also greenhouse gases, as well as the cause of other problems.

The greenhouse gases allow solar energy to pass through to the Earth's surface. Some energy enters food chains and is eventually transferred to the atmosphere as heat, which is radiated away from the Earth's surface.

Some heat energy escapes into space, but much of this is reflected back towards the Earth. Greenhouse gases keep our atmosphere at the temperatures that allow life to exist.

It is important to understand that the greenhouse effect is a natural process and without it the average temperature on the Earth would be about –17 °C. However, over the last 100 years, there has been a build-up of greenhouse gases.

Power stations, factories, domestic heating and transport use fossil fuels and release huge amounts of carbon dioxide into the atmosphere. As we have seen, deforestation has resulted in large areas of forest being removed. In South America, the trees cleared for farming are burned, which releases carbon dioxide into the atmosphere. Roots and other remaining tree parts are decomposed by microbes in the soil, producing even more carbon dioxide.

There has also been a significant increase in methane due to the expansion of rice cultivation and cattle rearing. Methane is released by cattle and also by bacteria in the anaerobic conditions found in flooded rice fields and natural wetlands. Rotting material in landfill sites and rubbish tips, as well as the extraction of oil and natural gas, are other sources of methane.

Of all the greenhouse gases the largest increase has been in carbon dioxide, which has risen by 10% in the last 30 years. However, methane and CFCs cannot be ignored as they are much more effective greenhouse gases than carbon dioxide.

Climate change

Human activities are causing an increase in the concentration of greenhouse gases so the atmosphere is getting warmer. This is causing the **enhanced greenhouse effect**. The surface of the Earth has warmed by 0.7 °C over the past century. There are fears that this warming is increasing.

If the temperature of the Arctic and Antarctic were to rise above 0 °C, the polar ice would start to melt. This would cause a rise in sea level and flooding of many low-lying areas, for example in Bangladesh. These areas include the sites of some of the capital cities of the world.

There could also be a change in wind patterns and the distribution of rainfall leading to more extreme weather. Some parts are expected to become very dry. Some of these are important agricultural areas, such as parts of the USA and Asia, so that warming of the climate could mean a massive reduction in the grain crops of Central Asia and North America. The pattern of the world's food distribution could be affected with economic and political consequences.

Measures to reduce the effects of climate change involve reducing carbon emissions. This may be done by encouraging public transport, using energy more efficiently, recycling and changing the diet of cattle to stop them releasing methane.

One method is to use fossil fuels more efficiently as happens in a combined heat and power plant in which the heat is used for domestic heating.

The greenhouse effect

Figure 19.5.3 shows in more detail how the greenhouse effect works.

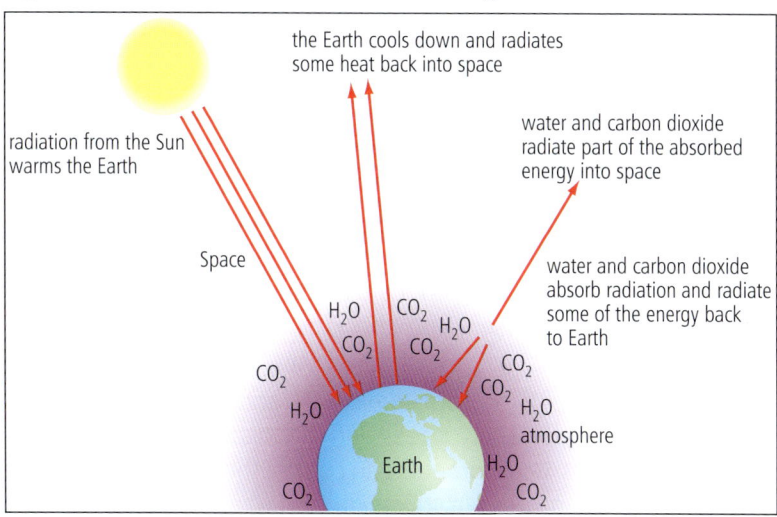

Figure 19.5.3 How the greenhouse effect works.

- Solar radiation passes through the atmosphere and warms the Earth's surface.
- The Earth radiates heat energy back into space. This is mainly infra-red radiation.
- Some of this heat energy is absorbed by the greenhouse gases.

This causes the air to warm up. Without carbon dioxide and water vapour, the heat energy would pass straight back out into space. Human activity is causing a large increase in the atmosphere of carbon dioxide and other greenhouse gases such as methane, which has a greater impact than carbon dioxide (see question 11 on page 267).

It is the increases in these gases in the atmosphere that are responsible for the enhanced greenhouse effect.

SUMMARY QUESTIONS

1 Copy and complete:

warms   absorbed   air   energy   space   radiates   infra-red   atmosphere

Solar _____ passes through the _____ and _____ the Earth's surface. The Earth _____ heat energy back into _____. This is mainly _____ radiation. Some of the heat energy is _____ by the greenhouse gases. This causes the _____ to warm up.

2 List the gases that contribute to the enhanced greenhouse effect and in each case give one human activity that acts as a source of the gas.

3 a Describe the likely consequences of global warming.

b Describe the ways in which people and their governments can reduce the emission of greenhouse gases and attempt to reduce the effects of global warming.

19.6 Sustainable resources

LEARNING OUTCOMES

- Define the term *sustainable resource*
- State that forests and fish stocks can be conserved and managed sustainably
- **S** Explain how forests and fish stocks can be conserved and managed sustainably

A **resource** is any substance, organism or source of energy that we take from the environment.

A **sustainable resource** is a resource that is renewed by the activity of organisms so that there should always be enough for us to take from the environment without it running out. Examples of these sustainable or renewable resources, that can be maintained, are timber harvested from forests and fish harvested from the sea and lakes.

Non-renewable resources include energy sources, such as fossil fuels and minerals, e.g., copper, zinc and lead ores. These resources cannot be replaced: once they are gone, they are gone forever.

Figure 19.6.1 In Haiti, the production of charcoal for cooking has contributed to the removal of almost all the tree cover from the country.

Figure 19.6.2 Sustainable forestry: the mature trees are ready to be felled for the timber trade. The young trees have another 20 years to grow.

Supplement

Sustainable timber production

Timber is a valuable resource as it used in many industries to make a great variety of products. Much of this timber comes from forests that are managed to ensure continuous production.

- As soon as trees are felled, they are replanted. The replanting is an example of restocking so that the resource can continue to be harvested at the same rate without becoming depleted. These trees tend to be softwood, coniferous trees, as in Figure 19.6.2.
- Publicity campaigns can educate timber companies, retailers and consumers to use timber that is certified as taken from sustainable sources.
- Not all timber is taken from commercial softwood forests. Some hardwoods, such as teak and mahogany, comes from tropical forests. Governments can control the use of this timber by issuing permits to logging companies and using quotas so there is no long-term damage to forests by loss of large, mature trees.
- Protected areas are established where felling is banned. These areas protect the biodiversity of tree species as well as all the other organisms that live in these ecosystems.
- National and international organisations campaign for the conservation of forests and also educate the public in the value of these ecosystems.

Sustainable fishing

Populations of fish are known as fish stocks. Many of these populations are now so severely depleted by overfishing that they are not sustainable. For example, stocks of the Atlantic cod collapsed in the 1990s.

Some species of fish and other sea creatures such as crustaceans and molluscs can be hunted sustainably. To continue to do this, the following steps need to be taken.

- Marine parks have been established where fishing is banned at all times. The Cape Rodney/Okakari Point Marine Reserve in New Zealand provides a refuge for fish and other marine life.
- Populations of fish are continually monitored by fisheries scientists to ensure that fish stocks are sustainable.
- Certain areas of the sea are declared off-limits for fishing during the breeding season. This ensures that a suitable number of young fish are recruited to the stock each year. The Cayman Islands in the Caribbean has a 'no take' zone to help conserve a fish known as the Nassau grouper. Thailand has banned fishing in the Andaman sea off the west coast of the country for three months of the year to allow fish and other marine life to breed. This has allowed fish stocks to increase.
- Quotas are issued by fishing authorities so that boats cannot take too many fish in each season.
- Some types of fishing are regulated to limit the decrease in fish stocks. A seine net is a very long net that completely surrounds a certain area and is operated from boats or from the shore. All the fish caught within the net are hauled in as shown in Figure 19.6.3. The use of these nets has been limited or banned in many areas of the sea.
- Fishing nets have to be a certain minimum size so that small fish can escape and survive to breed.
- Consumer awareness campaigns can educate people to buy types of fish that are known to come from sustainable stocks.
- Some stocks of fish can be restocked if it is possible to breed them in captivity and raise them in facilities known as hatcheries. Examples are found in Japan where shellfish, prawns and several species of fish including red sea bream and flounder, are released as young stages (fry) into the sea each year. Sturgeon to provide caviar are reared and released into the Caspian Sea by facilities in Iran.

Figure 19.6.3 One way to conserve fish stocks is to regulate the mesh size of nets.

Figure 19.6.4 Large factory ships like this one have taken so many fish from the oceans that stocks are not sustainable and some species are being driven to extinction.

KEY POINTS

1 A sustainable resource is one that can be removed from the environment without it running out. Examples are forests and fish stocks.
2 Forests can be conserved using education, protected areas, quotas and replanting.
S 3 Fish stocks can be conserved using education, closed seasons, protected areas, controlled net types and mesh sizes, quotas and monitoring of populations of fish.

SUMMARY QUESTIONS

1 Explain what is meant by a *sustainable resource*.

2 Explain why forests and fish stocks may be sustainable, but fossil fuels are not.

3 Research examples of conserving forests and fish stocks in your country or neighbouring countries.

S 4 Discuss the role of education in conserving forests and fish stocks.

A species becomes **extinct** if there are no individuals left alive anywhere in the world. Some species are kept alive only in captivity and they are described as 'extinct in the wild'. Extinction occurs all the time. However, the number of extinctions occurring now is far higher than in the past. This is because of all the human activities we describe in this Unit.

When the numbers of a species decrease to such a level that it is at risk of becoming extinct, the species is described as **endangered**.

We have already considered most of the causes of species becoming endangered and extinct. These are habitat destruction, climate change, pollution and overfishing. Others causes include:

- hunting for food, sport and to kill animals that are a threat to human life, crops or livestock
- competition with species that humans have moved from one part of the world to another
- predation by introduced or alien species.

Threats to the survival of species

Climate change is occurring as a result of changes to the composition of the atmosphere. Climate change was a feature of the past as the world cooled and heated at different times. During the last ice age, most of Europe, North America and much of Asia were covered in ice sheets. As the world warmed, the ice melted and organisms colonised the land that was exposed.

Now the world is warming even more, and plants and animals associated with tropical regions are extending their ranges away from the equator. Plants better able to survive in changing environments will outcompete other species. Examples of this are evident on mountains. Plants adapted to cold conditions are disappearing from lower altitudes and retreating towards mountain peaks. If this continues there will be nowhere left for them to go.

Habitat destruction was covered in Topic 19.2. The Atlantic forests of Brazil have largely been cut down. This has robbed many species, including the golden lion tamarin, of their habitats.

Deforestation is not the only cause of habitat loss. Wetlands are very important habitats as they provide ecological services for us and also feeding grounds for migrating birds. Many wetlands across the world have been drained to provide land for farming, housing, industrial development and leisure facilities.

Pollution is occurring on a global scale. Coral reefs are endangered ecosystems as a result of coastal pollution. Sea otters, like many marine animals, are at risk of extinction thanks to oil pollution. Oil sticks to fur and feathers and stops them providing insulation. At the same time, the animals ingest toxic chemicals that cause organ damage (Topic 19.2). Many sea otters died following the release of oil from the tanker Exxon Valdez in the Bay of Alaska in 1989.

Figure 19.7.1 The last thylacine (Tasmanian wolf) died in Hobart Zoo in 1936.

Figure 19.7.2 It is sometimes difficult to know when a species becomes extinct, but the Golden Toad, *Incilius periglenes*, was restricted to the Monteverde Reserve in Costa Rica and has not been seen there since 1989.

Figure 19.7.3 Cycads are ancient trees that are little changed since the Jurassic period over 100 million years ago. They are threatened in their natural habitats, but can grow safe in botanic gardens, as here in Sri Lanka.

Introduced species have been moved around the globe by humans. Sometimes we have done this deliberately by introducing animals for sport or to control pests:

- Almost half the small to medium-sized native marsupials in Australia have become extinct. This is due to competition with rabbits and predation by foxes introduced by European colonists.

- The small Indian mongoose was introduced to Jamaica from India in 1872 to control rats that were eating much of the sugar cane crop. They proved so successful that they were introduced to other parts of the Caribbean. However, as with many such introductions, the mongoose fed on other prey as well as the one it was intended to control.

- Plant species can become extinct because of overgrazing by introduced species especially goats. This has done untold damage to many islands.

Hunting has caused the extinction of some species. They have been hunted for food, for sport, for trade and as a deliberate policy to remove dangerous animals or pests. Humans have always hunted animals and harvested plants from the wild for food. With increasing populations many of these populations became unsustainable. About 15000 years ago people migrated from Asia into the Americas. As they moved southwards they destroyed many of the large mammal species.

People have hunted animals such as 'big cats' for sport; they have also collected beautiful and interesting plants for show. Animals were trapped and shot for their fur in the days before synthetic materials became available. In some places, animals are shot as a deliberate policy; this is the case with elephants that invade farmland. In Tasmania, people hunted the thylacine, or Tasmanian wolf, to extinction partly because it was believed to kill sheep.

Figure 19.7.4 A project to reintroduce the golden lion tamarin to its natural habitat in Brazil has involved captive breeding in many zoos.

Figure 19.7.5 Coral reef communities are threatened by pollution, climate change and overexploitation.

Supplement

There is very little genetic variation in small populations. This makes rare and endangered species at particular risk of becoming extinct. With few individuals left alive, many of the alleles of the genes in that species are no longer present. This reduces the chances of the population evolving in response to changes in the environment (Topic 17.3).

SUMMARY QUESTIONS

1 Define the following terms: **a** *extinction*, **b** *endangered species*.

2 Explain how the following threaten the survival of species:
   a hunting   **b** overfishing
   c habitat destruction   **d** climate change

3 Use examples to explain the possible effects that the introduction of species may have on food webs and ecosystems.

S 4 When a population decreases in size, variation within the population decreases. What are the implications for the survival of a species when this happens?

KEY POINTS

1 Organisms become extinct when there are no individuals left alive in the wild or in captivity. Endangered species are those that are at risk of extinction.

S 2 An endangered species is at risk of becoming extinct because its population is small with very little genetic variation that is needed to adapt to changing conditions.

3 Causes of extinction are climate change, habitat destruction, hunting, pollution and predation by introduced species or competition with them.

19.8 Captive breeding

Figure 19.8.1 The California condor is the largest North American land bird. It has been reintroduced after becoming extinct in the wild in 1987.

Figure 19.8.2 The Mexican wolf has benefited from artificial insemination. A recent census found 163 individuals in the wild in New Mexico and Arizona, compared to 131 in 2018.

Captive breeding

Captive breeding involves breeding animals in human-controlled environments such zoos and wildlife reserves. Sometimes, the process includes release of these animals into the wild if there is sufficient natural habitat to support and protect them. Captive breeding programmes promote biodiversity and have saved many species from extinction. However, captive breeding may also reduce genetic diversity through inbreeding.

Captive breeding has been used to help species that are threatened by the results of human activities such as habitat destruction, over-hunting or fishing, pollution, predation, disease and parasitism. Successful examples of captive breeding include the California condor, the black-footed ferret, the golden lion tamarin and the red fox – all of these have been reintroduced to the wild.

A captive breeding programme should aim for adequate genetic diversity. Breeders usually select individuals from different populations. Unfortunately, too often these populations have consisted of only a few individuals because of their threatened state. This means they are more at risk of the problems of inbreeding.

Supplement

Artificial insemination

Artificial insemination is the process of collecting semen containing living sperm from the male animal and placing it, via a fine plastic tube, into the uterus of the female. It has long been used in domestic animals such as cattle because enough sperm can be collected from the best bulls and used to fertilise many cows. This is an important tool in selective breeding. It also means cows can be inseminated without the need for a farmer to keep a bull. The technique has been adopted in the interests of conservation of endangered species.

Artificial insemination is used to reduce problems such as inbreeding and lack of interest in mating. Giant pandas, for example, lose interest in mating once they are captured. Female giant pandas release eggs once a year over a period of 48 to 72 hours. Artificial insemination has allowed scientists to overcome the problem of non-receptive females. AI has also been instrumental in saving such animals as the Mexican wolf, the Black-footed ferret and Magellanic penguin from extinction.

Cryopreservation is a technique that allows gene banks to preserve and store live sperm, eggs or embryos in frozen conditions. The Zoological Society of San Diego established a 'frozen zoo' to store tissue from the world's rarest and most endangered species using cryopreservation. So far, the gametes or embryos of more than 350 species including mammals, reptiles and birds have been preserved in this way.

In vitro fertilisation

In vitro fertilisation (IVF) is another technique that has been used in captive breeding, when other methods have been unsuccessful. *In vitro* means 'in glass'. Eggs and sperm are collected from the potential parents and mixed together in a laboratory dish where fertilisation occurs.

To increase the chances of success, the female animal is injected with hormones to stimulate the development of her eggs. This way several eggs can be collected from her ovaries. A small incision is made in the body wall and a fine plastic tube inserted to collect the eggs from the ovary surface. The eggs are kept in a solution that contains nutrients and oxygen.

Semen is collected from the male animal and is mixed with the eggs and left for up to 24 hours for fertilisation to occur. A technician checks that fertilisation has occurred and that the zygotes have started to divide into embryos. These are kept in a solution for a few days and then a scientist places them into the uterus of the female animal. If the procedure is successful, the embryo will develop into a fetus.

The last male Northern White rhino died in 2018 and just two females remain, but IVF techniques have enabled two embryos to be formed, in the hope of saving the species. Animals from the big cats (like cheetahs and lions) to tiny amphibians (like the Wyoming toad and the Puerto Rican toad) have benefited from IVF techniques to increase their numbers.

Figure 19.8.3 Cheetahs are the fastest land animal, capable of running 50 to 80 mph.

Figure 19.8.4 Can eggs extracted from these females be used in IVF with frozen sperm from the dead male, to save the Northern White Rhino from extinction?

Figure 19.8.5 In 2019, the first Puerto Rican crested toad was hatched via *in vitro* fertilisation.

SUMMARY QUESTIONS

1 Explain, using examples, how captive breeding has saved animals from extinction.

S 2 **a** Explain what is involved in the following techniques:

In vitro fertilisation (IVF) and cryopreservation

b i Explain why eggs to be used in IVF are kept in a solution of nutrients and oxygen.

ii Explain why embryos are kept for a few days before implanting them into the uterus.

3 Research how AI and IVF have helped in the survival of **two** of the following species: giant pandas, black-footed ferrets, Magellanic penguins and lions.

KEY POINTS

1 Captive breeding involves breeding animals in controlled environments, such as zoos.

S 2 Artificial insemination involves transferring semen from the male to the uterus of the female.

3 *In vitro* fertilisation is a technique in which fertilisation takes place in a Petri dish.

19.9 Conservation

Figure 19.9.1 Putting rings onto the legs of birds has given scientists information about how long they live and how far they travel. Some bird species migrate over thousands of miles.

Figure 19.9.2 Horton Plains National Park, Sri Lanka. Areas of wilderness like this should be conserved because they are the habitats of rare plants and animals, and also for the enjoyment of future generations.

Figure 19.9.3 The Arabian oryx, *Oryx leucoryx*, has been bred in zoos and reintroduced into the wild, sometimes successfully.

Endangered species can be conserved. This is never achieved simply by putting a barrier around an organism's habitat and leaving the area alone. Human activities affect everywhere on the planet, so it is important to manage the organism's habitat and closely monitor the changes in population.

Species monitoring can involve labelling, radio tagging and counting flocks of birds and herds of animals. Researchers in East Africa use aeroplanes to find herds of elephants and count them. Other techniques have to be used for forest elephants.

No species lives in isolation, so we have a duty to conserve ecosystems and habitats. Here are some ways in which this is done:

- National parks are large tracts of land set aside for wildlife but which may be occupied by people and are patrolled by wardens. Examples include the game parks of East Africa, such as Masai Mara in Kenya and Serengeti in Tanzania.
- Marine parks protect areas of the sea from damage by fishing and pollution. Goat Island Marine Reserve in New Zealand is an example.
- Rescuing endangered animals and breeding them in captivity and then returning them to the wild. Many species of *Partula* snails became extinct on Pacific islands during the 20th century. Some are now bred in captivity ready to be released back into their habitats. The Arabian oryx was bred in zoos and reintroduced to Oman where it had become nearly extinct.
- Growing endangered plants in botanical gardens and re-establishing them in the wild.
- Reducing habitat destruction – for example by issuing licences for logging in forests to prevent deforestation and protecting wetlands to prevent them being drained.
- Re-establishment of ecosystems where land has become degraded. A very long-term project is the re-establishment of tropical dry forest in Guanacaste National Park in Costa Rica – this may take up to 300 years to achieve!
- Preventing trade in endangered species. Despite an international ban on all species of Pangolin, huge numbers of them are still being illegally traded from Africa to Asia. In particular the animals' scales are sought after for use in folk medicine.
- Encouraging sustainable management of ecosystems. Trees removed from forests should be replaced by planting or allowing time for natural replacement from seeds.
- Seed banks are cold stores that conserve seeds of endangered or valuable species. Seed banks around the world hope to collect and store seeds from many species in case they become extinct in the wild. Their genes may be useful for crop improvement in the future or to produce valuable products such as medicinal drugs. Seeds are collected from plants in the wild and are put into long-term storage. Seeds of many species are dehydrated so they contain only 5% water and can thus survive being kept at –20 °C.

Removing water from seeds slows down their metabolism so that they remain viable for many years. However, seeds do not remain viable in seed banks forever. Some seeds from each sample are removed from storage, thawed and tested to see if they will germinate. This takes place every five years. Collections continue to be made if possible to 'top up' the bank for each species.

Increasing people's knowledge and understanding of the way in which their actions affect wildlife will help to ensure the success of conservation programmes.

Figure 19.9.4 These wardens in Papua New Guinea are conserving the eggs of leatherback turtles.

Supplement

Reasons for conservation

We should conserve ecosystems, habitats and species because:

- Ecosystems provide us with services such as treating waste, providing food and fuels, and giving us areas for recreation. They also provide us with useful substances such as medicines.

- Ecosystems help to maintain the balance of life on the planet – or example, nutrient cycles.

- Habitats support a wide variety of organisms that interact in ways we do not fully understand, often to continue life on this planet, for example by keeping pests and diseases in check.

- There are few foods, apart from some fish species, that we take directly from the wild in large quantities. But there are many that we take in smaller quantities, such as Brazil nuts. There may be many more plants and animals that we could utilise as food sources.

- Fuels – fossil fuels will not last forever. We still need timber to provide fuels, such as biomass fuels (Topic 19.6).

- Drugs – drugs for treating cancer have been discovered in plants.

- Genes – as a result of selective breeding there is very little genetic diversity in our three main staples – rice, wheat and maize. It is important to conserve any locally adapted varieties that exist and also any wild relatives that have genes we could use in the future. These plants can be kept in botanic gardens and their seeds in seed banks.

- Prevention of species becoming extinct, especially those threatened directly by human activities.

Figure 19.9.5 The Svalbard seed bank in northern Norway is cut into the Arctic permafrost.

KEY POINTS

1 Endangered species can be conserved by monitoring the sizes of their populations, protecting habitats, using captive breeding programmes and putting seeds into long term storage in seed banks.

2 Educating people of all ages about the importance and practices of conservation is essential for the survival of endangered species.

S 3 There are many reasons for conservation, such as reducing extinction, protecting vulnerable environments and maintaining ecosystem functions, such as nutrient cycling and providing resources such as food, drugs, fuel and genes.

SUMMARY QUESTIONS

1 Explain the meaning of each of the following:
   a endangered species   b species monitoring
   c captive breeding

2 State two species that are endangered in your region or country. Describe the steps being taken to conserve them.

3 Explain the part played by each of the following in conservation:
   a national parks   b zoos   c botanic gardens   d seed banks

S 4 Explain why it is necessary to conserve habitats and ecosystems.

Practice questions

1 Competition between crop plants and weeds for resources is one of the biggest potential causes of reduced yields. Which reduces competition between crop plants and weeds?

  A chemical fertilisers

  B herbicides

  C pesticides

  D selective breeding

  (Paper 1) [1]

2 The boll weevil (see diagram) is a notorious pest of cotton plants. What should be used to control the boll weevil?

  A fungicide

  B herbicide

  C insecticide

  D molluscicide

  (Paper 1) [1]

3 An ecosystem is all the:

  A organisms in an area

  B organisms in an area and the physical factors that influence them

  C physical factors that influence an organism in an area

  D plants and animals in an area and the interactions between them

  (Paper 1) [1]

4 What is the negative effect of the over-use of fertilisers on farmland?

  A flooding

  B pollution of freshwater

  C global warming

  D more photosynthesis

  (Paper 1) [1]

5 Which pair of gases contributes to the enhanced greenhouse effect?

  A carbon dioxide and methane

  B oxygen and carbon dioxide

  C sulfur dioxide and water vapour

  D water vapour and oxygen

  (Paper 2) [1]

6 Fish stocks are in serious decline in many major fishing areas. Which is *not* a method that could be used to make stocks sustainable?

  A decreasing the mesh size of nets

  B setting up exclusion zones and 'no-take' zones

  C reducing the time that fishing boats can stay at sea

  D issuing quotas for the maximum numbers of fish that can be caught.

  (Paper 2) [1]

7 When a population becomes endangered the genetic variation within the population decreases. This means that within the population there are fewer:

  A alleles

  B chromosomes

  C DNA base sequences

  D genes

  (Paper 2) [1]

8 Eutrophication may happen when untreated sewage or fertilisers enter bodies of water. The oxygen concentration in the water may decrease because:

  A decomposers are respiring aerobically

  B less oxygen dissolves in water from the air

  C more carbon dioxide is released by bacteria

  D there are more animals in the water

  (Paper 2) [1]

9 Humans have used a variety of methods to increase food production including the use of monocultures for growing crops.

  (a) (i) State what is meant by the term *monoculture*. [1]

  (ii) State **one** advantage and **one** disadvantage of using monocultures for the growing of crops. [2]

  (b) Describe **three** ways **other than** monocultures that humans use to increase the production of food. [6]

  (c) Agriculture is a source of non-biodegradable plastics. Describe the effects of these plastics on the environment if they are not collected for proper waste disposal. [3]

  (Paper 3)

10 When populations reach a certain size they are declared to be endangered.

(a) State what is meant by the term *endangered species*. [2]

(b) The Mercury Island tusked weta, *Motuweta isolata*, is a large carnivorous insect from New Zealand. It became endangered following the introduction of alien species and was rescued by captive breeding and a reintroduction programme.

Suggest:

(i) why this species of tusked weta did not survive the introduction of alien species to New Zealand [2]

(ii) how a captive breeding programme for such an insect would be carried out. [2]

(c) State three reasons, **other than** the introduction of alien species, that can lead to plant and animal species becoming endangered or extinct. [3]

(Paper 3)

11 The table shows how the concentrations of three gases have increased between 1890 and 1990 and the estimated concentrations in. It also shows how effective they are as greenhouse gases compared with carbon dioxide. The main sources of nitrous oxide are motor vehicles and fertilisers.

gas	concentration in the atmosphere in certain years / parts per million			relative effect as greenhouse gas compared with CO_2
	1890	1990	2030 (estimate)	
carbon dioxide (CO_2)	290	354	400–500	1
methane (CH_4)	0.9	1.7	2.2–2.5	30
nitrous oxide (N_2O)	0.28	0.3	0.33–0.35	160

(a) Calculate the percentage increase in the concentration of carbon dioxide in the atmosphere between 1890 and 1990. Show your working. [2]

(b) Explain why the concentrations of (i) carbon dioxide, and (ii) methane increased in the 20th century. [4]

(c) Explain how the gases shown in the table may cause global warming. [5]

(d) Suggest why it is important to know the relative effects of these gases as shown in the table. [3]

(e) Suggest **three** ways to prevent the concentrations of the gases in the table from reaching the levels estimated for 2030. [3]

(Paper 3)

12 The Asian elephant, *Elephas maximus*, is a large herbivorous animal. It is classified by the International Union for Conservation of Nature as an endangered animal. In 2008, researchers published the first estimate of the population in the Taman Negara National Park in the centre of Peninsular Malaysia. The researchers counted piles of dung to estimate the population. Their estimate was 631.

(a) Suggest why the researchers could not count the elephants, but had to count piles of dung instead. [3]

(b) Suggest why it is important for conservationists to have reliable estimates of the population sizes of endangered species. [3]

(c) It has been estimated that Asian elephant populations can increase at a rate of 2% per year if there are no limiting factors.

(i) List four factors that are likely to limit the population growth of Asian elephants in a National Park such as Taman Negara. [4]

(ii) Explain why it is important to conserve large herbivorous animals, such as the Asian elephant, but not allow the population in a National Park to increase too much. [5]

(d) Artificial insemination (AI) and *in vitro* fertilisation (IVF) are two methods of assisted reproductive technology that are used in captive breeding programmes for some species of mammal.

Explain the use of AI and IVF in captive breeding programmes. [5]

(Paper 4)

20.1 Microorganisms and biotechnology

LEARNING OUTCOMES

- State how bacteria are useful in biotechnology and genetic modification
- **s** Explain that there are few ethical concerns about the use of bacteria in biotechnology and genetic modification
- Explain why plasmids found in bacteria are useful in genetic modification
- Describe the role of anaerobic respiration in yeast in the production of ethanol for biofuels
- Describe the role of yeast in breadmaking

The use of microorganisms and biotechnology

Biotechnology is the use of microorganisms such as bacteria and fungi to make useful products or to carry out services for us. An example of such a service is making waste products harmless.

The use of both bacteria and fungi in various processes is widespread.

- Bacteria reproduce very quickly with a generation time (the time taken for numbers to double) that is often as little as 30 minutes.
- Unlike animals or plants, microbes can convert raw materials into the finished product very quickly, in hours rather than months or weeks.
- The use of bacteria means that food production can be independent of climate.
- Bacteria can produce complex proteins (like enzymes and antibiotics) that pass out into the surrounding medium and can be harvested. Enzymes made by microorganisms are used in the food industry.

Examples of enzymes that are produced by bacteria for use in the food industry are:

- amylase for breaking down starch in the production of glucose syrup
- pectinase for extracting juice from fruit (see page 271)
- sucrase for breaking down sucrose in making confectionery
- protease for making meat more tender.

Figure 20.1.1 Some of the products of traditional biotechnology using bacteria and fungi.

Figure 20.1.2 Many people have concerns about the welfare and health of chickens kept in these conditions.

Supplement

With the aid of genetic modification, scientists can quickly alter microbes and so modify products. In contrast, breeding new varieties of plants and animals can take a long time.

In addition to the single loop of DNA, bacterial cells also contain small circles of DNA called **plasmids**. Plasmids are easy to work with because they can replicate very quickly. Plasmids can be cut open by enzymes and a gene from another organism can be spliced into them. In nutrient media, the bacteria multiply rapidly, making many copies of the plasmid and the inserted gene (see page 273).

The use of bacteria and fungi in the manufacture of complex chemicals does not raise the same ethical concerns over their manipulation and welfare as would be the case in the use of other living organisms – for example, keeping chickens in battery farms (Figure 20.1.2).

Biofuels and breadmaking

Many microbes can respire successfully without oxygen. Yeasts can respire with or without oxygen.

When yeast respires without oxygen (anaerobically) it is called fermentation.

Glucose $\longrightarrow$ alcohol + carbon dioxide + energy released

Biofuels

Biofuels are fuels made from biological material. Sugar-rich products from sugar cane and maize can be fermented anaerobically with yeast to produce ethanol. Some countries have no oil supplies of their own and have developed a biofuel substitute. Brazil, for example, uses fermenters to produce ethanol. The raw material often used is sugar cane juice, which contains a lot of carbohydrates. Glucose from maize starch is another raw material. This is obtained by treating the starch with carbohydrase enzymes.

Ethanol is a good substitute for petrol because it has a high energy content and when burned does not produce toxic gases. Using ethanol as a fuel means that there is no overall increase in carbon dioxide in the atmosphere. As a result this fuel is **carbon neutral**.

Car engines need to be modified to be able to use pure ethanol. **Oil seed rape** is a crop that provides oil that is converted into fuel called **biodiesel**.

Bread-making

Baking is an example of using yeast to help produce a food. Bread is made from **dough**, a mixture of flour, water, salt, sugar and yeast. This mixture is kept at a warm temperature. The yeast starts to ferment the sugar, producing carbon dioxide gas and alcohol.

Bubbles of carbon dioxide are trapped inside the dough and make it rise. When the bread is baked in a hot oven the bubbles of the gas make the bread 'light' in texture. The heat causes the alcohol to evaporate, leaving behind the traditional taste of bread.

Figure 20.1.3 Some biofuels are made from waste material. This biobus runs on fuel made from used cooking oil and other waste products from the food industry.

Figure 20.1.4 Yeast produced carbon dioxide, which caused the bread dough to rise and gave these loaves their light texture.

SUMMARY QUESTIONS

1 List four ways in which bacteria are useful in biotechnology.

2 a Write a word equation for anaerobic respiration in yeast.

  b In bread-making, why is yeast mixed with sugar before it is added to flour and water?

  c What happens to the ethanol made by yeast in bread-making?

3 Make a table summarising the advantages and disadvantages of ethanol as a fuel for cars.

S 4 Explain the use of plasmids in genetic modification.

KEY POINTS

1 Bacteria are useful in biotechnology and genetic modification as they reproduce rapidly and make complex molecules.

S 2 Plasmids are used to transfer genes into bacteria.

3 There are few ethical concerns about the genetic modification of bacteria and their use in biotechnology.

4 Yeast carries out anaerobic respiration in the production of ethanol for biofuels.

5 Yeast carries out anaerobic respiration during bread-making.

20.2 Enzymes and biotechnology

LEARNING OUTCOMES

- Describe an experiment to investigate the use of pectinase in the production of fruit juice
- Describe an experiment to investigate the use of enzymes in biological washing powders
- **S** Explain the use of lactase in the production of lactose-free milk

Figure 20.2.1 Biological washing powder.

Figure 20.2.2 Granules of biological washing powder. Some of the granules are partially opened. Inside these granules are enzymes that act as cleansing agents.

Enzymes have become very important in industry.

Biological washing powders

Biological washing powders may contain one or more of the following types of enzyme:

- **proteases** – break down protein stains (e.g., blood, grass and egg)
- **lipases** – break down fats in grease stains (e.g., butter, lipstick and mayonnaise)
- **amylases** – break down starch (e.g., food stains containing starch)
- **cellulases** – break down cellulose fibres on the outside of cotton fabrics to remove the dirt attached to them.

The enzymes listed above have been modified so that they withstand the high temperatures and alkaline conditions required for some washing powders. During a washing cycle, the enzymes break down stains. The products of the reactions then dissolve or are suspended in water and are removed when the washing machine empties. For example, proteins are broken down into amino acids; starch and cellulose are broken down into glucose.

PRACTICAL

Comparing the action of biological and non-biological washing powder

1 Decide what sort of stain you are going to have (e.g., egg, soya sauce, mustard).
2 Decide what sort of fabric you are going to use (e.g., cotton, linen or wool). Also, decide what size of fabric to use and cut two pieces the same size.
3 Add the same amount of stain to each piece of fabric and let it dry before you put it into the beakers to 'wash'.
4 Weight out 5 g of each powder and dissolve each in 500 cm^3 of water at 30 °C in a beaker.
5 Add a piece of the same fabric with the same stain to each beaker.
6 Decide how often you are going to stir the 'washing' over a 10-minute period.
7 Decide how to compare the cleanliness of each fabric after washing. You could produce a scale of cleanliness.
8 Repeat your tests to see if you get the same results.

Safety: Wear gloves and do not let the washing powders come into contact with your skin to avoid a possible allergic reaction.

Extracting fruit juice

Pectinases are enzymes that break down pectins, which are molecules that act like a 'glue' in plant cell walls. Pectinases are used for extracting fruit juices and for softening vegetables.

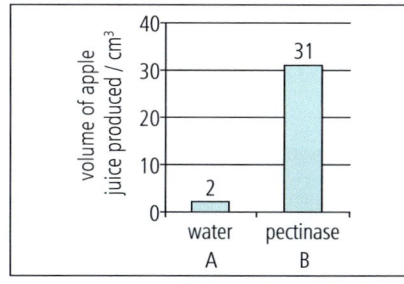

Figure 20.2.3 Extracting apple juice using pectinase.

PRACTICAL

Extracting juice from apples with pectinase

1 Cut an apple in half.

2 Chop each half into small pieces and put them into separate beakers.

3 Pour 10 cm³ of water over the apple in beaker A and 10 cm³ of pectinase solution over the apple in beaker B.

4 Place both beakers in a water bath at 40 °C for 20 minutes.

5 Filter the juice from each of the beakers of apple pieces (Figure 20.2.3).

- Explain the results shown in Figure 20.2.4.
- How does the pectinase release this juice from the apple cells?
- Why was 10 cm³ of water added to beaker A?

Figure 20.2.4 Sample results from the practical to extract juice from apples.

Supplement

The use of lactase in making lactose-free milk

Some people do not produce enough lactase and therefore have problems digesting and breaking down lactose, the sugar found in milk. They may get diarrhoea, wind and stomach cramps if they consume milk or milk products. Lactose is a complex sugar, similar to sucrose, that can be broken down into simple sugars by the enzyme lactase.

$$\text{lactose} \xrightarrow{\text{lactase}} \text{glucose} + \text{galactose}$$
(complex sugar)   (2 simple sugars)

You can investigate the effect of lactase in breaking down lactose by using test strips that detect the concentration of glucose. These are the test strips that are used to detect glucose in urine. You cannot use Benedict's test as all three sugars in the equation above are reducing sugars and give a positive result.

SUMMARY QUESTIONS

1 Describe what each of the following enzymes do.
   a amylase   b protease
   c lipase   d pectinase

2 Explain the advantages of using enzymes in biological washing powders.

3 Look at the practical box above. Explain:
   a why water was added to beaker A
   b the results in Figure 20.2.4
   c how pectinase releases juice from apple tissue.

S 4 a How is lactose-free milk produced?
   b Why is there a demand for lactose-free milk?
   c Outline an experiment to discover which is the best temperature to produce lactose-free milk.

KEY POINTS

1 Biological washing powders contain enzymes that work at low wash temperatures and in alkaline conditions.

2 Pectinase is used to extract fruit juice by breaking down the pectin in the cell walls in fruits such as apple.

S 3 Lactase breaks down lactose to produce lactose-free milk.

20.3 Fermenters

Supplement

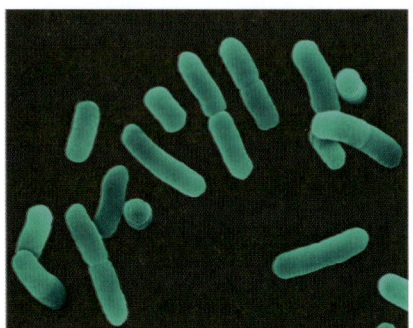

Figure 20.3.1 Lactic acid bacteria are used in the industrial production of yoghurt and cheese.

You may recognise the term *fermentation* as meaning respiration without oxygen (see page 140). However, in an industrial context it refers to any process that uses microorganisms to produce a useful product or to carry out a useful process for us. The organisms involved may respire with or without oxygen.

Industrial production of bacterial and yeast products

Large-scale industrial fermentation can be used to produce useful products from both bacteria and yeasts. The fermenter is inoculated with a culture of a suitable microorganism, which then grows under controlled conditions including the correct nutrients, constant temperature and constant pH. You should already know that useful genes can be spliced into the plasmids of bacteria. These GMO bacteria are transferred into crop plants to give beneficial properties. For example, herbicide resistance in oilseed rape and insect resistance in maize and cotton plants (see page 274).

Invertase is an enzyme extracted from yeast. It splits sucrose into glucose and fructose and is used commercially to produce soft centres and improve the shelf-life of confectionery.

Chymosin is an enzyme similar to rennet, but produced by genetically modified yeast. It has replaced rennet, which was extracted from the stomachs of calves and used to clump milk protein together in cheese and yoghurt production. Chinese scientists in Tokyo succeeded in growing the bacterium *Corynebacterium glutamicum* in fermenters and isolating crystals of **glutamate**. Monosodium glutamate (MSG) has since turned out to be an ideal seasoning, enhancing the original flavour of foods.

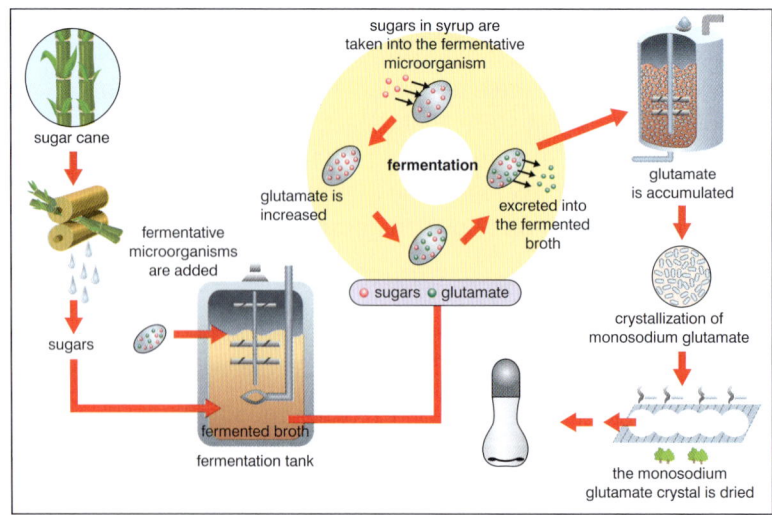

Figure 20.3.2 The production of monosodium glutamate by fermentation.

Industrial fermentation

Industrial fermenters are large tanks (Figures 20.3.3 and 20.3.4) that can hold as much as 500 000 dm³ of fermenting mixture. Conditions inside these huge tanks are carefully controlled.

- The **fermentation vessel** is made of stainless steel and is filled with a medium containing the required nutrients. These may include sugars and ammonium salts. To this, some of the bacteria or yeast is added.
- The microorganism grows well in the conditions inside the fermenter. Sugars provide energy for respiration and ammonium salts are used by the microorganism to make proteins and nucleic acids (DNA and RNA).
- A **stirrer** keeps the microorganisms suspended so they always have access to nutrients and oxygen. Stirring also helps to maintain an even temperature throughout the fermenter.
- An **air supply** provides oxygen for the aerobic respiration of the microbe. (Oxygen is supplied because without it, yeasts and fungi respire anaerobically to make alcohol. See page 140.)
- A **water-cooled jacket** removes the energy released by fermentation to give a constant temperature of 24 °C.
- **Probes** monitor the temperature and make sure the pH is constant at 6.5 by adding alkalis if necessary.

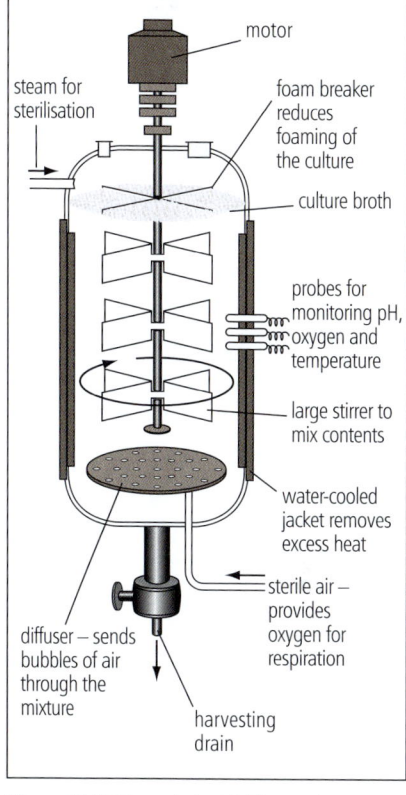

Figure 20.3.3 Large industrial fermenter.

SUMMARY QUESTIONS

1. Explain the function of each of the following in an industrial fermenter:
   a a stirrer   b an air supply   c a water-cooled jacket
   d temperature and pH probes

2. Explain how and why these factors change during a fermentation process:
   a temperature   b oxygen concentration   c pH

3. Explain what the following products are and why they are useful:
   a monosodium glutamate (MSG)   b chymosin
   c invertase

Figure 20.3.4 Industrial fermenters.

KEY POINTS

1. Microorganisms reproduce quickly in the ideal conditions found inside large, industrial fermenters. They can produce many useful bacterial and yeast products.

2. The ideal conditions for fermentation include the correct nutrients and a constant pH and temperature.

20.4 Genetic modification

LEARNING OUTCOMES

- Define the term *genetic modification*

- State that genetic modification can be used in the production of human medicines, herbicide-resistant and insect-resistant plants, and providing additional vitamins in crop plants

S • Outline how genetic modification is used to make human proteins, such as insulin

- Discuss the advantages and disadvantages of genetically modifying crops

Figure 20.4.1 Herbicide-resistant soya can be sprayed with herbicides to reduce the competition from weeds

KEY POINTS

1 Genetic modification involves taking a gene from one species and putting it into another species.

2 The human insulin gene can be transferred into a bacterial plasmid. The bacteria grow to produce lots of human insulin.

Genetic modification involves changing the genetic material of an organism by removing, changing or inserting individual genes. A gene is a section of DNA that codes for the production of a specific protein.

Genetic modification is the transfer of a gene from the DNA of one species to the DNA of another species. This is a process that can never be achieved by artificial selection, because only in rare cases would one species breed with another. Genetic modification allows the transfer of genes between totally unrelated species – for example, from humans to bacteria and from bacteria to plants.

Examples of genetic modification

Human medicines

Many human proteins, including insulin, human growth hormones and blood clotting agents, are produced by bacteria that have been genetically modified with the appropriate human genes. These proteins are mass produced and used as medicines.

Herbicide resistance in crop plants

A genetically modified variety of oilseed rape has been developed in which a gene is transferred into it from a soil bacterium. This gene makes oilseed rape resistant to the herbicide **glufosinate**. If applied to fields of normal oilseed rape, this herbicide would kill the weeds and the crop itself. But the genetically modified oilseed rape is not affected by the herbicide and so will continue to grow while the weeds are destroyed. A herbicide-resistant soya has also been developed (Figure 20.4.1).

Insect resistance in crop plants

Maize and cotton are both important cash crops, but both can be attacked by insects, reducing their yields. The major pests of maize are root worm larvae and stem-boring caterpillars. Caterpillars and cotton boll weevils are pests of cotton plants.

At the beginning of the 20th century, a soil bacterium was found that produced a toxin that kills certain caterpillars. In the 1990s, the gene for this toxin was isolated and transferred to maize and cotton plants. The genetically modified plants produce their own toxin that kills insect larvae that feed on it. This has less environmental impact than spraying pesticides because only the insects feeding on the crop plants are killed.

Additional vitamins in crop plants

Genetically modified varieties of rice have been developed. One that has genes from maize and a soil bacterium enables people to make vitamin A. The resultant rice is pale yellow in colour giving it its name, **Golden Rice**™. If this rice becomes available to grow on a large scale it will improve the diets of the people who eat it. They will be able to make vitamin A. This will reduce the high mortality rates as a result of poor immune systems.

Supplement

Industrial production of insulin

Genetic modification makes it possible to make insulin quickly and cheaply on a large scale (Figure 20.4.3). The human gene that codes for the production of insulin is identified. **Restriction enzymes** act as chemical scissors to cut the human insulin-making gene from the rest of the DNA.

A circular piece of DNA called a **plasmid** is removed from a bacterium. The same restriction enzymes are then used to cut open the plasmid. The two ends of the DNA of the insulin-making gene are an exact match with the two DNA ends of the plasmid. These are called 'sticky ends.'

Another enzyme called ligase is used to attach the sticky ends of the insulin-making gene to the sticky ends of the plasmid. The plasmid is now known as a **recombinant plasmid** and is inserted back into the bacterium, which now has the gene code for insulin. The bacteria multiply very rapidly inside an industrial fermenter and produce insulin.

Advantages and disadvantages of genetically modified crops

The benefits

- **Solving global hunger** – genetic modification could feed more people as the crops that are produced are able to tolerate extreme climate conditions and soils. Food production could be increased in marginal areas.

- **Environmentally friendly** – genetically modified crops such as soya, maize and rice can be resistant to insects, **herbicides** and diseases so there would be less use of pesticides. Also, genes that improve nitrogen uptake would mean less need for chemical fertilisers and lessen the environmental threat that they cause.

- **Consumer benefits** – genetically modified crops have already been produced with an improved flavour and better keeping qualities. They are easier to produce and require fewer additives.

The concerns

- **Environmental safety** – genetically modified plants might become successful weeds if pollen from crops that are resistant to weedkillers is transferred to other plants by insects or the wind.

- **Food safety** – new gene combinations may have effects that are so far unknown. They may result in harmful substances being produced.

- **Biodiversity** – increasing use of herbicides and plant breeding will reduce the number of plant varieties and wild relatives.

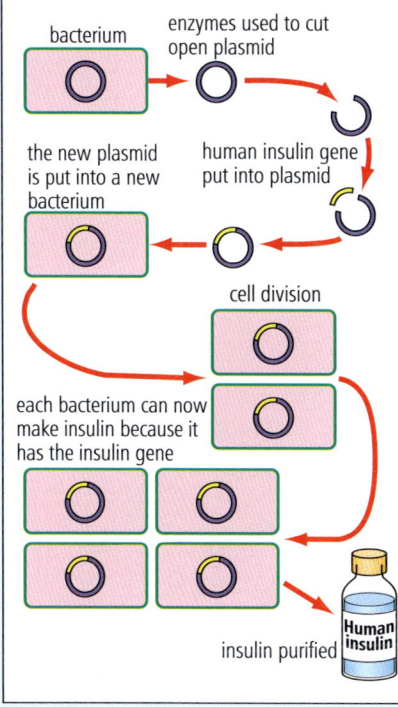

Figure 20.4.2 Human insulin is made by genetically engineered bacteria.

SUMMARY QUESTIONS

1 Explain what is meant by *genetic modification*.

2 Describe three different ways in which crops have been genetically modified to improve them.

S 3 What is a plasmid?

4 List the stages involved in producing bacteria that make human insulin.

5 Discuss the benefits of genetically modified crops and the concerns people have about their safety.

6 Research the types of human proteins, other than insulin, that are made by genetically modified organisms.

Practice questions

1 In order to transfer a feature from one species to another it is necessary to transfer:

   A a chromosome   B a gene

   C a genotype   D a nucleus

   (Paper 1)   [1]

2 Yeast is used in the production of some biofuels. This is because:

   A yeast cells are a good source of energy

   B yeast cells produce oils that are added to petrol (gasoline)

   C yeast cells respire to produce carbon dioxide

   D yeast cells respire without oxygen to produce alcohol

   (Paper 1)   [1]

3 A student is planning a laboratory investigation to compare the effectiveness of three washing powders at removing stains. Which combination of controlled factors influences the activity of the enzymes during the washing process?

   A size of container, frequency of stirring, type of washing powder

   B type of fabric, length of washing time, size of fabric

   C type of stain, size of fabric, final appearance of fabric

   D washing temperature, volume of water, pH of washing mixture

   (Paper 1)   [1]

4 Which substance causes bread dough to rise?

   A carbon dioxide   B ethanol

   C lactic acid   D oxygen

   (Paper 1)   [1]

5 Bacteria are used in many industrial processes. Which is a reason for this?

   A bacteria have complex requirements from their environment

   B bacteria have many enzymes so can carry out complex sets of reactions to make useful products

   C bacteria reproduce slowly under optimum conditions

   D there are many safety concerns with modifying the genes of bacteria compared with modifying the genes of crop plants and livestock

   (Paper 2)   [1]

6 Which is the *best* reason for controlling temperature in a fermenter that contains bacteria?

   A at high temperatures enzymes denature

   B at high temperatures the rate of fermentation increases too much

   C at low temperatures bacteria reproduce too fast

   D at low temperatures there is no diffusion of nutrients into bacteria

   (Paper 2)   [1]

7 The following processes occur in the genetic modification of bacteria to produce a human protein:

   1 a gene is isolated from a human cell

   2 a plasmid is cut open by restriction enzymes

   3 bacteria divide by binary fission

   4 bacteria are grown in a fermenter to produce human protein

   5 recombinant plasmids are taken up by bacteria

   6 bacteria make multiple copies of the recombinant plasmid

   7 plasmids and human genes combine to form recombinant plasmids

   Which sequence is the correct order of events?

   A 2, 1, 5, 7, 6, 3, 4   B 2, 1, 7, 6, 4, 5, 3

   C 1, 2, 3, 7, 6, 5, 4   D 1, 2, 7, 5, 6, 3, 4

   (Paper 2)   [1]

8 An advantage of growing herbicide-resistant soya is that:

   A less herbicide needs to be sprayed on the crop

   B there is less competition between weeds and soya plants

   C there is less risk of disease attacking the soya plants

   D the soya plants grow faster

   (Paper 2)   [1]

9 A student investigated the effect of pectinase on the extraction of apple juice. Several apples were cut up and crushed to make a pulp. The pulp was divided into two samples, A and B. Sample A was mixed with pectinase powder. Sample B was used as a control.

(a) Suggest:

(i) what should be added to sample B as it is being used as a control, [1]

(ii) three variables that the student should keep constant. [3]

(b) The student obtained 35 cm³ of juice from sample A and 15 cm³ from sample B. She noticed that juice from sample A looked more appealing than the juice from sample B.

(i) Explain why there was much more juice from sample A. [3]

(ii) Suggest why the student thought that the juice from sample A was more appealing. [2]

(c) State three ways in which the student could improve the investigation to find out how effective pectinase is at increasing the yield of fruit juice. [3]

(Paper 6)

10 Many organisms can be genetically modified so that they can be used in biotechnological processes, but bacteria are one of the most common.

(a) Define the term *genetic modification*. [2]

(b) State the advantages of using bacteria in industrial processes. [2]

(c) State three examples of genetic modification. [3]

(Paper 3)

11 Look at the diagram of the fermenter on page 273. *Penicillium* is a fungus that can be grown in fermenters to produce an antibiotic.

(a) State **three** features that are shared by all organisms that are classified in the kingdom Fungus. [3]

(b) Name the antibiotic produced by *Penicillium*. [1]

(c) State five conditions that are maintained inside the fermenter. [5]

(d) For each condition, describe how it is maintained and explain why it must be maintained inside the fermenter. [5 + 5]

(e) Suggest how useful products are obtained from the fermenter shown on page 273. [2]

(Paper 4)

12 Human growth hormone (HGH) promotes growth and is given to young people as a treatment for reduced growth. It is a protein made of 188 amino acids.

(a) (i) State how many bases in DNA code for this protein. [1]

(ii) Explain your answer. [2]

Genetically modified bacteria make human growth hormone and a similar hormone found in cattle. HGH is used to treat children who do not grow properly. There may be a genetic reason for this or they may have received radiation treatment for leukaemia or have had a brain tumour.

(b) Outline how bacteria are genetically modified to produce a human protein, such as HGH. [5]

(c) Suggest the advantages of using genetic modification to produce HGH. [3]

(Paper 4)

13 Lactose is a complex sugar similar in structure to maltose. Lactose is also known as milk sugar as it is the main carbohydrate in milk.

(a) State the chemical elements present in lactose. [1]

(b) Suggest how lactose is digested in the human digestive system. [3]

(c) Some people cannot digest lactose and they often feel ill after drinking milk or eating some milk products. Explain how it is possible to make milk that does not contain any lactose. [2]

(Paper 4)

Practical investigations

Introduction

In your practical examination you will either carry out a practical investigation (Paper 5) or be provided with information about one or two practical investigations (Paper 6).

During your course you should do as much practical work as possible to help you develop the skills that are tested in these papers. These skills are:

- use techniques, apparatus, and materials safely

- follow a sequence of instructions

- plan experiments and investigations including selecting suitable techniques, apparatus and materials

- make and record observations, measurements and estimates

- interpret and evaluate experimental observations and data

- evaluate methods and suggest possible improvements (including the selection of appropriate techniques, apparatus and materials).

(See your syllabus for more details).

Here we will guide you through the procedures involved in carrying out a practical investigation in the laboratory.

At the beginning of the question there is usually a statement telling you what you are going to investigate. We will use the following enzyme practical as an example:

You will investigate the effect of temperature on the activity of a protease.

This statement is followed by a series of instructions and questions. Before you read the instructions have a look at the apparatus and materials that have been provided for you. This will give a good idea of what you are going to do. In this practical the substrate for the enzyme is the protein in a solution made from milk powder. As the enzyme breaks down the protein the solution changes from cloudy to clear.

When you read the sequence of instructions, imagine yourself carrying them out so that you have a mental picture of what you are going to do. You can underline or highlight the instructions if that helps. After reading the instructions, read through the questions and find out whether you will put your results into a table that has been drawn for you or whether you have to design and draw your own table in a blank space.

Before you carry out the first instruction, identify the independent variable and the dependent variable for the investigation.

- The **independent variable** is the variable that you are going to change. You will often be told the values of this variable in the instructions, e.g. the temperatures chosen for the investigation are 20 °C, 30 °C and 40 °C.

- The **dependent variable** is the variable that you are going to measure to judge the effect of changing the independent variable. You do not know the values of this variable before you start. You are carrying out the investigation to find out! In the investigation the dependent variable is the time taken for the milk solution to clear.

When you follow the instructions work carefully and work safely.

One of the first things you may need to do is measure out some liquids. In our investigation, you may use a disposable pipette for small volumes of the enzyme solution and a plastic syringe or a measuring cylinder to measure out the milk powder solution.

You should always wash out the piece of apparatus with the solution you are going to dispense. This makes sure that there is no water left in the apparatus to dilute the solution. If you are using a syringe then put the nozzle inside the liquid and move the plunger in and out several times without lifting the syringe out of the liquid. This helps to prevent air bubbles forming inside the syringe. Then fill the syringe to a level *above* the volume you wish to dispense, looking carefully at the scale (graduations) on the syringe. Push the syringe plunger down towards the nozzle and stop when the rim at the end of plunger is over the correct mark on the scale. At this point the whole syringe, including the nozzle, holds the correct volume of liquid you wish to dispense.

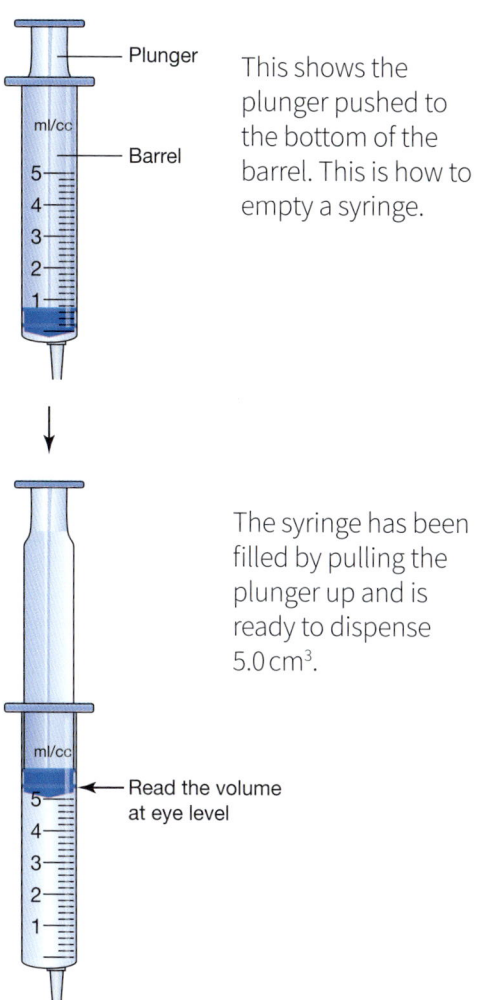

This shows the plunger pushed to the bottom of the barrel. This is how to empty a syringe.

The syringe has been filled by pulling the plunger up and is ready to dispense 5.0 cm³.

You may use measuring cylinders to dispense liquids. Always read the volume by using the method shown below.

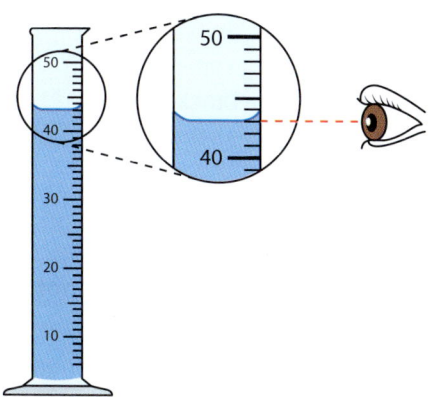

As you complete each step in the instructions put a tick by the side of it.

In an investigation with an enzyme, such as a protease, you may have to put the enzyme solution and the substrate solution into a water bath for a while before mixing them together. This is to get them both to the same temperature before mixing, otherwise the temperature will change during the reaction, which means there will be an uncontrolled variable.

When you have completed setting up the apparatus and materials, you should be ready to take results. Here is a typical instruction for this stage.

- Put the contents of tube **A** into tube **B** and immediately start a timer.

Here you have to do two things at the same time and it may be impossible to start the timer at *exactly the same moment* that you add the liquids together. A way to solve the problem is to start the timer before adding the liquids and noting the time. Record this time in your table or elsewhere. The next instruction is likely to be:

- As soon as the milk solution clears stop the timer and record the time.

This is not just 'the time'. This should be recorded in a table as 'time taken for milk solution to clear' or 'time for solution to clear'.

If you have to make your own table for results, then do this before starting to collect data so you can write the results directly into the table and not have to copy them up. The table must be constructed carefully. Here are the rules to follow when making tables for results.

> **Rules for making tables for results**
>
> - All the data from your investigation should be presented in one table.
>
> - Use a pencil and a ruler to draw up a table with columns and rows. The table should have borders (top, bottom and sides).
>
> - It is usual to put the independent variable on the left in the first column and the dependent variable in a column to the right of the independent variable. In our investigation *temperature* should
>
> (*Continued*)

(Continued)

go in the left hand column and *time for solution to go clear* should go in the right hand column.

- Column headings should be informative with the correct units; for example 'time' on its own is not informative enough; 'time for solution to go clear' is informative.

- Units must be shown in the column headings and not after each number in the table. Units are usually shown after a solidus (/) or can be in brackets:
  three examples: time / s, length / mm and volume of gas (cm³)

- The data should be recorded to the same degree of precision. For example, if one measurement of length in a column of data is given to the nearest millimetre, then all the lengths in that column should be given to the nearest millimetre.

- If appropriate, a table should have an informative title. A good way to write this is: 'A table to show ......'

Interpreting

After you have collected your data you may have to process your results by:

- calculating rates, e.g. by dividing volume of gas produced by the length of time taken for collection (with units of cm³ per s)
- calculating percentages or percentage changes
- calculating means if you have made two or more readings for each value of the independent variable.

You may have to use the results you have collected or the processed results in other questions.

In Papers 5 and 6 you will be given data to display graphically. The two most common ways to present data are as line graphs and bar charts.

Here are the rules to follow when drawing line graphs.

Rules for drawing line graphs

- Make the best use of the space available, using over half of the length and width of the grid provided. Often a grid is provided so that your graph will fit perfectly into the space provided.

(Continued)

(Continued)

- The independent variable is on the *x*-axis and the dependent variable is on the *y*-axis. This means taking the first column from your table and using it to make the *x*-axis and taking the second column and using it to make the *y*-axis.

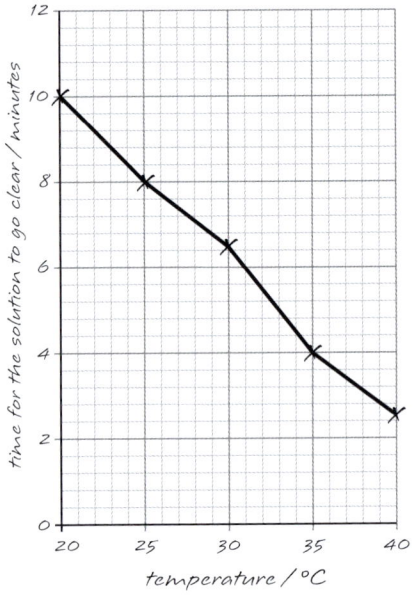

- Axes should be correctly scaled so that points are easy to plot and data can be read easily from the graph.

- Use a sensible scale, for example a 20 mm square on the graph paper can represent 1, 2, 4, 5 or 10 units. Make sure that the numbers on the axes increase uniformly, e.g. 0, 1, 2, 3, 4 or 0, 10, 20, 30, 40 and go up (e.g. 0 → 40 °C) and not down. An exception to this is if you have positive and negative numbers to plot – see question 2(b) on page 290 and question 10 on page 37 for two examples.

- Axes should be labelled in the same way as column headings in a table. Units should be shown as for table headings, separated from the label by a solidus (/) or a bracket, e.g. time for the solution to clear / s *or* volume of gas (cm³).

- Points should be plotted on the grid with a sharp pencil (**not** a pen) and must be visible. Use a fine saltire (x) or a dot in a circle. Do **not** use a thick blob.

(Continued)

(Continued)

You now have to draw a line on your graph. Lines should be drawn with a sharp pencil (not a pen) and should be continuous without any gaps.

Often it is obvious that the data points fall either on a straight line or on a smooth curve. If so, draw the line from the first to the last plotted point using a ruler (if straight) or freehand if a curve. If the points are not quite on a straight line or a curve you can draw a **line of best fit**. Lines and curves of best fit should show an even distribution of points on either side of the line along its whole length. For example, you may have two points above the line and two points below.

Sometimes it is not possible to be sure if the line should be straight or a smooth curve. In this case join the points with straight lines made with a ruler. The lines should go through the centre of each plotted point.

If appropriate, a graph should have an informative title. A good way to write this is: 'A graph to show ......'

You can find plenty of examples of graphs throughout your textbook. Now you have read these rules take a look at the graphs in Answers 1 and 2 on page 289. Read the comments on the graphs.

In an investigation your independent variable may not be a continuous variable, such as temperature or pH. For example, you may carry out an investigation to find out how yeast respires when provided with different sugars. There is no continuous scale that you can draw for three of four different types of sugar, so you cannot use a line graph. Instead you will have to plot your result as a bar chart. Here are the rules for drawing bar charts.

Rules for drawing bar charts

- Use at least half the grid provided, do not make the chart too small.
- Draw the chart in pencil.
- Bar charts are made of blocks of equal width, which do **not** touch.

(Continued)

(Continued)

- The intervals between the blocks on the *x*-axis should be the same.
- The *y*-axis should be properly scaled as in a line graph.
- The y-axis should be labelled and units included.
- The lines or blocks can be arranged in any order, but it can help to make comparisons if they are arranged in descending order of size (see example).
- Each block should be identified by writing a label underneath. There is no need to shade the blocks or colour code them.

Here is an example.

Type of sugar	Rate of respiration / bubbles per minute
Glucose	16
Fructose	13
Sucrose	12

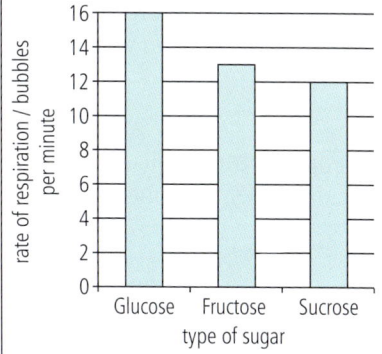

After you have completed your graphs you may be asked to do some more processing by:

- calculating a rate using the gradient on a line graph
- finding values intermediate between the plotted points on a line graph
- predicting a value beyond the range for the independent variable - this is known as extrapolation

You may be asked to describe the results. This means putting into words what you can see in the figures in the table or can see in the line on a graph. When describing, you need to use words such as: increase, decrease, maximum, minimum, peak and constant. Always give some of the data from the table or the graph to illustrate the points that you make. You may also have to make a conclusion. To do this you are only expected to use the information given in the question. You are not expected to use your knowledge of the topic you have investigated to explain the results.

Evaluating

You also will be asked to evaluate the results and the procedure. A common way to ask about this is to ask about errors. An error is **not** a mistake that you have made in the procedure, such as dropping a test tube. It is an inbuilt fault with the procedure. For example, with the apparatus provided it may not have been possible to keep the temperature in a water bath constant. You may then be asked to suggest improvements – in which case you need to think of a way to overcome the error. 'Do the experiment with someone else' is not a suitable answer – you have to think of something suitable for working on your own. For example, water baths can be kept constant by adding hot, or cold, water as soon as the temperature starts to change or you can suggest the use of a thermostatically controlled water bath.

Planning

On both Paper 5 and Paper 6, there will usually be a planning question for six marks. The question may ask you to extend the investigation you carried out in Paper 5 or read about in Paper 6. It may involve a new technique that you have not come across in your course. You can use this book to read about all the practicals listed in the syllabus. However, the question may ask you to plan something totally unexpected. Practise writing plans as part of your course. Each time you write a plan think about:

- the independent variable and the dependent variable
- the variables that should be controlled and how they should be controlled
- an appropriate range for the independent variable and the number of intermediate values (e.g. 10 °C to 80 °C for the range, and 20 °C, 30 °C, 40 °C, 50 °C, 60 °C and 70 °C as the intermediate values)
- suitable apparatus to use in the method
- the method to follow as a sequence of steps as on page 283
- the collection and recording of the results of the investigation
- a safety hazard for the experiment and an appropriate precautions for the hazard

Make sure your plans cover all these points.

EXAM TIP

There are planning questions in the Summary Questions for 6.2, 17.1 and 17.2. Try those questions and compare your answers with those given on the support website.

1 Ascorbic acid (vitamin C) is produced in animal and plant tissues from glucose. Ascorbic acid is an antioxidant, required in small quantities by organisms to prevent damage to key molecules.

The concentration of ascorbic acid in plant tissues can be estimated by using a modification of the test for starch in which iodine forms a coloured complex with starch molecules.

Ascorbic acid reacts with iodine so that it does not form the coloured complex with starch.

A student investigated the effect of using five different concentrations of ascorbic acid. The student was provided with six test tubes, a $0.5\,cm^3$ starch solution, an iodine solution and five solutions of ascorbic acid as follows:

A 1.00%

B 0.50%

C 0.25%

D 0.10%

E 0.05%

The student used three $1\,cm^3$ syringes, labelled 1, 2 and 3, in this method.

Step 1 Use syringe 1 to add $0.5\,cm^3$ starch solution to the test tube labelled **A**.

Step 2 Use syringe 2 to add $0.5\,cm^3$ of ascorbic acid solution **A** to the same test tube. Shake the test tube thoroughly.

Step 3 Use syringe 3 to add iodine solution to the mixture in the test tube drop by drop. Shake the test tube after the addition of each drop.

Step 4 Continue to add iodine solution until a blue colour appears and remains in the mixture after shaking. If the solution does not remain blue after the addition of $1\,cm^3$ iodine solution, refill the syringe with iodine solution and continue.

Step 5 When the blue colour remains, record the volume of iodine solution that has been added.

Step 6 Repeat steps 1 to 5 with ascorbic acid solutions **B**, **C**, **D** and **E**.

Step 7 Repeat steps 1 to 5 but use **water** instead of an ascorbic acid solution.

EXAM TIP

Before you start Paper 5 or 6, look through the questions and find the 6-mark planning question. You may need to use information from the investigation in question 1 or 2 when writing your plan.

EXAM TIP

There is a mark scheme for the two questions in this section. See the support website.

EXAM TIP

Practise writing your own methods for investigations like this – with numbered steps. This will help you follow instructions in Paper 5 and write answers to the planning question in Papers 5 and 6.

The student recorded the results in a notebook.

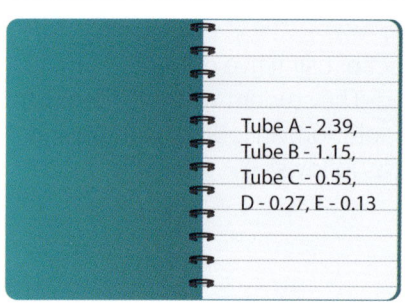

Tube A - 2.39,
Tube B - 1.15,
Tube C - 0.55,
D - 0.27, E - 0.13

(a) Make a table to record the student's results. [4]

Answer 1

concentration of ascorbic acid / %	volume of iodine solution / cm³
0.05	0.13
0.10	0.27
0.25	0.55
0.50	1.15
1.00	2.39

Answer 2

tube	ascorbic acid	volume of iodine
A	1%	2.39 cm³
B	0.5%	1.15 cm³
C	0.25%	0.55 cm³
D	0.1%	0.27 cm³
E	0.5%	0.10 cm³

Commentary

Tables should have columns and rows drawn with a pencil and ruler. The columns should have header rows with informative headings. Units should never be put into the individual cells within the table as in Answer 2. They should be put in the headings. Use a solidus (/) as in Answer 1 or put brackets around the units. Figures should be shown consistently. As the concentrations are given to two decimal places in the question paper, they should be shown in the same way in the results table. It is usual to arrange the data by increasing value of the independent variable as in Answer 1. Note that Answer 2 has a copying error.

(b) (i) State why three separate syringes were used instead of just one syringe. [1]

Answer 1

To avoid contamination.

Answer 2

This saves having to wash out the syringe each time between adding starch, ascorbic acid and iodine solution. Any water left in the syringe will dilute one or more of the solutions.

Commentary

As you read the method used in the investigation think about the purpose of each step. This will help you to anticipate questions asking you to explain a particular step. The question asks about avoiding contamination of one of the solutions with another. In an enzyme experiment you would not want to contaminate the solution of the substrate with any enzyme solution because otherwise the reaction will start too early. 'To avoid contamination' needs further explanation to be sure of gaining the mark, for example, 'To avoid contaminating the ascorbic acid solution with starch'.

(ii) Explain why the blue colour does not appear immediately on the addition of iodine solution to test tubes A to E. [2]

Answer 1

The iodine is diluted by the contents of the tubes.

Answer 2

The iodine reacts with ascorbic acid as soon as it is added.

Commentary

Make sure you understand how the experiment 'works' as you read the information provided. Read the introduction to the question slowly and carefully and then read it again. Write notes on the exam paper if that helps. Here the iodine solution that is added in the first few drops reacts with the ascorbic acid and therefore cannot bind with the starch, so no blue colour is seen. Eventually, all the ascorbic acid in the test tube has reacted so there is none left to react with the iodine. Now the iodine is able to combine with starch to give a blue colour.

(iii) Explain why the procedure was repeated with water (step 7). [1]

Answer 1

To make sure that there was a colour change as soon as the iodine solution was added to the starch solution.

Answer 2

To find out how much iodine solution has to be added to see any colour change when it combines with starch. This would be the same volume for all the test tubes.

Commentary

As it says in Answer 2, this is to find the minimum volume of iodine solution needed to see any colour change. This should be less than all the results obtained for the ascorbic acid concentrations. If this happens, then the student knows the method is suitable for measuring ascorbic acid concentrations. You may be tempted to write down 'for a control' in answer to this question. This always should be qualified with an explanation of why the control is necessary to be certain of gaining a mark. Here 'for a control' would not gain the mark.

(c) Identify **one** hazard of this procedure **and** describe how the risk of this hazard could be reduced. [2]

Answer 1

Liquid may splash into the eyes when using a syringe. Wear eye protection.

Answer 2

Wear safety glasses

Commentary

Safety precautions are an essential part of all experimental work. Be prepared to state a hazard of any investigation that you carry out and give a suitable precaution. Answer 1 does this and is the better answer. Answer 2 answers only part of the question.

(d) (i) Give **one** reason why a syringe was used for adding the iodine solution and not a dropping pipette. [1]

Answer 1

A syringe is more accurate.

Answer 2

Dropping pipettes do not have scales on them so it is impossible to tell how much iodine solution is added other than counting drops.

Commentary

Answer 2 identifies the problem with using dropping pipettes. To say that they are not accurate is not enough to gain a mark. It is better to say that as there is no scale on the pipette it is not possible to read off the volume of iodine added. We know from the student's results that the 1 cm³ syringe had divisions that were 0.01 cm³ apart.

(ii) Suggest how the method could be improved to reduce the chance of any errors when repeated by someone else. [1]

Answer 1

The shaking of the test tubes after adding ascorbic acid should be the same each time. The instructions could say that the test tubes must be shaken three times.

Answer 2

The test tubes could be stirred and not shaken. A glass rod could be used to stir the contents of the test tube for five seconds after adding a drop of iodine solution.

Commentary

Both Answers are correct ways to gain the mark. Instructions should be as detailed as possible so that everyone implements the steps in exactly the same way. You should think about this when writing your answer to the planning question.

(e) Some plants are rich sources of ascorbic acid. Many also store reducing sugars, such as glucose and fructose, in their fruits. A student wished to compare the relative quantities of reducing sugars in the fruits of five different species of citrus tree.

Plan an investigation to compare the reducing sugar content of fruits from five different species of citrus tree. [6]

Answer 1

1   The independent variable is the species of citrus tree.

2   Peel each fruit and put into a juicer. Put the same volume of juice in five different labelled beakers.

3   Put 2.0 cm³ of the first fruit juice into a labelled test tube and add 2 cm³ of Benedict's solution.

4   Put the test tube into a boiling water bath and watch for a colour change.

5   When there is no further colour change remove the test tube.

6   Repeat steps 3 to 5 in exactly the same way with the other fruit juices.

7   Compare the colours (dependent variable). Red and orange means the fruits have lots of reducing sugar. Green means they have very little.

8   Wear gloves and safety goggles when using the boiling water bath to avoid splashes getting into the eyes.

Answer 2

Use the fruits to make some fruit juices by squeezing them. Use a pipette to put the same volume of fruit juice into 5 different test-tubes. Put the same volume of Benedict's solution into each

EXAM TIP

Read the planning task very carefully. Make sure that you choose a method that will answer the question and do not rely simply on recalling practicals you have done before.

EXAM TIP

Notice the practical details in Answer 1. The Practical sections throughout the book include practical details like these.

test tube. Boil the fruit juices with the Benedict's solution and write down the time when they begin to change colour. The fruit juice that changes colour first has the most reducing sugar.

Commentary

It is a good idea to set out your answer to the planning question in the same way as the method given at the beginning of the question as a series of numbered steps. You do not have to write 'step' each time, numbers alone are enough. Look how much easier it is to follow Answer 1, compared with Answer 2. You can always write things like 'repeat step 3….' if necessary.

Note also that Answer 1 states the volumes of fruit juice and Benedict's solution to use. It is unlikely that two different people could follow the instructions in Answer 2 and get the same results. The instructions are too ambiguous. Note that in Answer 2 the IV and DV have not been identified and there is no safety hazard and precaution given.

2   A student used the apparatus shown in the diagram to investigate the effect of increasing the carbon dioxide concentration on the rate of photosynthesis in an aquatic plant.

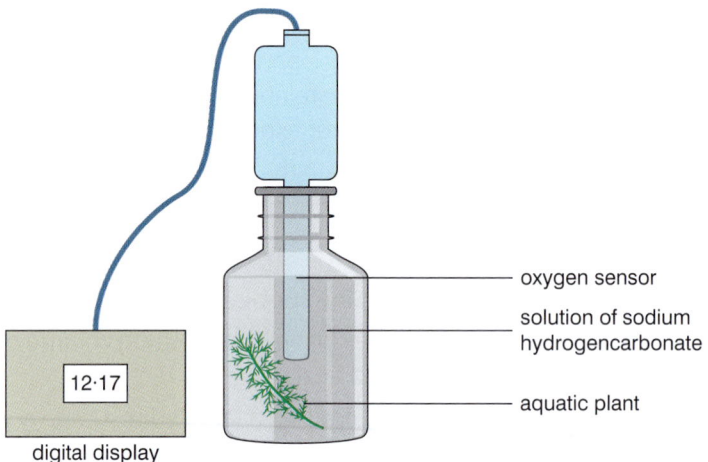

digital display

oxygen sensor

solution of sodium hydrogencarbonate

aquatic plant

The student used five different pieces of the aquatic plant and kept them in specimen bottles containing different concentrations of sodium hydrogencarbonate solution. The results of the investigation are in the table.

piece of aquatic plant	concentration of sodium hydrogencarbonate solution / g dm⁻³	dissolved oxygen concentration / mg dm⁻³		percentage change in dissolved oxygen concentration
		start	end	
P	0	12.07	11.38	– 5.72
Q	20	13.09	15.73	20.17
R	40	12.13	16.19	33.46
S	60	12.92	17.32	34.06
T	80	14.17	18.56	30.98

(a) Plot a graph to show the effect of increasing the concentration of sodium hydrogencarbonate solution on the percentage change in dissolved oxygen.

Draw a best fit line on your graph. [5]

Answer 1

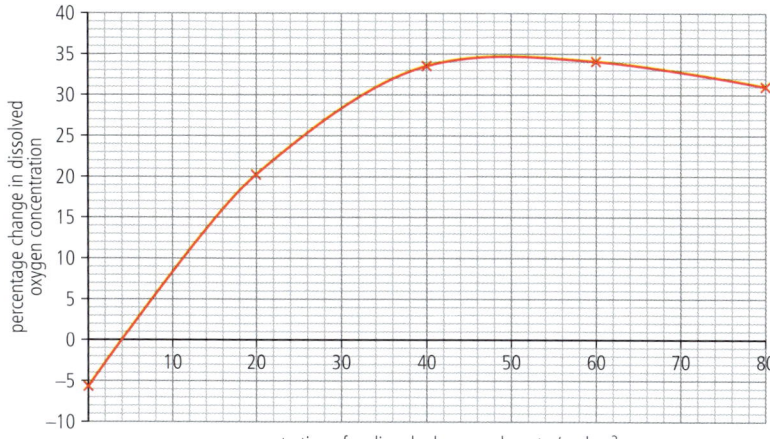

Answer 2

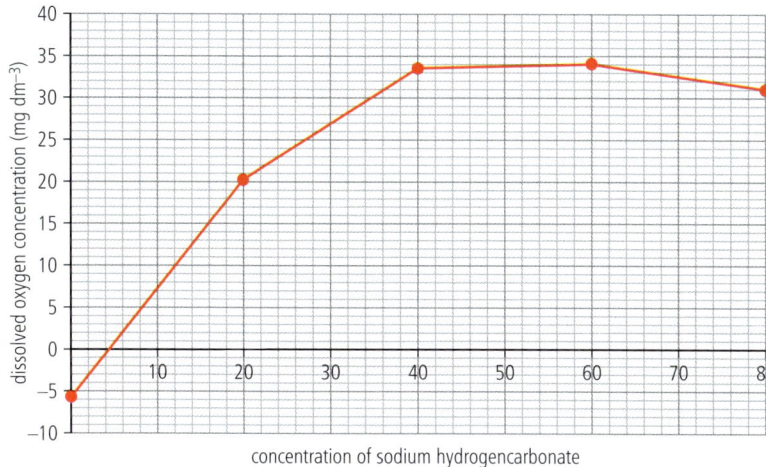

EXAM TIP

Here are some tips for drawing line graphs:

- Choose scales that make use of the whole grid or at least half the space provided.

- Use evenly spaced scales that increase in 1s, 2s, 4s, 5s or 10s. Do **not** use scales that increase in increments of 3 or 6.

- Use a sharp pencil to plot the points and draw the line. Use crosses rather than dots.

- Make a single line without any breaks and do not continue the line beyond the last plotted point.

- Label the axes and add the units. Axis labels are the same as the table headings.

Commentary

Choose scales for the x- and y-axes that make use of the whole grid. If you cannot do this then make sure to use **at least** half the distance along the x-axis and half the distance along the y-axis. Note the position of 0,0 on the y-axis. As there are positive and negative percentage changes this cannot be at the bottom of the y-axis as is more usual. You will plot this type of graph when you investigate the effect of different salt or sugar solutions on plant tissue when studying osmosis (see question 10 on page 37).

Answer 1 gains full marks. Answer 2 has made several errors. The plots are too big and the lines between the points are straight lines rather than a smooth curve. The question asked for a best fit line. The label for the vertical axis is incorrect. No unit for the independent variable has been included.

(b) Describe the pattern shown on your graph.   [3]

Answer 1

The dissolved oxygen concentration decreased when the plant was kept in the lowest concentration of carbon dioxide. This means that oxygen dissolved in the water was used up by the plant. As the carbon dioxide concentration increased the oxygen concentration increased as the plant was carrying out photosynthesis and so releasing oxygen. At high concentration ($80 \, g \, dm^{-3}$) the percentage change decreased compared with $40 \, g \, dm^{-3}$ and $60 \, g \, dm^{-3}$ as less oxygen was released. It decreased from 34% to 30%.

Answer 2

The dissolved oxygen concentration in the tubes around the aquatic plant increases as the concentration of sodium hydrogencarbonate increases. It reaches a peak at 60 and then decreases.

Commentary

Look carefully at any graph you have drawn and the mark allocation. The trend may be very simple if there is a straight line relationship. Here, it is more complex and you need to think of three different points to make. The graph clearly shows an increase in the concentration of the oxygen released up to a peak and then a decrease. If there are figures you can include in your answer, then use them as in Answer 1. Always remember to add the units from both axes. Answer 2 has included one figure but not given any units.

(c) Suggest why the student used a sensor to monitor changes in oxygen concentration in the water surrounding the aquatic plant rather than count bubbles of oxygen released by the plant.   [3]

Answer 1

When counting bubbles it is difficult to make sure you have counted all the bubbles. Most of the bubbles come from the cut end of the stem, but others can come from elsewhere on the plant. Some of the oxygen produced stays in solution. The concentration of oxygen can be read from a meter meaning there won't be any errors when counting the bubbles.

Answer 2

The student does not have to watch the experiment all the time counting bubbles. He might be interrupted and lose count.

Commentary

Counting bubbles is not a good way to take results for the reasons given. Similarly, when you take results by judging when the end point of a reaction occurs you might not stop timing at exactly the same point each time, especially if you have not had much practice. Experiments can be improved by having some apparatus to take results rather than relying on the experimenter's judgement. You may be asked to suggest a way to improve an investigation and if so, then think of some type of apparatus to use.

(d) State two variables that must be kept constant throughout this investigation. [2]

Answer 1

Light, temperature and pH

Answer 2

Light intensity (the number of lamps shining at the water plant). The temperature of the water (use a thermometer to monitor the temperature).

Commentary

This question is asking about the factors that influence photosynthesis. Looking at the diagram should give you a few clues as to the variables that should be kept constant. Question papers often have such clues. Answer 1 has not stated that the light *intensity* should be kept constant. It also includes three answers, not two as requested. Only the first two answers given will be marked, so this answer gains just one mark.

(e) Other students evaluated the investigation by stating that the results were of poor quality because there were no repeat readings included in the table.

Explain why the student should have included repeat readings in the investigation. [2]

> **EXAM TIP**
>
> Taking results and making observations can often be improved by using a smartphone to make video recordings. The results can be taken from the playback.

Answer 1

The whole investigation should be repeated with fresh plant material. This would show whether the results can be repeated with other pieces of plant. The results for each concentration of sodium hydrogencarbonate solution can used to calculate the mean average.

Answer 2

If repeats were included the student could tell whether the results are accurate and the trend is correct. Also whether any (or all) of the results are anomalous.

Commentary

Answer 1 states what is meant by a replicate. Replicates are included to make sure that the results obtained with one piece of plant can be repeated when tested on others. In school science it is usual to state that a minimum of three replicates should be planned so that any anomalous results can be identified. If all three results for each value of the independent variable are very similar then the experimenter can have confidence in the results.

(f) The photograph shows some cells in a leaf of the aquatic plant. The small dark structures in each cell are chloroplasts. The magnification of the photo is × 450.

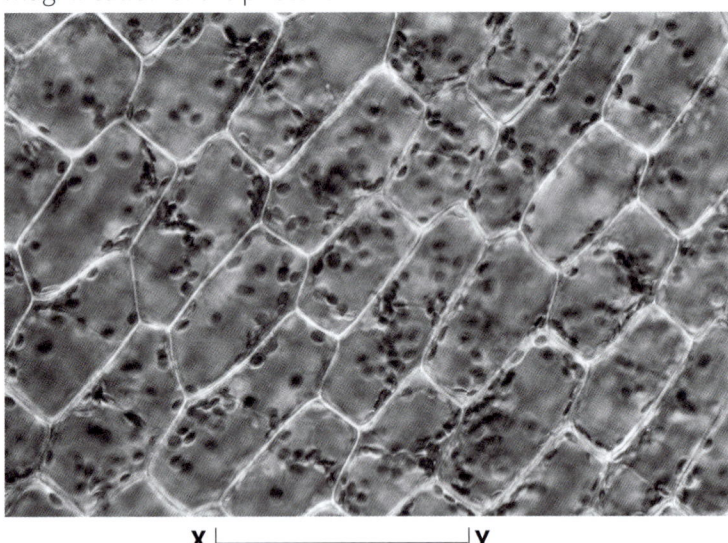

X |———————————————| Y

(i) Make a drawing of three adjacent cells from the photograph. Do not make the cells you draw smaller than those in the photograph.

Draw a line on the photograph around the cells that you have drawn.

Do not label your drawing. [4]

Answer 1

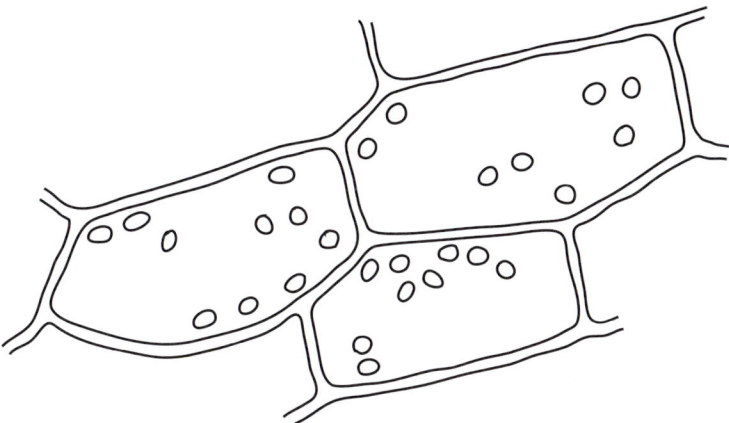

Answer 2

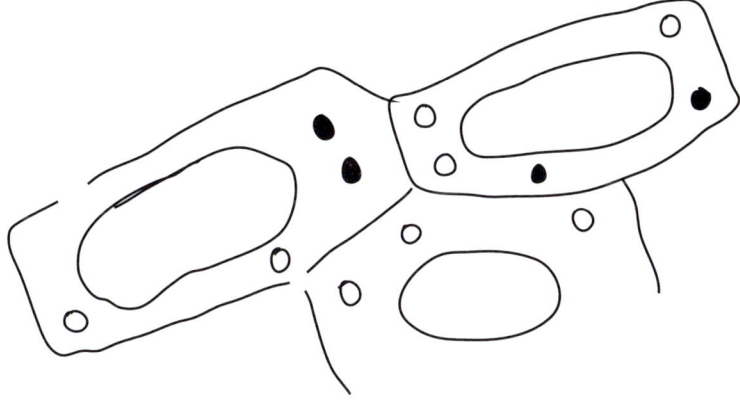

Commentary

Answer 1 gains full marks. The drawing should fill at least half the space provided on the exam paper and the instruction to draw three adjacent cells has been followed. The lines should be clear and continuous. Chloroplasts have been drawn. Two lines have been used to show the thickness of the cell walls around each cell. No shading has been used and there are no labels.

Answer 2 gains one mark for making a drawing that fills half the space available. The lines are not continuous as there are three gaps in the lines used to draw around the cells, shading has been used, the thickness of the cell walls has not been shown. There are only two complete cells, and vacuoles have been drawn in although they are not visible in the photo.

(ii) Measure the length of the line X–Y.

Use the formula to calculate the actual length of the cell:

$$\text{actual length} = \frac{\text{length of line X–Y}}{\text{magnification}}$$

Show your working and give your answer in millimetres and to two significant figures. [3]

Answer 1

Length of line X–Y = 32 mm

Working

Actual length = 32 mm / 450

= 0.07111 mm

Answer = 0.071 mm (2 sig. figs.)

Answer 2

Length of line X–Y = 3

Working

Actual length = 3 / 450

= 0.00666

Answer = 0.006 mm

Commentary

When you measure anything always give the units. There are several common errors in Answer 2. The correct length of the line is 3.2 cm not 3 cm. The working for finding the actual length is carried out correctly (and would gain a mark), but the answer (0.006) is the measurement in centimetres and not in millimetres. The answer is given to one significant figure (the '6') and not two significant figures as stated in the question.

EXAM TIP

Always make and record measurements in mm and not in cm. In Papers 5 and 6 you are expected to use millimetres. In Paper 4 you may be asked to convert from millimetres to micrometres. A common error is to measure in centimetres and multiply by 1000 to change to micrometres, instead of multiplying by 10 000.

Exam skills and techniques practice

Sample questions with candidate/student answers (in blue) and examiner's comments.

- Questions 1 and 2 are sample questions for Paper 3.
- Questions 3 and 4 are sample questions for Paper 4.
- Questions highlighted in green are on the supplement content of the syllabus.

1 Figure 1 shows a cell from the root of a plant.

a Complete the table that shows the names of the structures in the plant cell and their functions.

label	cell structure	function
A	cell membrane ✗	supports cell
B	mitochondria ✓	aerobic respiration
C	nucleous ✓	controls the cell activities ✓
D	vacuole ✓	stores water and ions
E	ribosomes	making proteins ✓

[6]

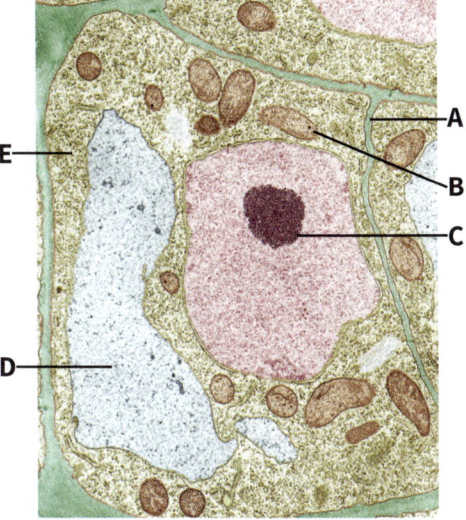

Figure 1

Exam advice:

The student has made one error here. Structure **A** is the cell wall, not the cell membrane. The nucleus (**C**) is not spelt correctly but the meaning is clear so a mark is awarded.

Exam technique:

Look very carefully at any stimulus material given in a question. In this case, it is a photograph that shows the detailed structure of a plant cell with the features that you should be able to identify from images like this. As you look at the stimulus material make some notes on the exam paper before looking at the questions. Here, for example, you could write identifications by the labels on Figure 1.

b State the names of **two** cell structures **shown in Figure 1** that are **not** present in the cells of bacteria. [2]

Nucleus (C) ✓ and vacuole (D) ✓

Exam advice:

The student has answered the question correctly giving structures that are visible in Figure 1 but are **not** found in bacterial cells. It is not necessary to include the letters and if they were incorrect, they would probably be ignored because the question asks only for the names.

Exam technique:

Always read questions carefully; read at least twice looking for important words, especially those that are in bold type, such as **not**.

c State **one** structure present in palisade mesophyll cells that is **not** present in root cells like the cell in Figure 1. [1]

Chloroplasts ✓

Exam advice:

The student has spelt the name of these cell structures correctly. Learn the correct spelling as well as the functions of the cell structures listed in Topic 2.1 and in the glossary (see pages 18 and 292).

Exam technique:

Palisade mesophyll cells are in leaves and are the site of most of the photosynthesis that occurs in plants. These cells are full of chloroplasts to absorb light and make sugars. These facts should come to mind when reading the question.

d Leaves stop growing once they reach their maximum size. Roots continue to grow throughout the life of a plant.

State how new cells are produced in the roots of plants. [1]

Old cells divide into two to form two new cells ✓.

Exam advice:

This is the answer that is required in Paper 3. The same question in Paper 4 would require the term *mitosis* to be used in the answer.

Exam technique:

The command word for this question is *state*, so a short answer is all that is required. It is important to say that each cell divides into two new cells in answer to this question. 'A cell divides' would not be sufficient to gain the mark because a cell could divide into more than two cells.

Exam question comments:

This student gained 9 out of a possible 10 marks. Notice how the answers use the appropriate terms, but a mark is lost for a common error – confusing the cell membrane with the cell wall.

[Total: 10]

2 a Complete the sentence to describe what is meant by an *adaptive feature*.

An adaptive feature is an ......*genetic*...... feature that helps an organism to survive and ......*breed* ✓...... in its environment.

[2]

Exam advice:

The correct term for the first space is *inherited* or *hereditary*. The expected answer for the second space is *reproduce*, but 'breed' is an acceptable alternative because it also means to produce offspring.

Exam technique:

Make sure that answers you give to sentence completion questions read correctly. In this answer, 'an genetic' does not read correctly.

b Explain the difference between pollination and fertilisation in flowering plants. [3]

Pollination is the transfer of pollen grains. ⌀ *Fertilisation is when the pollen grains fertilise the egg.* ✗

Exam advice:

Always give full answers. In the first sentence, the student needs to say that 'Pollination is the transfer of pollen grains from anthers to stigma' to gain the first mark. In the second answer, the student should have written 'Fertilisation in flowering plants involves the fusion of a pollen nucleus with the nucleus in an ovule'.

Exam technique:

Make sure that you learn the full definitions of biological terms. Use the glossary to help you learn these definitions. Also avoid giving a definition that includes the word being defined as the student has done when writing about fertilisation. The student should have used the word *fuse* or *fusion* as in the fusion of a sperm and an egg in animals.

The photographs in Figure 2 show plants of the species *Babiana ringens* that grows in South Africa. The plant is unusual in that its red flowers grow from the base of the plant rather than on stems up in the air (Figure 2a).

The flowers are pollinated by malachite sunbirds. Sunbirds perch on the long stalks that grow from the plants while they visit the flowers to feed (Figure 2b).

The researchers investigated the importance of the stalks by comparing the success of pollination in two populations of *B. ringens*. The researchers removed the stalks from some of the plants in two populations (**A** and **B**) that were 5 km apart. The birds had to stand on the ground to feed on flowers without stalks (Figure 2c).

Figure 2

c Suggest **three** features of the flowers of *B. ringens* that
attract pollinators. [3]

> Three features of the flowers of **B. ringens** that attract
> pollinators are that they have large flowers ✓, produce nectar
> ✓ and have colourful petals ✓.

Exam advice:

These are features that you should associate with insect
pollination, but the photographs suggest that they also attract
birds. You may know about humming birds that visit flowers to
feed on nectar. There is no point in rewriting the question. All that
is required are the three features, which you can give as a list.

Exam technique:

The command word is *suggest* because bird pollination is not
something that you will have learnt about. You are being asked
to apply your knowledge to a new example. There will be many
questions that test this skill in Papers 3 and 4.

d The researchers tested the prediction that the long stalks
provide a perch for the birds while they feed on the flowers.
They counted the number of seeds produced by each
plant in the two populations and then calculated the mean
number of seeds per plant.

The results are shown in the table.

plants of *B. ringens*	mean number of seeds per plant	
	population A	**population B**
perch present	120	70
perch removed	86	27

Use all the information provided to explain the results shown
in the table. [4]

> The seeds are formed from fertilised ovules. The plants that
> have perches have produced more seeds than those without
> ✓ because they have been pollinated and the ⬡ ovules are
> fertilised. They were pollinated because more birds visited the
> flowers ✓. Perhaps it was easier for them to feed on the flowers
> while on a perch rather than standing on the ground ✓.

Exam advice:

The student has made the link between the birds as pollinators and fertilisation of the ovules to form seeds. The omission mark ⌢ shows where the student should have said that *more* ovules have been fertilised. The student makes the reasonable suggestion that the presence of a perch results in more birds visiting the flowers.

Exam technique:

Look carefully at the data in the table and make some notes before you start writing your answer. There are four marks, so make at least four different points.

Exam question comments:

The student gained 7 marks out of 12. Marks were lost because the answer to **(b)** was not precise enough, the data in the table was not used in **(d)** to qualify the comparison between the seed production in the plants with and without perches, and only two suggestions were made to explain the difference.

[Total: 12]

3 Figure 3 shows three stages during one heartbeat. The black arrows show the flow of blood through the heart and adjacent blood vessels.

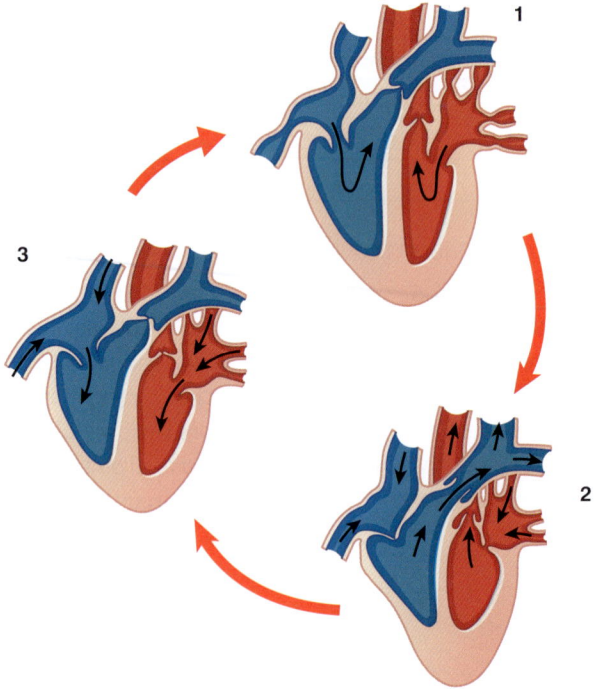

Figure 3

a Use the information in Figure 3 to explain how the internal structure of the heart makes sure that blood flows in one direction. [6]

There are valves between the atrium and the ventricle on each side of the heart. There are also valves between the ventricle and the artery (aorta and pulmonary artery) on each side. These valves make sure that blood flows in one direction through the heart. The blood flows into the right atrium and into the left atrium, then flows into the ventricles and is pumped out into the pulmonary artery and the aorta.

Exam advice:

The four valves in the heart are responsible for making sure blood travels in one direction. The question asks for an explanation, but the student has only identified the position of the valves and not how they function as required by the question. The student has clearly misinterpreted the question as the second part of the answer describes the pathway shown by the arrows. No marks would be gained by this answer.

Exam technique:

This question has six marks and requires a lengthy answer. The command word here is *explain*, which requires answers showing a good understanding of how the valves work to make sure blood flows in one direction through the heart and prevents backflow. The answer requires detail about the opening and closing of the valves. As they work in the same way on the right and the left, it is only necessary to describe one side in your answer but make sure you say it is the same on both sides. See page 100 [Topic 9.2].

b Two people measured their heart rates during vigorous exercise. They each used a heart monitor to record their heart rate. They exercised for four minutes. The graph shows their heart rate at the end of each minute.

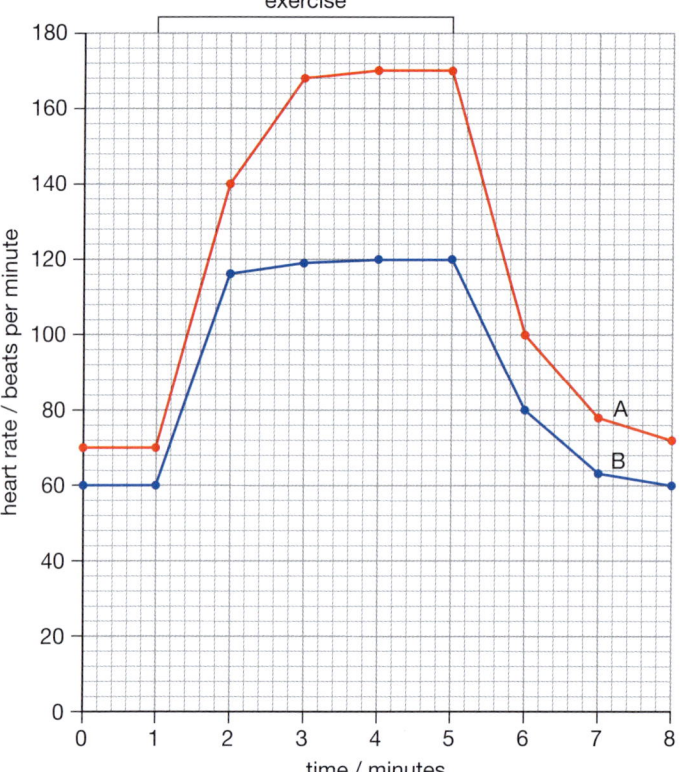

Figure 4

i Calculate the percentage increase in person **B**'s heart rate during the first minute of exercise. Show your working. **[2]**

heart rate at 1 minute = 60 bpm

heart rate at 2 minutes = 116 bpm

increase in heart rate = 56 bpm

$\frac{56}{116}$ x 100 ✓ = 48.28 = 48% ✓

ii Describe how the changes in person **B**'s heart rate differ from those of person **A**. [3]

The heart rate has not increased as quickly ✓ and has not increased to such a high rate ✓. The heart rate at 2 minutes was 116 bpm and at 4 minutes was only 120 bpm.

iii Explain why the heart rates of both students increased. [5]

The heart rates have increased to 120 bpm and 170 bpm because the heart needs to pump more blood into the body ⌒ because the students are exercising. The blood carries oxygen that is needed by the muscles so that they can respire more ✓ to provide more energy ✓.

Exam advice:

The student has written 'pump more blood into the body', which is very vague. It would be better to say that 'more blood is pumped out by the heart every minute. Increasing the heart rate is one way to achieve this. The other is to increase the volume of blood pumped out every time the heart beats. Most of the extra blood pumped out by the heart during vigorous exercise goes to the muscles.' The student is correct in referring to oxygen, but blood provides oxygen to muscles all the time. It's important here to say that 'more oxygen is provided to muscles so that they can respire faster.'

Exam technique:

This question requires knowledge of respiration from Unit 12, and of the heart and exercise from Unit 9. Always be aware that questions can test your knowledge of more than one Unit at a time. In your revision, you should look for links between the different Units.

iv Explain why the heart rate does not decrease to the value at rest immediately the people stop exercising. [4]

Blood needs to flow to the liver to remove the oxygen debt ✔. During exercise, lactic acid is produced in the muscles. This is transported in the blood to the liver where it is broken down in respiration ✔. Oxygen is needed for this and so more blood is needed to supply the extra oxygen needed ✔ and to take away the extra carbon dioxide produced ✔.

Exam advice:

The student has realised that this question is asking about the oxygen debt even though it is not mentioned in the question. This means that the student has recalled all the relevant information about this topic.

Exam technique:

When you read through a question, think of the learning outcome for the syllabus that is being tested or the relevant pages from your textbook. The student has made two links here – between high heart rate supplying more oxygen and the lactic acid that is broken down in the liver (see page 141 [Topic 12.2]).

Exam question comments:

The student gained 10 marks out of 20. Notice in part **(a)** how important it is to use the stimulus material given in the questions. Also it is important to plan out an answer to a question part with 6 marks so that you write at least six different points and organise them in a logical sequence. In this case, the diagrams could have helped the student plan an answer.

[Total: 20]

4 During the outbreak of coronavirus disease 2019 (Covid-19) scientists investigated the effects of a variety of drugs as possible treatments for the disease.

    a Explain what is meant by the term *drug* in this context.   [2]

> In the treatment of the disease it's a substance that is taken by the body ✓ that kills the disease ⌒.

Exam advice:

It is not the disease that is killed, but the organism that causes the disease. The student should have written 'disease-causing organism' or 'pathogen'.

Exam technique:

In this case, the definition given in the syllabus would only gain one mark as the question says 'in this context', which refers to the treatment of the disease. Also note that there are two marks. Always read every word of the question and look at the mark allocation.

    b The virus that causes Covid-19 is similar in structure to the human immunodeficiency virus (HIV) as they are surrounded by a membrane that is formed from the human cells that it invades.

      Figure 5 is a diagram of the virus that causes Covid-19.

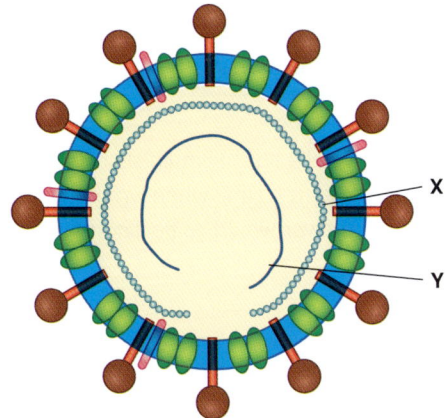

X

Y

Figure 5

Name the parts of the virus labelled X and Y. [2]

X Protein coat ✓

Y Genetic material ✓

Exam advice:

The genetic material in HIV and the coronavirus that causes Covid-19 is RNA. Some viruses have DNA as their genetic material. You are not expected to know which type of nucleic acid is found in any specific type of virus. 'Genetic material' is sufficient to gain the mark for identifying Y.

Exam technique:

Always double check that you have written the answers in the right places. If you write the correct answers in the wrong place (e.g., 'X is genetic material') then no marks are awarded.

c As Covid-19 was a new disease, no vaccination existed for it.

Other research groups developed potential vaccines for Covid-19.

Explain the advantage of using a vaccine to control the spread of a disease such as Covid-19. [4]

Vaccines give active immunity ✓ that lasts a long time ✓. This is because there are cells that remain in the body that remember the virus and respond quickly the next time it enters the body. This means that people who have been vaccinated can't get the disease and can't pass it on ✓.

Exam advice:

The student should identify the 'cells that remain' as memory cells, which is an important technical term to use when writing about immunity. The last part of the answer would be better if the term *transmission* was used.

Exam technique:

This question could prompt answers that describe active immunity or describe how vaccines work to produce active immunity. These would not answer the question completely as they do not explain the *advantage* of using vaccination.

d Apart from drugs, another potential treatment that was trialled in 2020 was the use of plasma obtained from blood donors who had recovered from Covid-19.

i Blood collected from blood donors is treated with a substance that prevents blood clotting.

Describe what happens to blood to cause it to clot. [2]

Fibrinogen in the blood plasma changes to fibrin ✓. Fibrin is fibrous and forms a mesh that traps blood cells forming a blood clot ✓.

Exam advice:

Fibrinogen and fibrin are proteins. The student could have included that information in the answer, but it is not required to gain full marks.

Exam technique:

Notice that the first sentence alone would not be enough to gain full marks for this question. The syllabus statement refers to the 'mesh' so you should include it in your answer.

ii Explain why scientists decided to trial a treatment using blood plasma from people who had recovered from Covid-19. [3]

The blood plasma contains the antibodies that the people have produced against the virus ✓. When injected, the antibodies will fight ✗ the virus and destroy them. This is an example of passive immunity ✓ but it only lasts for a short-time and is not permanent.

Exam advice:

The student has made the link between blood plasma and antibodies and this is the 'key' that unlocks this question as the student also realises that this is passive immunity.

Exam technique:

Do not use the word 'fight' when writing about the immune system. It would be better here to say that the antibodies fit onto antigens on the surface of the virus (see Figure 5) and stop the virus entering host cells. There are no marks for giving *disadvantages* of passive immunity. Note that Figure 5 shows the antigens that are like spikes on the surface of the virus. Always look carefully at drawings and diagrams for clues when you read through a question before starting your answers.

Exam question comments:

The student gained 10 marks out of 13. Notice that the question covered both active and passive immunity. It is important to understand the difference between them in terms of how and where the antibodies are formed. Use the word 'specific' when writing about vaccines and antibodies. A typical question is to explain why the vaccine for one disease does not provide protection against a different disease. The 'spikes' on the virus in Figure 5 act as antigens. The antibodies that combine with the spikes have a specific shape that only combines with these spikes.

[Total: 13]

Glossary

This glossary provides definitions of some of the important biological terms used in this book. The underlined words are key terms as described or defined in the syllabus, which you are required to know for the examinations. You may, however, be expected to explain terms other than these.

A

Absorption movement of digested food molecules through the wall of the intestine into the blood or lymph.

Accommodation the adjustment of the shape of the lens so as to focus light onto the retina.

Active immunity the body produces its own antibodies in response to an antigen, either on the surface of a pathogen or in a vaccine.

Active site part of the surface of an enzyme molecule into which the substrate fits.

Active transport the movement of molecules or ions in or out of a cell through the cell membrane from a region of their lower concentration to a region of their higher concentration against a concentration gradient, using energy released from respiration.

Adaptation the process by which populations become more suited to their environment over many generations as a result of natural selection.

Adaptive feature an inherited feature that helps an organism to survive and reproduce in its environment.

Adrenaline a hormone produced by the adrenal glands that prepares the body for emergencies and stress, for example by increasing the glucose concentration of the blood.

Aerobic respiration the chemical reactions that occur in cells that use oxygen to break down nutrient molecules, such as glucose, to release energy.

AIDS Acquired immunodeficiency syndrome – a collection of diseases that result from a weakening of the body's immune system following infection by HIV.

Allele any of two or more alternative forms of a gene.

Alveolus a tiny air sac in the lungs where exchange of gases between the air and blood occurs.

Amino acid a molecule made up of carbon, hydrogen, oxygen and nitrogen. Amino acids link together by chemical bonds to form protein molecules. Some amino acids also contain sulfur.

Amniotic fluid the liquid that is contained within the amnion and which protects the fetus from mechanical damage.

Amniotic sac the membrane that surrounds the developing fetus in the uterus.

Amylase the enzyme that breaks down starch.

Anaerobic respiration the chemical reactions that occur in cells that break down nutrient molecules, such as glucose, to release energy without using oxygen.

Anaemia a disorder of haemoglobin in the red blood cells which means that the blood cannot carry sufficient oxygen.

Antagonistic muscles a pair of muscles which brings about movement at a joint. When one contracts, the other relaxes, e.g., the biceps and triceps, which move the lower arm, also the radial and circular muscles in the iris, which control the width of the pupil.

Anther the part of a flower where pollen grains are produced.

Antibiotic a drug that is taken to kill or stop the growth of bacteria.

Antibody a protein released by lymphocytes to protect against pathogens.

Antigen any substance that stimulates lymphocytes to produce antibodies.

Antitoxin a chemical released by lymphocytes that neutralises the poisonous waste products (toxins) produced by bacteria.

Arteriole type of blood vessel between artery and capillaries. Contraction and relaxation of muscle in the wall controls flow of blood into capillaries. See vasoconstriction and vasodilation.

Artery a blood vessel through which blood travels away from the heart.

Arthropods the phylum that consists of animals that have jointed legs and a protective exoskeleton.

Artificial insemination placing semen into the uterus or oviduct as a means of treating infertility.

Artificial selection the selection of plants and animals for breeding because of their useful characteristics, e.g., high crop yield. Also called selective breeding.

Asexual reproduction reproduction without the formation of gametes. New individuals are genetically identical (clones) of the parent.

Assimilation the movement of digested food molecules into the cells of the body where they are used, becoming part of the cells or respired to release energy.

Atherosclerosis a narrowing of the arteries caused by deposits of cholesterol in the internal walls of arteries which slows down the rate of blood flow.

Atrium a chamber of the heart that receives blood from veins and pumps it to a ventricle (plural atria).

Auxin a plant growth hormone which controls cell elongation.

B

Bacterium a type of microorganism consisting of single cells. Each bacterial cell has cytoplasm surrounded by a cell membrane and cell wall, but has no nucleus.

Balanced diet a diet that provides sufficient energy for a person's needs and all the food nutrients in the correct proportions.

Bile an alkaline fluid made in the liver and stored in the gall bladder. It is released into the small intestine through the bile duct to help with the digestion and absorption of fats.

Glossary

Binary fission a type of asexual reproduction in which one cell divides into two. Bacteria reproduce by binary fission.

Binomial system internationally agreed system for naming organisms in which the scientific name is made up of two parts consisting of the genus name and a name that applies only to one species in the genus. The scientific name for humans is *Homo sapiens*.

Biodegradable something that can be broken down by biological processes. Some plastics are biodegradable.

Biodiversity the number of different species and the variety in types of species in an ecosystem.

Biomass the mass of living material in a particular area.

Biotechnology the use of organisms to make useful products, such as foods and medicines, or to carry out useful services, such as making waste products harmless.

Bladder a muscular sac that stores urine and passes it out through the urethra.

Blind spot part of the retina where there are no receptors.

Brain the part of the central nervous system that coordinates most activities of the body.

Bronchiole a small branch of a bronchus which ends in alveoli (air sacs).

Bronchus one of the two tubes that branch off the trachea and pass into the lungs (plural: bronchi).

C

Capillary the smallest blood vessel with walls only one cell thick. Substances are exchanged through capillary walls between blood and the fluid that surrounds all cells (tissue fluid).

Carbohydrase an enzyme that digests carbohydrates to simple sugars, e.g., amylase digests starch to sugars.

Carbohydrate a class of food substance that provides energy, e.g., starch and glucose. Composed of carbon, hydrogen and oxygen.

Carbon cycle the flow of carbon compounds through plants, animals, decomposers and their environment.

Carbon dioxide the gas produced as a waste product during respiration and absorbed by plants and used during photosynthesis to make simple sugars.

Carnivore an animal that gains its energy by eating other animals – a meat-eater.

Carrier (1) an individual who has a recessive allele for a genetic disease and can transmit it to the next generation; (2) a person who is infected by the pathogen for an infectious disease, does not have the symptoms of the disease, but is able to transmit it to others.

Catalyst a substance that increases the rate of a chemical reaction and is not changed by the reaction.

Cell structural and functional unit of living organisms. All organisms are composed of cells. Viruses are not.

Cell membrane the boundary of the cell which controls the materials that pass into and out of it.

Cell sap the liquid that fills the vacuole of a plant cell.

Cell wall the outer layer of a plant cell made of cellulose, which supports the cell and gives it shape.

Cellulose a complex carbohydrate that makes up the cell walls of plant cells.

Central nervous system brain and spinal cord.

Chemical digestion the breakdown of large, insoluble molecules into small, soluble molecules.

Chlorophyll the green pigment found in many plant cells that absorbs light for photosynthesis.

Chloroplast small structure containing chlorophyll that is found in some types of plant cell, e.g., palisade cells in leaves. Carries out photosynthesis.

Cholera an intestinal disease caused by a bacterium.

Cholesterol a lipid-based chemical made in the liver and found in the blood. High levels of cholesterol in the blood are linked to an increased risk of atherosclerosis and heart disease.

Chromosome a thread-like structure made up of DNA. Chromosomes are found inside the nucleus and contain genetic information in the form of genes.

Cilium a tiny process found on cells that line some tubular organs, e.g., airways and oviducts (plural: cilia). In the airways, they beat to move dust and microbes out of the lungs and up to the throat.

Ciliary muscle a muscle in the eye that controls the shape of the lens during focussing.

Circulatory system the organ system made up of blood vessels and the heart that transports blood. Mammals have a double circulation with blood passing through the heart twice in one circuit of the body.

Classify to sort living organisms into groups according to features they have in common.

Clone an organism that is genetically identical to its parent.

Clotting a series of chemical reactions that cause blood cells to stick together. At a wound this stops the loss of blood and results in the formation of a scab.

Codominance the existence of two alleles for a particular characteristic where neither is dominant over the other and both are expressed in heterozygous individuals.

Colon the part of the alimentary canal between the small intestine and the rectum where the absorption of water occurs.

Community all the animals, plants and microorganisms that are found in a particular ecosystem.

Competition contest between organisms for resources such as food, water and mates.

Concentration gradient the difference in concentration of a substance between two places, e.g., either side of a cell membrane, between air in the alveolus and blood in the lungs.

Cone a sensory cell in the retina of the eye that responds to light of high intensity and detects colour.

Constipation a condition where compacted faeces are difficult to pass out of the body.

Consumer an organism that gains its energy by feeding on other organisms.

Continuous variation variation in a feature that shows a range of phenotypes between two extremes with many intermediates, e.g., human height.

Coronary arteries arteries that branch from the aorta to supply oxygenated blood to heart muscle.

Cornea the transparent layer at the front of the eye which helps to refract light rays onto the retina.

Coronary heart disease heart disease caused by blockage of coronary arteries that supply heart muscle with blood.

Cotyledon part of the embryo of a flowering plant – a seed leaf. In many plants cotyledons are a store of energy and nutrients for the embryo.

Cross-pollination the transfer of pollen grains from the anther of a flower to the stigma of a flower on a different plant of the same species.

Cuticle the waxy covering of the epidermis in plant stems and leaves that reduces the loss of water by transpiration.

Cytoplasm jelly-like contents of the cell not including the nucleus.

D

Deamination the process, which takes place in the liver, where the nitrogen-containing part of amino acids is removed to form ammonia. Ammonia is then converted into urea.

Decay the breakdown of dead organisms and waste material by decomposers.

Decomposers microorganisms, mainly bacteria and fungi that gain their energy by breaking down dead organisms and waste material.

Deficiency disease a condition when an important nutrient, such as a vitamin or mineral, is missing from the diet and results in a disease.

Deforestation the removal of trees by humans in order to exploit the land.

Denitrification conversion of nitrate ions into nitrogen gas by bacteria.

Denitrifying bacteria bacteria which live in anaerobic conditions, such as water-logged soil, and convert nitrate ions into nitrogen gas.

Development an increase in complexity as an embryo grows and gains new tissues, organs and organ systems.

Diabetes a medical condition in which the blood glucose concentration is not controlled. Type 1 diabetes is caused by a failure of the pancreas to secrete insulin.

Dialysis the use of a partially permeable membrane to separate substances.

Diaphragm a sheet of muscular and fibrous tissue that separates the thorax from the abdomen. Its movements cause air to flow in and out of the lungs.

Diarrhoea a condition where the faeces released are loose and watery. A symptom of diseases, such as cholera.

Dicotyledon a type of flowering plant with an embryo that has two cotyledons and a net-like arrangement of veins in its leaves. (Also known as eudicotyledon.)

Diffusion the net movement of molecules or ions, from a place with a higher concentration to a place with a lower concentration down a concentration gradient as a result of random movement.

Digestion the breakdown of large, insoluble food molecules into small, water-soluble molecules using mechanical and chemical processes.

Digestive system the organ system that breaks down food and absorbs it into the blood.

Diploid nucleus a nucleus containing two sets of chromosomes (e.g., in body cells).

Discontinuous variation limited number of phenotypes for a feature with no intermediates, e.g., human blood groups.

DNA the molecule that forms the genetic material. The sequence of bases in DNA codes for the sequence of amino acids in proteins.

Dominant an allele that is expressed if it is present (e.g., T or G).

Drug any substance taken into the body that modifies or affects chemical reactions in the body.

Duodenum the first part of the small intestine.

E

Ecosystem all the living organisms (the community) in a place, and the interactions between them and their physical environment.

Egestion the passing out of food that has not been digested or absorbed, as faeces, through the anus.

Embryo the early stage of an animal or plant as it develops from a fertilised egg.

Emulsification the breakdown of large fat globules into many tiny globules.

Endangered species any species at very high risk of becoming extinct in the wild.

Energy transfer the transfer of energy from one trophic (feeding) level to the next.

Enhanced greenhouse effect increase in concentration of greenhouse gases in the atmosphere leading to global warming.

Enzymes proteins that function as biological catalysts to increase the rate of all metabolic reactions, e.g., those of chemical digestion and respiration.

Eutrophication the enrichment of waters with plant nutrients that can stimulate growth of algae and plants.

Evolution the process in which inherited features change in populations of organisms over time.

Excretion the removal from organisms of waste products of metabolism and substances in excess of requirements.

Extinct when a species no longer exists on the Earth.

F

Fats lipid molecules that consist of glycerol and three fatty acids. Fats are energy rich as they contain a high proportion of carbon and hydrogen atoms.

Fatty acids molecules that combine with glycerol to form fats.

Fermentation (1) anaerobic respiration in which glucose is converted into ethanol and carbon dioxide; (2) an industrial process in which microorganisms are used to make a useful product.

Glossary

Fermenter a large container in which fungi or bacteria are grown under sterile, controlled conditions.

Fertilisation the fusion of the nuclei of gametes, e.g., sperm and egg.

Fertilisers chemicals that provide plant nutrients and are put on the land to increase the growth of a crop and produce a higher yield.

Fetus a stage during development of a mammal when all the major organs are recognisable.

Fibre indigestible plant material, mainly cellulose and lignin that provides bulk to assist the passage of food through the gut by peristalsis.

Flaccid a plant cell that has lost water by osmosis so that the cell contents no longer push outwards against the cell wall is described as flaccid.

Food chain the feeding relationships in a community showing the flow of energy and nutrients from one organism to the next beginning with the producer.

Food web a network of interconnected food chains showing the energy flow through an ecosystem.

G

Gall bladder a sac in the liver which stores bile before it is released down the bile duct into the small intestine.

Gametes sex cells with the haploid number of chromosomes.

Gas exchange system organ system comprising trachea, bronchi and lungs that involves moving air into the alveoli where oxygen and carbon dioxide are exchanged with the blood.

Gene a length of DNA, found on a chromosome, that codes for a protein.

Gene mutation a random change in the base sequence of DNA that gives rise to new alleles of genes.

Genetic modification changing the genetic material of an organism by removing, changing or inserting individual genes.

Genotype the genetic makeup of an organism in terms of the alleles it possesses (e.g., Tt or GG).

Genus a group of species with similar characteristics, e.g., the lion, *Panthera leo* and the tiger, *Panthera tigris* are in the same genus.

Glucagon a hormone produced by the pancreas that stimulates the liver to convert glycogen to glucose and so increase the concentration of glucose in the blood.

Glycogen a complex carbohydrate found as an energy store in the liver and muscles.

Goblet cells mucus-secreting cells.

Gravitropism a response in which parts of a plant grow towards or away from gravity.

Greenhouse effect the Earth is kept warm because carbon dioxide and other greenhouse gases in the atmosphere reduce the escape of heat (thermal) energy into space. Greenhouse gases radiate heat towards the Earth and maintain the temperature higher than it would be otherwise.

Growth a permanent increase in size and dry mass by an increase in cell number or cell size or both.

H

Haemoglobin the red pigment in red blood cells that combines with oxygen to form oxyhaemoglobin.

Haploid nucleus a nucleus containing a single set of unpaired chromosomes (e.g., nucleus in sperm and egg).

Hepatic portal vein the vein through which absorbed food travels from the small intestine to the liver.

Herbicide a chemical that kills weeds.

Herbivore an animal that obtains all of its energy by feeding on plants.

Heterozygous a genotype where the two alleles of a gene are different.

HIV (human immunodeficiency virus) the virus that causes AIDS. HIV attacks and destroys lymphocytes reducing the body's ability to defend itself against disease.

Homeostasis maintenance of a constant internal environment. This involves controlling factors, such as temperature and the concentration of glucose in the blood.

Homozygous a genotype where both alleles of a gene are identical.

Hormone a chemical messenger produced by an endocrine gland that is transported in the blood and alters the activity of one or more specific target organs.

Hydrophyte a plant that is adapted to living in water.

Hypha (plural: hyphae) thin thread-like structure that is part of the body (mycelium) of a mould fungus.

I

Ileum the part of the small intestine between the duodenum and the colon, the major function of which is the absorption of digested food.

Immune system tissues, cells and chemicals that act together to give a defence against pathogens.

Immunity protection against disease provided by the immune system, including lymphocytes and antibodies.

Implantation the embedding of an embryo into the lining of the uterus.

Ingestion the process of taking food into the body.

Inheritance the transmission of genetic information from one generation to the next.

Inherited disorders disorders caused by dominant or recessive alleles that are passed from one generation to the next.

Insulin a hormone produced by the pancreas that stimulates the liver and muscles to store glucose as glycogen so causing a reduction in the concentration of glucose in the blood.

Intercostal muscles muscles between the ribs; external intercostal muscles contract during inspiration; internal intercostal muscles contract during expiration.

***In vitro* fertilisation (IVF)** fertilisation occurs outside the body by mixing sperm and eggs in a laboratory dish. The resulting embryo is placed into the uterus.

Iris the coloured part of the eye around the pupil. It alters the size of the pupil and so controls the amount of light entering the eye.

K

Kidney the organ that filters waste chemicals out of the blood and controls the water and salt levels in the body.

L

Lactase the enzyme that breaks down lactose (milk sugar).

Lacteal a lymph capillary found inside a villus, which contains absorbed fats.

Lactic acid the chemical produced in muscles when glucose is respired anaerobically. Some bacteria also produce lactic acid during anaerobic respiration.

Large intestine the final parts of the alimentary canal that absorb water and store faeces.

Leaf the plant organ that absorbs light energy to convert into chemical energy during photosynthesis.

Lens part of the eye that focuses light rays onto the retina.

Limiting factor the factor that is in the shortest supply and restricts processes, such as the rate of photosynthesis or the rate of growth.

Lipase the enzyme that digests fats to fatty acids and glycerol.

Liver the organ in the abdomen that produces bile, breaks down amino acids into ammonia, converts ammonia into urea and stores glucose in the form of glycogen.'

Lymph the fluid formed when tissue fluid drains into lymph vessels.

Lymphatic system system of thin-walled vessels that transport lymph and lymph nodes where white blood cells are found.

M

Magnesium the element needed for the synthesis of chlorophyll. It is absorbed by the roots in the form of magnesium ions.

Malnutrition the condition caused by eating an unbalanced diet. Undernutrition occurs when a diet is deficient in one or more food types. Overnutrition can lead to obesity and coronary heart disease.

Maltase the enzyme that breaks down maltose to glucose.

Meiosis a type of division of the nucleus to reduce the chromosome number by half. Diploid nuclei give rise to haploid nuclei. Also called a reduction division.

Memory cells lymphocytes produced during the first invasion of a pathogen or to a vaccine; these cells provide a fast immune response to infection.

Menstrual cycle the sequence of events, that occurs in a woman, where the lining of the uterus thickens in order to receive a fertilised egg after ovulation. The cycle is controlled by hormones from the pituitary gland and the ovaries.

Menstruation the breakdown of the soft lining of the uterus, discharging blood and cells through the vagina.

Metabolism all the chemical reactions that take place inside a living organism.

Microorganism (microbe) microscopic organisms, such as bacteria, fungi and viruses.

Mineral ions inorganic nutrients such as iron and calcium that are needed for a balanced diet. Also required by plants to make compounds, such as amino acids.

Mitochondrion cell structure where aerobic respiration occurs.

Mitosis a type of division of the nucleus that gives rise to genetically identical cells in which the chromosome number is maintained by the exact duplication of chromosomes.

Monocotyledon a type of flowering plant with an embryo with one cotyledon and parallel veins in its leaves.

Monohybrid inheritance the inheritance of one gene with two or more alleles.

Motor (effector) neurone a neurone that transmits impulses away from the brain or spinal cord to effector organs, e.g., muscles and glands.

Movement an action by an organism or part of an organism causing a change of position or place.

Mucus a slimy, sticky substance that traps dust and some microbes in the air pathways. It also acts as a lubricant to help the passage of food along the gut.

Muscle tissue that is capable of contracting and relaxing to bring about movement of bones at joints. Muscle tissue also moves food along our gut and keeps our heart beating.

Mutation a change in a gene or in a chromosome.

Mycelium a network of thin threads (hyphae) that make up the body of a fungus.

N

Natural selection factors such as competition and predation affect the survival of a species. As a result, the individuals adapted to survive have a greater chance to pass on their alleles to the next generation.

Negative feedback the mechanism used in homeostasis for maintaining near constant conditions in the body.

Nerve a bundle of nerve cells (neurones) which pass from the central nervous system to a certain part of the body.

Neurone a nerve cell.

Nitrification conversion of ammonium ions to nitrate ions by bacteria.

Nitrifying bacteria bacteria found in the soil that convert ammonium ions into nitrite ions or nitrite ions into nitrate ions.

Nitrogen-fixing bacteria bacteria found both in the soil and in the roots of legume plants that convert atmospheric nitrogen into nitrogen-containing compounds such as amino acids.

Nitrogen an element needed for healthy plant growth, taken up by plant roots in the form of nitrate ions. Chemical fertilisers often supply nitrate ions.

Nitrogen cycle the flow of nitrogen compounds through plants, animals, decomposers and their environment.

Non-biodegradable materials that will not break down in the environment, e.g., many plastics.

Nucleus the part of the cell which contains genetic information in the form of chromosomes. The nucleus controls the activities of the cell. Nuclei may be haploid or diploid.

Nutrition the taking in of materials for energy, growth and development.

O

Oesophagus the muscular tube connecting the mouth with the stomach. Food passes down the oesophagus by a wave of muscular contraction called peristalsis.

Oestrogen a hormone secreted by the ovaries that stimulates the development of secondary sexual characteristics in females and helps to control the menstrual cycle.

Optic nerve Nerve from the retina in the eye to the brain.

Organ a number of tissues working together to carry out a function in the body.

Organ system a number of different organs that work together to carry out functions for the body.

Osmoregulation the control of the water content of the body.

Osmosis the net movement of water molecules through a partially permeable membrane, from a region of higher water potential (dilute solution) to a region of lower water potential (concentrated solution).

Ovary female sex organ where ova or eggs are produced.

Ovulation the release of an ovum (egg) from the ovary.

Ovule the structure inside the ovary of a plant that contains the female gamete. After fertilisation an ovule develops into a seed.

Ovum the female sex cell (gamete) produced in the ovary. Also called the egg.

Oxygen debt the extra oxygen that is needed by the body to respire lactic acid produced during anaerobic respiration.

P

Palisade mesophyll tissue in leaves that contain many chloroplasts and is the main site of photosynthesis.

Pancreas the organ in the abdomen that produces digestive enzymes and makes insulin and glucagon that regulate the concentration of glucose in the blood.

Partially permeable membrane a membrane that allows small molecules to pass through, but does not allow large molecules to pass through.

Passive immunity providing antibodies from another person or an animal; the person receiving the antibodies does not make the antibodies himself or herself.

Pathogen an organism that causes a disease.

Pectinase an enzyme that breaks down cell walls of plants.

Peripheral nervous system nerves that arise from the brain and spinal cord and go to all the organs of the body.

Peristalsis a wave of muscular contraction that squeezes food down the oesophagus to the stomach.

Pesticide a chemical that kills pests.

Phagocyte a type of white blood cell that ingests and destroys pathogens.

Phagocytosis the ingestion of food particles into a cell.

Phenotype the physical or other features of an organism due to both its genotype and its environment.

Phloem the plant tissue that transports sugars and amino acids.

Photosynthesis process by which plants manufacture carbohydrates from raw materials using energy from light.

Phototropism a growth response of a plant to the direction of light. a response in which parts of a plant grow towards or away from the direction of a light source.

Physical digestion the breakdown of food into smaller pieces without chemical change to the food molecules.

Pituitary gland a gland at the base of the brain that secretes hormones to control the activity of other organs, e.g., kidneys, testes and ovaries.

Placenta the organ that connects a mammalian embryo to its mother and through which it receives food and oxygen and removes carbon dioxide and chemical waste.

Plaque fatty deposit in the lining of an artery.

Plasma the liquid part of the blood which transports dissolved foods, urea, carbon dioxide and hormones.

Plasmid small piece of circular DNA used as a vector in genetic engineering.

Plasmolysis the separation of the cell membrane from the cell wall of a plant cell when water leaves the cell by osmosis.

Platelets small pieces of cells that release substances that cause blood to clot.

Pollination the transfer of pollen from the anther of a flower to the stigma.

Pollution the release by humans of materials or energy that will harm the environment.

Population a group of individuals of the same species living in the same habitat at the same time.

Population growth growth in numbers of a population. In absence of limiting factors growth is usually exponential.

Predators animals that hunt and kill other animals for their food.

Prey animals that are hunted and killed for food by predators.

Producer an organism that makes its own organic nutrients, usually using energy from sunlight through photosynthesis.

Progesterone a hormone secreted by the ovaries and by the placenta that maintains the lining of the uterus during the second half of the menstrual cycle and during pregnancy.

Protease an enzyme that breaks down proteins to amino acids.

Proteins compounds made up of amino acids, which are needed for growth and repair of tissues in the body. They contain the elements carbon, hydrogen, oxygen, nitrogen and sulfur.

Puberty the age at which secondary sexual characteristics appear in boys and girls.

Pulse when the left ventricle of the heart contracts, it forces blood out of the heart along the arteries. The arteries swell and this can be felt as a pulse in various parts of the body, such as the wrist.

Pupil the hole through which light enters the eye. The size of the pupil is controlled by the iris.

Pyramid of biomass a way to show biomass at each trophic level in an ecosystem. The area of each horizontal bar is proportional to the mass of living material at each trophic level.

Pyramid of numbers a way to show numbers of organisms at each trophic level in an ecosystem. The area of each horizontal bar is proportional to the number of individuals at each trophic level.

Q

Quota a limited number or quantity of a resource that can be taken from the environment, e.g., timber and fish.

R

Receptor cells or organs that are sensitive to a stimulus.

Recessive an allele that is expressed only when there is no dominant allele of the gene present (e.g., **a** or **b**).

Rectum the last part of the large intestine where faeces are stored before passing out through the anus.

Recycle to convert material back into a form that can be useful to humans.

Red-green colour blindness a sex-linked characteristic in humans. More commonly found in males.

Reflex action an automatic, rapid response to a stimulus which is often protective.

Reflex arc the arrangement of neurones that controls a reflex action.

Relay (connector) neurone a neurone in the central nervous system that transmits impulses from sensory to motor neurones.

Reproduction processes that make more of the same kind of organism.

Respiration a complex series of reactions, taking place in all living cells, that break down nutrient molecules to release energy for metabolism.

Response the reaction of an organism to a particular stimulus.

Restriction enzyme an enzyme used in genetic engineering to cut DNA at specific places.

Retina the part of the eye that contains light sensitive cells.

Ribosome small cell structure where proteins are made.

Rod a sensory cell in the retina of the eye that responds to light of low intensity.

Root the organ that anchors plants into the ground and absorbs water and mineral ions.

Root hairs specialised cells in a root that provide a large surface area for the absorption of water and mineral ions.

Rough endoplasmic reticulum system of membranes inside cells where ribosomes are attached and proteins are made.

S

Sclerotic the tough, white outer layer of the eye which keeps it in shape.

Secretion the release of useful substances by cells.

Selective breeding see artificial selection.

Self-pollination the transfer of pollen grains from the anther of a flower to the stigma of the same flower or different flower on the same plant.

Sense organ a receptor organ that is sensitive to a particular stimulus or stimuli, e.g., touch, light, sound, temperature and chemicals.

Sensitivity the ability to detect changes in the environment (stimuli) and make responses.

Sensory neurone a neurone that transmits impulses from receptors to the brain or spinal cord.

Sex chromosomes the pair of chromosomes that determine a person's sex. In humans, XX is female and XY is male.

Sex-linked characteristic a feature in which the gene responsible is located on a sex chromosome (usually the X chromosome) making it more common in one sex than in the other.

Sexual reproduction the process involving fusion of haploid nuclei to form a diploid zygote and the production of genetically dissimilar offspring.

Sexually transmitted infection a disease that is transmitted through body fluids during sexual activity, e.g., HIV.

Single circulation in fish, blood flows through the heart once in a complete circulation of the body.

Small intestine the region of alimentary canal that completes digestion and absorbs digested food.

Solute a substance that dissolves in a solvent.

Solvent a liquid that dissolves substances.

Species a group of organisms with similar characteristics which are capable of interbreeding and producing fertile offspring.

Sperm the male sex cell or gamete.

Starch an insoluble carbohydrate made from glucose molecules. An energy store in plant cells and an important component of the human diet.

Stem the organ that supports the leaves, flowers and fruits of a plant. It contains xylem tissue to transport water and mineral ions and phloem tissue to transport sugars and amino acids.

Stem cells unspecialised cells that divide by mitosis to produce daughter cells that can become specialised for specific functions.

Stimulus (plural: stimuli) a change in the environment that is detected by a sense organ.

Stoma (plural: stomata) a small hole in the epidermis of leaves that allows gases to diffuse in and out. The size of the hole is controlled by guard cells.

Stomach a muscular sac at the end of the oesophagus that mixes food with gastric juice. The chemical digestion of protein begins in the stomach.

Suspensory ligaments fibres that hold the lens in the eye in place and alter its shape when the ciliary muscles contract and relax.

Sustainable resource a resource, e.g., timber or fish, which is produced as rapidly as it is removed from the environment so that it does not run out.

Sweat glands coiled glands found in the dermis of the skin which secrete sweat to lose heat by evaporation.

Synapse a gap between two neurones across which a chemical transmitter is released to stimulate an impulse in the second neurone.

T

Test cross a genetic cross used to find out the genotype of an organism.

Testis (plural: testes) the male sex organs where sperm cells and testosterone are produced.

Glossary

Testosterone the hormone, produced by the testes, that stimulates the development of secondary sexual characteristics in males and the development of sperm cells.

Tissue a group of similar cells that act together to perform the same function.

Tissue fluid the fluid that bathes the cells when plasma and some white blood cells pass out of the capillaries.

Toxin a poisonous substance. Bacteria release toxins as waste products which bring about the symptoms of a disease, e.g., high fever.

Translocation the movement of sucrose and amino acids in the phloem from sources to sinks.

Transmissible disease a disease in which a pathogen can be passed from one host to another.

Transpiration the evaporation of water from the surfaces of mesophyll cells in leaves followed by diffusion of water vapour through stomata into the atmosphere.

Trophic level the position of an organism in a food chain, food web, pyramids of numbers, biomass or energy. For example, herbivores are primary consumers and form the second trophic level.

Tropism a growth response by part of a plant to a stimulus, e.g., light or gravity.

Thrombosis a blood clot that occurs in a vein or an artery.

Tuber a swollen stem of a potato for storage and asexual reproduction.

Turgor the pressure exerted by a plant cell onto its cell wall. Turgid cells are firm because they are full of water; this can support plant organs such as stems.

U

Umbilical cord the cord connecting the placenta to the fetus. It contains two umbilical arteries and a vein.

Urea a waste product formed from excess amino acids in the liver. It is filtered out of the blood in the kidneys and passed out in the urine.

Ureter a tube through which urine passes from the kidney to the bladder.

Urethra a tube that passes urine from the bladder to the outside at intervals.

Urinary system organ system that filters blood, produces and stores urine.

Urine an excretory fluid produced in the kidneys and stored in the bladder. It contains excess water, urea, and excess ions.

Uterus the womb; a muscular chamber with a soft lining in which the fetus develops.

V

Vaccine a preparation containing mild or dead pathogenic microorganisms or their antigens to stimulate the production of antibodies and give immunity.

Vaccination giving a vaccine to provide immunity.

Vacuole fluid-filled sac containing cell sap present in most plant cells.

Vagina a tube leading from the uterus to the outside. It receives the erect penis during sexual intercourse and is the birth canal.

Variation differences between individuals in features that may show continuous variation or discontinuous variation.

Vasoconstriction contraction of muscles in arterioles to reduce blood flow through capillaries.

Vasodilation relaxation of muscles in arterioles that increases blood flow through capillaries.

Vein a blood vessel through which blood travels towards the heart.

Ventricle the lower, more muscular chamber of the heart. In mammals, the right ventricle pumps blood to the lungs and the left ventricle pumps blood to the rest of the body.

Vertebrate an animal with a backbone.

Vesicles small cell structures surrounded by membrane that contain substances made by cells, such as enzymes or neurotransmitters.

Villus (plural: villi) tiny, finger-like projection from the wall of the small intestine. Villi increase the surface area of the small intestine for absorption of digested food.

Virus a microorganism that cannot be seen using the light microscope. Smaller than bacteria, they consist of nucleic acid (e.g., DNA or RNA) surrounded by a protein coat. They can reproduce only inside living cells.

Vitamin a micronutrient, needed in small amounts in the diet. If not enough is provided in the diet a deficiency disease can result.

W

Water potential the tendency for water molecules to move by diffusion. Water diffuses from an area of higher water potential to an area of lower water potential.

White blood cells they form part of the body's defence system. Lymphocytes release antibodies and antitoxins to combat pathogens. Phagocytes ingest and kill pathogens.

Wilting loss of turgor causes plant leaves to droop if water is lacking.

X

Xerophyte a plant adapted to dry conditions.

Xylem the plant tissue that transports water and mineral ions from roots to leaves, flowers and fruits.

Y

Yeast a single-celled fungus that ferments sugar to produce ethanol and carbon dioxide.

Yield the quantity of product from a crop or from a fermentation process.

Z

Zygote a fertilised egg.

Index

Index

Index